Green Analytical Chemistry

Green Analytical Chemistry

Current Status and Future Perspectives in Sample Preparation

Edited by

Marcello Locatelli
University "G. D'Annunzio" of Chieti-Pescara, Department of Pharmacy, Chieti, Italy

Savaş Kaya
Sivas Cumhuriyet University, Faculty of Science, Department of Chemistry, Sivas, Turkey

Elsevier
Radarweg 29, PO Box 211, 1000 AE Amsterdam, Netherlands
125 London Wall, London EC2Y 5AS, United Kingdom
50 Hampshire Street, 5th Floor, Cambridge, MA 02139, United States

Copyright © 2025 Elsevier Inc. All rights are reserved, including those for text and data mining, AI training, and similar technologies.

Publisher's note: Elsevier takes a neutral position with respect to territorial disputes or jurisdictional claims in its published content, including in maps and institutional affiliations.

No part of this publication may be reproduced or transmitted in any form or by any means, electronic or mechanical, including photocopying, recording, or any information storage and retrieval system, without permission in writing from the publisher. Details on how to seek permission, further information about the Publisher's permissions policies and our arrangements with organizations such as the Copyright Clearance Center and the Copyright Licensing Agency, can be found at our website: www.elsevier.com/permissions.

This book and the individual contributions contained in it are protected under copyright by the Publisher (other than as may be noted herein).

Notices
Knowledge and best practice in this field are constantly changing. As new research and experience broaden our understanding, changes in research methods, professional practices, or medical treatment may become necessary.

Practitioners and researchers must always rely on their own experience and knowledge in evaluating and using any information, methods, compounds, or experiments described herein. In using such information or methods they should be mindful of their own safety and the safety of others, including parties for whom they have a professional responsibility.

To the fullest extent of the law, neither the Publisher nor the authors, contributors, or editors, assume any liability for any injury and/or damage to persons or property as a matter of products liability, negligence or otherwise, or from any use or operation of any methods, products, instructions, or ideas contained in the material herein.

ISBN: 978-0-443-16122-3

For information on all Elsevier publications visit our website at https://www.elsevier.com/books-and-journals

Publisher: Candice Janco
Acquisitions Editor: Gabriela Capille
Editorial Project Manager: Debarati Roy
Production Project Manager: Nadhiya Sekar
Cover Designer: Greg Harris

Typeset by TNQ Technologies

Contents

Contributors xi
About the editors xv
Acknowledgments xvii

1. **Green chemistry and green analytical chemistry**
   Miryam Perrucci, Vincenzo De Laurenzi, Enrico Dainese, Marcello Locatelli, Halil I. Ulusoy, Abuzar Kabir, Imran Ali and Fotouh R. Mansour

1. Introduction	1
2. Green chemistry	2
3. Green analytical chemistry	6
4. Conclusions	12
Declaration of interests	13
Author contributions	13
Funding	13
Acknowledgments	13
References	13

2. **Principal concepts and guidelines for GAC applied in sample preparation**
   Naeem Ullah and Mustafa Tuzen

1. Introduction	15
2. Green chemistry	17
2.1 Green chemistry metrics	19
3. Green analytical chemistry	20
3.1 Background of green analytical chemistry	21
4. Key steps used in sample reparation	24
5. The principles of green sample preparation in the context of GAC	28
6. The main components of green analytical methods: Core principles of GAC	30
6.1 Sample collection	31
6.2 Analytical process and quantification techniques	32
6.3 Reagents safety and secure operation	33
6.4 Waste generation	34

7.	Conclusion	34
	Acknowledgments	35
	References	35
	Further reading	42

3. Conceptual density functional theory–based applications in extraction studies
Savaş Kaya

1.	Introduction	43
2.	Hard and soft acid-base principle	46
3.	Hard acids	48
4.	Soft acids	48
5.	Hard bases	48
6.	Soft bases	49
7.	Maximization and minimization in stable states of hardness, polarizability, electrophilicity, and magnetizability	50
8.	Conceptual DFT-based applications in extraction studies	52
9.	Conclusion	56
	References	56

4. Sorbent-based extraction procedures
Mohammad Reza Afshar Mogaddam, Sarina Beiramzadeh, Mohammad Nazari Koloujeh, Aysan Changizi Kecheklou, Mir Mahdi Daghi, Mir Ali Farajzadeh and Mustafa Tuzen

1.	Introduction	59
	1.1 Solid phase extraction	64
	1.2 Solid phase microextraction	70
	1.3 Fiber solid phase extraction	80
	1.4 In-tube solid phase microextraction	81
	1.5 Matrix-dispersed solid phase extraction	83
	1.6 Sorbent-based dynamic extraction	85
	1.7 Stir bar sorptive extraction	87
	1.8 Rotating disk solid phase extraction	89
	1.9 Magnetic solid phase extraction	92
	1.10 Thin film microextraction	95
	1.11 Dispersive solid phase extraction	99
	1.12 Packed sorbent microextraction	101
	1.13 Micro solid phase extraction	103
	1.14 Fabric phase sorptive extraction	104
	1.15 Air-assisted solid phase extraction	107
	1.16 Summary and outlook	109
	Abbreviations	109
	References	110

5. **Membrane-based extraction techniques**
   Abuzar Kabir and Basit Olayanju

   1. Introduction — 119
   2. Membrane-based solid-phase microextraction and associated techniques — 121
      - 2.1 Fabric phase sorptive extraction — 121
      - 2.2 Capsule phase microextraction — 127
      - 2.3 Micro solid phase extraction — 128
      - 2.4 Carbon nanomaterial-based solid phase extraction — 130
      - 2.5 Biofluid sampler — 130
      - 2.6 Membrane-protected molecularly imprinted materials — 132
      - 2.7 Membrane-protected solid phase microextraction — 133
      - 2.8 Thin film microextraction — 135
   3. Membrane-based liquid phase microextraction — 135
      - 3.1 Porous membrane–based liquid phase microextraction techniques — 136
      - 3.2 Nonporous membranes — 138
      - 3.3 Hollow-fiber based liquid phase microextraction — 138
      - 3.4 Electromembrane extraction — 138
      - 3.5 Bulk membrane extraction — 139
   4. Conclusions — 139
   References — 140

6. **Introduction and overview of applications related to green solvents used in sample preparation**
   Seçkin Fesliyan, Hameed Ul Haq, Nail Altunay, Mustafa Tuzen, Ebaa Adnan Azooz and Zainab Hassan Muhamad

   1. Introduction — 145
   2. Amphiphilic solvents — 146
   3. Supramolecular solvents — 149
   4. Hydrophilicity switchable solvent — 152
   5. Ionic liquids — 157
   6. Deep eutectic solvents — 165
   7. Conclusions and prospects for the future — 169
   References — 175
   Further readings — 188

7. **Advancements and innovations in solvent-based extraction techniques**
   Muhammad Farooque Lanjwani, Muhammad Yar Khuhawar, Mustafa Tuzen, Seçkin Fesliyan and Nail Altunay

   1. Introduction — 189
   2. Dispersive liquid liquid microextraction — 190
      - 2.1 Innovations in the use of extracting solvents — 192

viii Contents

       2.2 Ionic liquids (ILs)   193
       2.3 Supramolecular solvents (SUPRASs)   197
       2.4 Reverse micelles based SUPRAS   198
       2.5 Deep eutectic solvents (DES) in DLLME   201
       2.6 Switchable solvents for micro extraction   203
  3. **Pressurized fluid extraction (PFE)**   209
       3.1 Instrumentation and principles of pressurized fluid extraction   210
       3.2 Solvent used in PFE   210
  4. **Dispersive solid phase extraction**   211
       4.1 Applications of dispersive solid phase extraction (DLPE)   213
  5. **Micellar-assisted extraction or microvawe assisted extraction**   214
       5.1 Applications of micelle-assisted extractions   215
  6. **Supercritical fluid extraction (SFE)**   215
       6.1 Properties of supercritical fluid   218
       6.2 Solvent used in SFE   218
       6.3 Instrumentation used in SFE   219
       6.4 Applications of supercritical fluid   220
  7. **Liquid liquid extraction (LLE)**   221
       7.1 Principle of LLE   221
       7.2 The history of LLE   221
       7.3 Solvents used in LLE   222
  8. **Liquid phase microextraction (LPME)**   223
       8.1 Specific extraction media in SPME   224
       8.2 Solvent used in LPME   225
  9. **Conclusion**   226
 10. **Future recommendations**   226
       References   226

8. Solvent free extraction procedures

Moumita Saha, Rahul Makhija and Vivek Asati

  1. **Introduction**   247
       1.1 Principles of solvent-free extraction   248
  2. **Methods of solvent-free extraction**   248
       2.1 Grinding   248
       2.2 Pressing   248
       2.3 Gas-assisted mechanical methods   248
       2.4 Subcritical water extraction   249
       2.5 Pressurized liquid extraction (PLE)   249
       2.6 Supercritical fluid extraction   250
       2.7 Microwave-assisted extraction (MAE)   251
       2.8 Ultrasound-assisted extraction   251
       2.9 Ohmic heating extraction   252
      2.10 Ionic liquids   252
      2.11 Biocompatible extractions   253
      2.12 Solid phase microextraction (SPME)   253

	3.	Advantages	254
	4.	Considerations	255
	5.	Applications	256
	6.	Conclusion	256
		References	257

9. How to evaluate the greenness and whiteness of analytical procedures?

Ebaa Adnan Azooz, Farah Abdulraouf Semysim,
Estabraq Hassan Badder Al-Muhanna,
Mohammad Reza Afshar Mogaddam and Mustafa Tuzen

	1.	Introduction	263
	2.	A brief history of greenness evaluation tools in analytical chemistry	265
	3.	Classification of greenness evaluation tools	267
	4.	Summary of the chosen greenness evaluation tools	276
		4.1 National environmental method index (NEMI)	276
		4.2 The developed NEMI tool	278
		4.3 The Eco-scale tool	278
		4.4 Analytical eco-scale tool	282
		4.5 Green analytical procedure index (GAPI) tool	287
		4.6 Complex green analytical procedure index (ComplexGAPI)	287
		4.7 Analytical method greenness score (AMGS)	294
		4.8 Hexagon-CALIFICAMET scale	296
		4.9 The analytical GREEness (AGREE) scale	306
		4.10 The AGREEprep scale	314
		4.11 White analytical chemistry (WAC) scale	322
	5.	Applications of the studied greenness evaluation methods' publications	325
	6.	Conclusions	326
	7.	Visions for the future	326
		Abbreviations	348
		References	349

10. The CUPRAC method, its modifications and applications serving green chemistry

Reşat Apak, Mustafa Bener, Saliha Esin Çelik,
Burcu Bekdeşer and Furkan Burak Şen

	1.	Importance of green analytical chemistry	357
	2.	Oxidative stress and antioxidants	359
	3.	CUPRAC method and its advantages	360
		3.1 Main advantages of the CUPRAC assay	362
	4.	Modifications and applications of CUPRAC method serving green chemistry	364

4.1	Optical sensors based on CUPRAC method	364
4.2	CUPRAC microplate- and flow injection-based methods	368
4.3	Online HPLC-CUPRAC methods	370
4.4	Electroanalytical CUPRAC methods	372
4.5	Green solvents used in the CUPRAC method	374
4.6	Quencher CUPRAC methods	376
References		378

11. Conclusion and future perspectives

Miryam Perrucci, Vincenzo De Laurenzi, Marcello Locatelli, Halil I. Ulusoy, Abuzar Kabir, Imran Ali, Fotouh R. Mansour and Savas Kaya

Declaration of interests	387
Author contributions	387
Funding	387
Acknowledgments	387
References	387

Index   389

Contributors

Mohammad Reza Afshar Mogaddam, Food and Drug Safety Research Center, Tabriz University of Medical Sciences, Tabriz, Iran; Pharmaceutical Analysis Research Center, Tabriz University of Medical Sciences, Tabriz, Iran

Estabraq Hassan Badder Al-Muhanna, Radiology Department, College of Medical Technology, The Islamic University, Najaf, Iraq; Department of Pathological Analyses Sciences, Faulty of Science, Jabir Ibn Hayyan University for Medical and Pharmaceutical, Najaf, Iraq

Imran Ali, Department of Chemistry, Jamia Millia Islamia (Central University), Jamia Nagar, New Delhi, India

Nail Altunay, Sivas Cumhuriyet University, Faculty of Science, Department of Chemistry, Sivas, Turkey

Reşat Apak, Istanbul University-Cerrahpaşa, Faculty of Engineering, Department of Chemistry, Avcilar, Istanbul, Türkiye

Vivek Asati, Department of Pharmaceutical Chemistry, ISF College of Pharmacy, Moga, Punjab, India

Ebaa Adnan Azooz, The Gifted Students' School in Al-Najaf, Ministry of Education, Najaf, Iraq; Medical Laboratory Technology Department, College of Medical Technology, The Islamic University, Najaf, Iraq

Sarina Beiramzadeh, Food and Drug Safety Research Center, Tabriz University of Medical Sciences, Tabriz, Iran; Faculty of Chemistry, Tabriz Branch, Islamic Azad University, Tabriz, Iran

Burcu Bekdeşer, Istanbul University-Cerrahpaşa, Faculty of Engineering, Department of Chemistry, Avcilar, Istanbul, Türkiye

Mustafa Bener, Istanbul University, Faculty of Science, Department of Chemistry, Division of Analytical Chemistry, Fatih, Istanbul, Türkiye

Aysan Changizi Kecheklou, Food and Drug Safety Research Center, Tabriz University of Medical Sciences, Tabriz, Iran; Analytical Chemistry Department, Faculty of Chemistry, University of Tabriz, Tabriz, Iran

Mir Mahdi Daghi, Food and Drug Safety Research Center, Tabriz University of Medical Sciences, Tabriz, Iran; Faculty of Chemistry, Tabriz Branch, Islamic Azad University, Tabriz, Iran

Enrico Dainese, University of Teramo, Department of Biosciences and Agro-Food and Environmental Technologies, Teramo, Italy

Vincenzo De Laurenzi, Department of Innovative Technologies in Medicine & Dentistry, University "Gabriele d'Annunzio" of Chieti-Pescara, Chieti, Italy; Center for Advanced Studies and Technology (CAST), University "Gabriele d'Annunzio" of Chieti−Pescara, Chieti, Italy

Mir Ali Farajzadeh, Analytical Chemistry Department, Faculty of Chemistry, University of Tabriz, Tabriz, Iran; Engineering Faculty, Near East University, Nicosia, North Cyprus, Turkey

Seçkin Fesliyan, Sivas Cumhuriyet University, Faculty of Science, Department of Chemistry, Sivas, Turkey

Hameed Ul Haq, Gdansk University of Technology, Faculty of Civil and Environmental Engineering, Department of Sanitary Engineering, Gdansk, Poland

Abuzar Kabir, Department of Chemistry and Biochemistry, Florida International University, Miami, FL, United States; International Forensic Research Institute, Department of Chemistry and Biochemistry, Florida International University, Miami, FL, United States

Savaş Kaya, Department of Chemistry, Faculty of Science, Sivas Cumhuriyet University, Sivas, Turkey

Muhammad Yar Khuhawar, Dr M. A. Kazi Institute of Chemistry, University of Sindh, Jamshoro, Sindh, Pakistan

Muhammad Farooque Lanjwani, Tokat Gaziosmanpaşa University, Faculty of Science and Arts, Chemistry Department, Tokat, Turkey; Dr M. A. Kazi Institute of Chemistry, University of Sindh, Jamshoro, Sindh, Pakistan

Marcello Locatelli, University "Gabriele d'Annunzio" of Chieti-Pescara, Department of Pharmacy, Chieti, Italy

Rahul Makhija, Department of Pharmaceutical Analysis, ISF College of Pharmacy, Moga, Punjab, India

Fotouh R. Mansour, Department of Pharmaceutical Analytical Chemistry, Faculty of Pharmacy, Tanta University, Tanta, Egypt

Zainab Hassan Muhamad, Department of Computer Science, The Gifted Students School, Gifted Guardianship Committee, Ministry of Education, Najaf, Iraq

Mohammad Nazari Koloujeh, Department of Chemistry, Faculty of Science, University of Maragheh, Maragheh, Iran

Basit Olayanju, Department of Chemistry and Biochemistry, Florida International University, Miami, FL, United States

Miryam Perrucci, University of Teramo, Department of Biosciences and Agro-Food and Environmental Technologies, Teramo, Italy; Department of Innovative Technologies in Medicine & Dentistry, University "Gabriele d'Annunzio" of Chieti-Pescara, Chieti, Italy

Moumita Saha, Department of Pharmaceutical Analysis, ISF College of Pharmacy, Moga, Punjab, India; Department of Pharmaceutical Quality Assurance, Manipal College of Pharmaceutical Sciences MAHE, Manipal, Karnataka, India

Farah Abdulraouf Semysim, Department of Chemistry, The Al-Mutafawiqat High School in Najaf, Ministry of Education, Najaf, Iraq

Furkan Burak Şen, Istanbul University, Faculty of Science, Department of Chemistry, Division of Analytical Chemistry, Fatih, Istanbul, Türkiye

Saliha Esin Çelik, Istanbul University-Cerrahpaşa, Faculty of Engineering, Department of Chemistry, Avcilar, Istanbul, Türkiye

Mustafa Tuzen, Tokat Gaziosmanpaşa University, Faculty of Science and Arts, Chemistry Department, Tokat, Turkey

Naeem Ullah, Tokat Gaziosmanpaşa University, Faculty of Science and Arts, Chemistry Department, Tokat, Turkey; Department of Chemistry, University of Turbat, Balochistan, Pakistan

Halil I. Ulusoy, Department of Analytical Chemistry, Faculty of Pharmacy, Cumhuriyet University, Sivas, Turkey

About the editors

Marcello Locatelli is an Associate Professor of Analytical Chemistry, and his research activity is aimed at the development and validation of chromatographic methods for the qualitative and quantitative determination of biologically active molecules in human and animal, cosmetics, food, and environmental complex matrices. These procedures have been applied to different analytes and drug associations, also finding applications in clinical and preclinical studies, to characterize new delivery systems of the active principle to improve their pharmacological properties. In the development of the methods, predictive models and chemometrics are applied both for the optimization of extraction protocols and for final data processing. Particular attention is given to innovative (micro)-extraction techniques and new instrumental configurations for the quantitative analysis of complex matrices.

Savaş Kaya is an Associate Professor of Inorganic Chemistry at Sivas Cumhuriyet University, Health Services Vocational School, Department of Pharmacy, 58140, Sivas, Turkey. He was born in 1989 and obtained his doctorate degree in 2017 in the field of theoretical inorganic chemistry. He does research in theoretical chemistry, computational chemistry, materials science, corrosion science, physical inorganic chemistry, and coordination chemistry. Savaş Kaya has published more than 80 papers in international journals, indexed SCI, and SCI expanded. He is the editor of the books entitled *Conceptual Density Functional Theory and Its Application in the Chemical Domain* and *Corrosion Science: Theoretical and Practical Applications*. He is the author of 10 book chapters. Recently, he introduced the Kaya chemical reactivity approach and Kaya combined reactivity descriptor and proposed the "Nucleophilicity Equalization Principle."

Acknowledgments

The Editors would like to first thank Elsevier for the opportunity to publish this book, which aims to represent a small contribution in the panorama of Green Analytical Chemistry (GAC) and Green Sample Preparation (GSP).

Thanks go to all the contributors who made it possible, thanks to their very high professionalism, to bring this editorial project to completion.

Chapter 1

Green chemistry and green analytical chemistry

Miryam Perrucci[1,2], Vincenzo De Laurenzi[2,3], Enrico Dainese[1], Marcello Locatelli[4], Halil I. Ulusoy[5], Abuzar Kabir[6], Imran Ali[7] and Fotouh R. Mansour[8]

[1]*University of Teramo, Department of Biosciences and Agro-Food and Environmental Technologies, Teramo, Italy;* [2]*Department of Innovative Technologies in Medicine & Dentistry, University "Gabriele d'Annunzio" of Chieti-Pescara, Chieti, Italy;* [3]*Center for Advanced Studies and Technology (CAST), University "Gabriele d'Annunzio" of Chieti—Pescara, Chieti, Italy;* [4]*University "Gabriele d'Annunzio" of Chieti-Pescara, Department of Pharmacy, Chieti, Italy;* [5]*Department of Analytical Chemistry, Faculty of Pharmacy, Cumhuriyet University, Sivas, Turkey;* [6]*International Forensic Research Institute, Department of Chemistry and Biochemistry, Florida Interna-tional University, Miami, FL, United States;* [7]*Department of Chemistry, Jamia Millia Islamia (Central University), Jamia Nagar, New Delhi, India;* [8]*Department of Pharmaceutical Analytical Chemistry, Faculty of Pharmacy, Tanta University, Tanta, Egypt*

1. Introduction

Nowadays, it is normal to read the term "green chemistry" (GC) in each field that includes pharmaceutical, analysis and environmental chemistry. However, this term and the idea around it were born in last century, due to increasing industrialization, that from a part growing quality of life, but on the other hand the impact on environment was worst day by day [1,2].

Due to these evidences, Cathcart in 1990 used for the first time the well-known term "green chemistry", and in 1996, the term appeared for the first time in a paper, thanks to Anastas and Warner [3,4]. The maximum goal of this sector is to minimize pollution favor to environmental and human health [4,5]. How visible from Fig. 1.1, between 2010 and 2023 there was an important increase of following GC principles. GC was born for a sustainable society, starting from character, arriving to nature and production process of chemicals. A chemical process produces much waste and, often, it cause toxicity and bioaccumulation. Therefore, GC's main goal is designing for a green future [4–6].

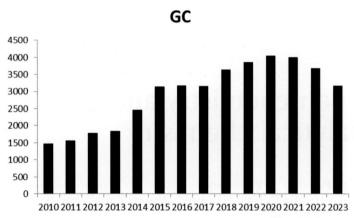

FIGURE 1.1 Trend over time of interest toward green chemistry principles.

Subsequently, many areas of chemistry followed GC principles sectorializing versus their type of research. Thus, green analytical chemistry (GAC) principles were born after 10 years from GC [7]. In addition, interest for GAC is increased during years, and certainly, it will continue to growth, as notable from Fig. 1.2.

With the aim of calculate, 'control' and evaluate the greenness of a procedure, various method were studied, as index or metric in different fields and with procedures of logarithmic scale independent each other.

2. Green chemistry

The definition of GC is well reported on numerous scientific paper and on the official site of this wide argument, and is *"the design of chemical products and*

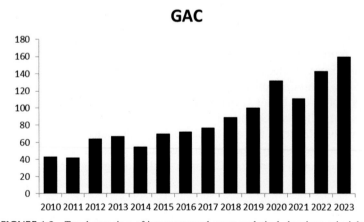

FIGURE 1.2 Trend over time of interest toward green analytical chemistry principles.

Green chemistry and green analytical chemistry **Chapter | 1** 3

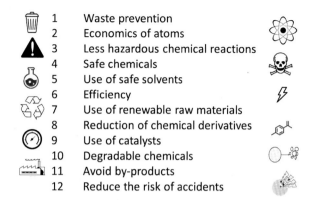

FIGURE 1.3 A schematic representation of 12 principles of GC.

processes that reduce or eliminate the use or generation of hazardous substances. GC applies across the life cycle of a chemical product, including its design, manufacture, use, and ultimate disposal" [8]. Practically the main aims are listed in Fig. 1.3.

Firstly, there is the necessity to identify the meaning of waste, and it is what is formed during a chemical process, except the final product of interest. Roger Sheldon proposed a calculating factor dividing the kilos of waste by kg of final products. More the value is close to zero, more the process is green [1,9].

About first point in each moment of a chemical cycle of a new (but not only for the newest ones) product (synthesis, analysis, characterization etc.) it is better to avoid or minimize waste. Examples are use of microextraction instead conventional ones, analysis should have short course etc. this is the first point, because waste can be in different forms and, consequently, it can have different effects, depending for example from nature, toxicity and how it can accumulate [1,4,5,8].

The second point is regarding economics of atoms, a concept introduced by Trost in 1990 called Atomy Economy (AE), that would maximize the possibility of using raw materials, so the final product can contain the same atoms of the initial step, minimizing in this way waste [6]. Thus, that means that there is necessity of planning synthesis so the final product will contain as much of reagents as possible and connect with the first point (avoid waste) [8,10].

The third and fourth points were born from the increasing interest about choosing safer chemicals and solvents, to avoid risks for environmental and human health, parallel to understand the important impact of some reagents and solvents and the possibility of replacing them with greenness ones [6,8,11].

Continuing this list of principles, reducing amount of solvents consist in reducing waste, because major part of waste came from solvents and, often, they are toxic and/or hazardous. Therefore, an important challenge in this part is to choose new and greener type of this class, as ionic liquids, deep eutectic solvents, natural deep eutectic solvents, water [6,8,11].

Another problem in this period is, certainly, the scarcity of petroleum stock, so renewable energy has been introduced (solar power, wind power, etc.) as possible options [6]. An example of reduction of energy consumption during a synthesis can be conducting at room temperature the experiment [8].

The use of renewable raw materials is preferred to non-renewable, of course. Example di renewable materials are chitin (a polymer present in exoskeleton of arthropods used to obtain chitosan, that has biomedical application, antimicrobial activity, water purification), glycerol, cellulose, lignin (from paper industry, used as additive or for production of vanillin, dimethylsulfoxyde) [6].

It is usual to read protocols with derivatization of drugs or component during a synthesis, for blocking groups or to make visible a drug to detector during an analysis process. When it's not necessary it's better to avoid derivatization process, because they often use hazardous solvents and increase quantity of waste [6,8].

Often, waste come from using stoichiometric reagents, instead, now, it is preferable to use as selective as possible of catalytic reagents [6].

An important problem that turns out with globalization is persistence, bioaccumulation, or incomplete degradation of chemical products into sea, ground. Thus, with the aim of avoid or reduce these phenomena, chemicals should decompose in innocuous products, that do not accumulate in sea and ground and consequently in human food [6,8].

A big goal during analysis is the possibility of analyzing a product without a sample preparation step. Many times this process is characterized by solid-liquid extraction, liquid-liquid extraction so process that require an important amount of solvents. On the other hand, it is also important in situ monitoring of a reaction, so without preparative step. In this way, the possible formation of intermediate products can be monitored quickly and saving energy [6].

This possible formation of metabolites can cause accident, as leaks or explosion, that can destroy both working and surrounding place [5,7,11].

Concluding, it is visible how GC can be useful in all part of a chemical pre- and post-production, how Anastas et al. defined *"all stages of chemical life-cycle"* [6]. Obviously, these 12 principles represent a possibility of various advantages and big goals that scientists can integrated in their everyday works.

Based these abovementioned principles, GC wants to minimize danger, consequently also risk will reduce, cause of the equation reported in Eq. (1.1) [11].

$$Risk = Danger \times Exposure \tag{1.1}$$

FIGURE 1.4 Schematic representation of RAMP.

Where risk represents the probability of damage resulting from exposure to a peril (hazardous chemical). From this, a new trend was born, following also GC, called RAMP, where the acronym is explain in Fig. 1.4 [10].

Following all of these principles phases of an experiment must be:

1. Design of experiment;
2. Recognize hazardous chemicals;
3. Assess risks;
4. Minimize risks;
5. Prepare for emergencies.

Point 3 of this list calculates risk's level by relating the exposure (low medium and high) to type of chemicals used, based on dangerousness of the latter (see Fig. 1.5) [10].

The first step to minimize and/or eliminate risks is eliminate hazardous chemicals, for example using room temperature and atmospheric pressure,

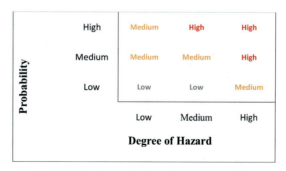

FIGURE 1.5 Example of calculate assessing risks.

instead extremely conditions. Considering the ninth point of GC principles, also using catalyst, instead of stoichiometric [6,10].

If elimination is not possible, you can consider a substitution of hazardous chemicals, with a non-flammable one, for example. Important is to consider the place of work and that it is working well and that it is optimal for this type of experiment (for example, the hood is working and has the right flow rate). In some lab, additionally to normal lab's law (as no eating in lab, no food storage in the fridge lab with chemicals), there are other Standard Operating Procedures (SOP), that are specifically for the type of lab's work and with the aim to understand well them, there is an appropriate course. Concluding, it is fundamental to use Personal Protective Equipment (PPE), both ones, and specifically for the type of work to do [10].

About the substitution of solvents, often it is easy to read that using buffer, water solutions, but after using them, column and chromatographic system need a long cleaning process, waste and consumption of energy. Additionally, buffer solutions cannot stay so long, so they need to prepare again with consequently consumption of solvent, buffer powder and time [1]. In Fig. 1.6 there is a schematic representation of solvents (from top to bottom) respectively less toxic and more toxic, provided by Prat and colleagues [1,12].

3. Green analytical chemistry

Later, GAC emerged, starting from GC's principles that were not all applicable in analytical chemistry, because more focused on synthesis. Only four of the

FIGURE 1.6 Solvents classification.

abovementioned principles are applied in analytical chemistry and are the first, the fifth, the sixth and the eighth, concerning respectively minimizing waste, using safer solvents, designing energy efficiency and reducing derivatization [13]. Therefore, new 12 principles focused on analytical chemistry were born (and represented in Fig. 1.7). Stepping back, analytical chemistry is defined as *"the study of the separation, identification, and quantification of the chemical components of natural and artificial materials"* [14].

In this contest, the key goal of analytical chemistry is continuing in researching coupled these 12 principles. Thus, reduction or elimination of hazardous chemical substances, minimization of consumption of energy and waste, in this way operator's safe will be increase [13]. The first principle wants to avoid pre-treatment and, when possible, do directly analysis. Consequently, minimal sample size and low number of samples represent the second principle of GAC, preferring micro and nano-scale. According the third principle, location in which will be measurement has to be more close possible. It is fundamental to decrease number of analytical steps. Automation of analytical procedures means minor exposure and automation is so important than miniaturization. About sixth principle, derivatization should be avoided, because to type of solvents used in it and for the number of additional steps. Obviously, miniaturization and following GAC principles will generate minor volume of waste that represents another important point of GAC, parallel to the possibility to analyze multi analyte in 1 h. Consequentially, also consumption of energy will be lower. Reagents obtained from a process, it is preferred to come from renewable source, and toxic reagents should be

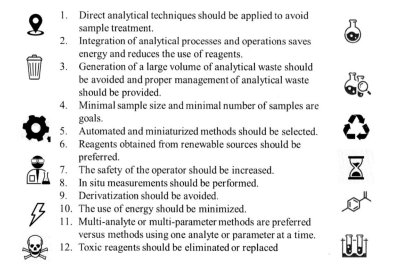

1. Direct analytical techniques should be applied to avoid sample treatment.
2. Integration of analytical processes and operations saves energy and reduces the use of reagents.
3. Generation of a large volume of analytical waste should be avoided and proper management of analytical waste should be provided.
4. Minimal sample size and minimal number of samples are goals.
5. Automated and miniaturized methods should be selected.
6. Reagents obtained from renewable sources should be preferred.
7. The safety of the operator should be increased.
8. In situ measurements should be performed.
9. Derivatization should be avoided.
10. The use of energy should be minimized.
11. Multi-analyte or multi-parameter methods are preferred versus methods using one analyte or parameter at a time.
12. Toxic reagents should be eliminated or replaced

FIGURE 1.7 A schematic representation of 12 principles of GAC.

minimized or eliminated, favor to more green solvents or reagents, some examples of greener solvents can be ionic liquids or supercritical fluids. Following these entire principles operator will work in safety [7].

Nowak and colleagues affirmed that GAC has to accompany the 'traditional' analytical chemistry, as logical consequence of the last mentioned [15]. Many researcher called "3R" approach (reduction, replacement, recycling) with the aim of minimize the use of hazardous solvents, or replace them with greener ones [13]. Instead, Koel and colleagues proposed a different approach called "4S", Specific methods, smaller dimensions, simpler methods and Statistics [16]. These approaches are focused on instrumental analysis, which is the main aim of analytical chemistry that have to study and process without replacing sensitivity and selectivity [13]. During past, the most essential focuses in analytical chemistry were sensitivity, selectivity, robustness, accuracy and precision, but now the situation is changed, following GAC principles, and more attention is given to reducing amount of solvents, samples and saving energy [13]. Keith and colleagues proposed four criteria for establish the greenness of an analytical process, and they affirmed that a method is less green if one of these parameters can be marked off [13,17] (see Fig. 1.8).

Thus, one of the major trends of last years is 'counting' how green is a procedure, and to do this are born metrics that judge the works through different points and scales, for example NEMI, Analytical Eco-Scale, GAPI and AGREE [7,18].

NEMI was one of the first analytical metrics to be studied, it's a simple tool that can provide information about greenness of an analytical procedure, but not quantitative. It supplies information about if reagents used are listed as persistent, bio-accumulative, or toxic, if reagents produce hazardous waste, if

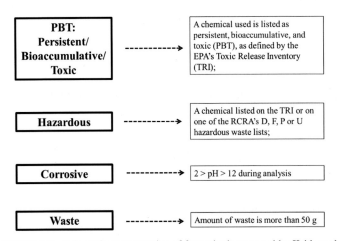

FIGURE 1.8 Schematic representation of four criteria proposed by Keith et al.

Green chemistry and green analytical chemistry Chapter | 1 9

FIGURE 1.9 Schematic representation of NEMI.

sample and procedure are corrosive (2 < pH < 12) and if quantity of waste is below 50 g [19]. In Fig. 1.9 there is an example of NEMI.

Galuszka and colleagues developed analytical Eco-Scale in 2012 and it works subtracting points from a total core of 100, which represents the "ideal" green analysis, more close is to 100, more green is the analysis [18–20]. Some parameters are considered for subtracting penalty points: solvents and reagents, energy consumption and waste [21].

About first point, it is based on pictograms on solvents/chemicals container, each pictogram has one penalty point (PPs). If chemical is characterized by the word "danger", number of pictograms are multiplied for two. Instead, if there is the word "warning" pictograms are multiplied for one. Additionally, if quantity of reagents/solvents is less than 10 mL or 10 g, score is multiplied by 1, if it's between 10 and 100 g or mL by two, and more than 100 mL or g it must be multiplied by 3. In Fig. 1.10, there is a simple representation of the calculation. A simple example can be Acetonitrile, that has 2

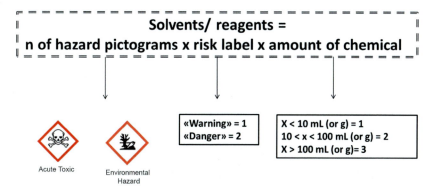

FIGURE 1.10 Schematic representation of assigning penalty points of the first point of analytical eco-scale.

TABLE 1.1 Examples of some common instrument with energy used and penalty points.

Instrument	Energy used	PPs
FTIR, spectrofluorometry, UV-VIS, sonicator	≤0.1 kWh	0
GC, ICP- MS, LC	≤1.5 kWh	1
NMR, GC-MS	≥1.5 kWh	2

pictograms and the write "Danger", so total is 4, that has to be multiplied for the quantity of solvent used [18].

About energy consumption the calculate is the same of the above mentioned, ≤0.1 kWh per sample is 0 penalty points, ≤1.5 kWh is 1 penalty points, and for ≥1.5 kWh are 2 points, as visible for some examples in Table 1.1.

The last parameter considered can be divided in two parts: the first one about amount of waste, and the second is regarding waste's processing.

Amount of waste (w):

- w < 1 mL (or g) = 1
- 1 < w < 10 mL (or g) = 3
- w > 10 mL (or g) = 5

Based on amount of waste, other PPs can be given, if it will be subjected to recycling 0 PPs, degradation 1 PPs, passivation 2 PPs, no treatment 3 PPs [18].

Green Analytical Procedure Index: GAPI.

This metric index was introduced in 2018, practically it is a pictogram with five pentagrams, and each pentagram describes a process of analytical methodology. They are colored with green, yellow and red, based on the gravity of the impact. It is easy to understand.

- *First pentagram*: sample collection, transport, storage:
    1. Sample collection: green for in-line collection; yellow for on-line and at-line; red for off-line.
    2. Sample preservation to preserve its integrity: no sample preservation is green, yellow for chemical or physical preservation; red for both, chemical and physical preservation.
    3. Sample transportation: green for no transport; yellow for need transport.
    4. Sample storage: green for no storage; yellow for normal storage conditions; red for special storage conditions.

- *Second pentagram*: type of method.
  5. Sample preparation: green for no preparation; yellow for easy step as filtration, decantation; red for extraction step.
- *Third pentagram*: different step of sample preparation.
  6. Scale of extraction: green for nano-scale; yellow for micro-scale; red for macro-scale.
  7. Nature of solvents in extraction step: green for solventless or solvent-free; yellow for green solvents; red for hazardous solvents.
  8. Additional treatment: green for no additional treatment; yellow for simple treatment as clean up; red for derivatization.
- *Fourth pentagram*: amount of solvents/reagents.
  9. Volume of solvents: $V < 10$ mL (or g) green field; yellow for $10 < V < 100$ mL (or g) and red for $V > 100$ mL (or g).
  10. Hazardous solvents: green for 0 or max 1 solvents classified toxic; yellow for 2 or 3; red for more than 4 toxic solvents.
  11. Safety hazard of the solvents: green if flammability or instability NFPA is 0 or 1; yellow if NFPA flammability or instability score is 2 or 3, and red if NFPA flammability or instability score is 4.
- *Fifth pentagram*: energy consumption
  12. Energy consumption by instrument per sample: green for <0.1 kWh; yellow for <1.5 kWh; red for >1.5 kWh.
  13. Occupational hazards: green for hermetic sealing; red for release of vapors or gasses into the atmosphere.
  14. Amount of waste (w): $w < 1$ mL (or g) green; $1 < w < 10$ mL (or g) yellow and red for amount of $w >$ than 10 mL (or g).
  15. Treatment of waste: green for recycling waste; yellow for degradation and red for no treating of waste.
- The *circle* in the second pentagram is used to indicate that the analysis is for quantification and qualification. If there is not the circle, it means that the analysis is only for qualification.

A simple example of GAPI pictogram is illustrated in Fig. 1.11.

Another metric tool used is AGREE, that was developed in 2020 and it results to be the most inclusive covering all 12 principles of GAC. It comprehends all of 12 GAC principles, giving different weight to each point, easy to understand thanks to color of each part of pictogram and easy to design with an online software [22]. In the central part of the pictogram, there is the total score for the method considered, as illustrated in Fig. 1.12.

12 Green Analytical Chemistry

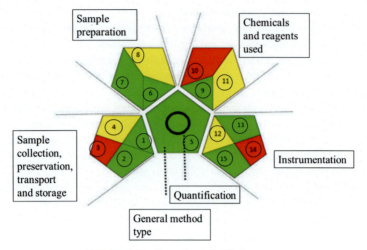

FIGURE 1.11 Example of GAPI pictogram.

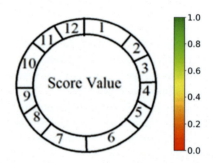

FIGURE 1.12 Schematic representation of AGREE.

4. Conclusions

How reported in this chapter, it is easy to understand the importance and the impact of GC and GAC principles. So many scientists have reported in their researches as key words also GC and GAC, highlighting the need to protect the environment and the operator. There are some principles in common between GC and GAC, as minimizing hazardous solvents and waste that are fundamental aspect in many field from synthesis to validation of analytical methods. In addition, the choice of solvents is fundamental in a chemistry process, because this is the cardinal point of many area of interest.

To understand the greenness of an analytical process many index were born and they are used in many validation method, in this case also the type of instrument used can be subject of criterion. For sure, interest versus GC and GAC will increase during years, and a big challenge can be getting as close as possible at the perfect idea of Green, for each field of chemistry. We are, also, sure that new index more precise and refined will born, and, consequently, we will try them and follow the idea of GAC.

Declaration of interests

The authors declare that they have no competing financial interests or personal relationships that could have appeared to influence the work reported in this paper.

Author contributions

All Authors contributed equally to Conceptualization; Investigation; Project administration; Resources; Supervision; Roles/Writing—original draft; and Writing—review & editing.

Funding

This research did not receive any specific grant from funding agencies in the public, commercial, or not-for-profit sectors.

Acknowledgments

This article is based upon the work from the Sample Preparation Study Group and Network, supported by the Division of Analytical Chemistry of the European Chemical Society.

References

[1] B.A. de Marco, B.S. Rechelo, E.G. Tótoli, A.C. Kogawa, H.R.N. Salgado, Evolution of green chemistry and its multidimensional impacts: a review, Saudi Pharmaceutical Journal 27 (2019) 1−8, https://doi.org/10.1016/j.jsps.2018.07.011.

[2] M. Tobiszewski, A. Mechlińska, B. Zygmunt, J. Namieśnik, Green analytical chemistry in sample preparation for determination of trace organic pollutants, TrAC, Trends in Analytical Chemistry 28 (2009) 943−951, https://doi.org/10.1016/j.trac.2009.06.001.

[3] C. Cathcart, Green chemistry in the Emerald Isle, Chemistry and industry 21 (1990) 684−687.

[4] P.T. Anastas, T.C. Williamson, Green Chemistry: Designing Chemistry for the Environment, vol 626, American Chemical Society Publications, 1998, pp. 1−6, https://doi.org/10.1021/bk-1996-0626.

[5] P.T. Anastas, N. Eghbali, Green chemistry: principles and practice, Chemical Society Reviews 39 (2010) 301−312, https://doi.org/10.1039/B918763B.

[6] J.B. Zimmerman, P.T. Anastas, H.C. Erythropel, W. Leitner, Designing for a green chemistry future, Science 367 (2020) 397−400, https://doi.org/10.1126/science.aay3060.

[7] M. Locatelli, A. Kabir, M. Perrucci, S. Ulusoy, H.I. Ulusoy, I. Ali, Green profile tools: current status and future perspectives, Advances in Sample Preparation 6 (2023) 1−15, https://doi.org/10.1016/j.sampre.2023.100068.

[8] EPA, Basics of Green Chemistry, https://www.epa.gov/greenchemistry/basics-green-chemistry, (Accessed 12 January 2023).
[9] R.A. Sheldon, Catalysis: the key to waste minimization, Journal of Chemical Technology & Biotechnology: International Research in Process, Environmental and Clean Technology. 68 (1997) 381−388, https://doi.org/10.1002/(SICI)1097-4660(199704)68:4<381::AID-JCTB620>3.0.CO;2-3.
[10] D.C. Finster, RAMP: a safety tool for chemists and chemistry students, Journal of Chemical Education 98 (2020) 19−24, https://doi.org/10.1021/acs.jchemed.0c00142.
[11] L. Heine, M.H. Whittaker, How chemical hazard assessment in consumer products drives green chemistry, Handbook of Green Chemistry 11 (2010) 231−280.
[12] D. Prat, J. Hayler, A. Wells, A survey of solvent selection guides, Green Chemistry 16 (2014) 4546−4551, https://doi.org/10.1039/C4GC01149J.
[13] M. Koel, Do we need green analytical chemistry? Green Chemistry 18 (2016) 923−931, https://doi.org/10.1039/c5gc02156a.
[14] A. Gałuszka, Z. Migaszewski, J. Namieśnik, The 12 principles of green analytical chemistry and the significance mnemonic of green analytical practices, TrAC, Trends in Analytical Chemistry 50 (2013) 78−84, https://doi.org/10.1016/j.trac.2013.04.010.
[15] P.M. Nowak, R. Wietecha-Posłuszny, J. Pawliszyn, White analytical chemistry: an approach to reconcile the principles of green analytical chemistry and functionality, TrAC, Trends in Analytical Chemistry 138 (2021) 1−10, https://doi.org/10.1016/j.trac.2021.116223.
[16] M. Koel, B.M. Kaljurand, Design in analytical chemistry, Critical Reviews in Analytical Chemistry 42 (2012) 192−195, https://doi.org/10.1080/10408347.2011.645378.
[17] L.H. Keith, L.U. Gron, J.L. Young, Green analytical methodologies, Chemical Reviews 107 (2007) 2695−2708, https://doi.org/10.1021/cr068359e.
[18] M. Sajid, J. Płotka-Wasylka, Green analytical chemistry metrics: a review, Talanta 238 (2022) 1−11, https://doi.org/10.1016/j.talanta.2021.123046, 123046.
[19] M. Shi, X. Zheng, N. Zhang, Y. Guo, M. Liu, L. Yin, Overview of sixteen green analytical chemistry metrics for evaluation of the greenness of analytical methods, TrAC, Trends in Analytical Chemistry (2023), https://doi.org/10.1016/j.trac.2023.117211, 117211.
[20] A. Gałuszka, Z.M. Migaszewski, P. Konieczka, J. Namieśnik, Analytical eco-scale for assessing the greenness of analytical procedures, TrAC, Trends in Analytical Chemistry 37 (2012) 61−72, https://doi.org/10.1016/j.trac.2012.03.013.
[21] L.P. Kowtharapu, N.K. Katari, S.K. Muchakayala, V.M. Marisetti, Green metric tools for analytical methods assessment critical review, case studies and crucify, TrAC, Trends in Analytical Chemistry 166 (2023), https://doi.org/10.1016/j.trac.2023.117196, 117196.
[22] F. Pena-Pereira, W. Wojnowski, M. Tobiszewski, AGREE—analytical GREEnness metric approach and software, Analytical Chemistry 92 (2020) 10076−10082, https://doi.org/10.1021/acs.analchem.0c01887.

Chapter 2

Principal concepts and guidelines for GAC applied in sample preparation

Naeem Ullah[1,2] and Mustafa Tuzen[1]

[1]*Tokat Gaziosmanpaşa University, Faculty of Science and Arts, Chemistry Department, Tokat, Turkey;* [2]*Department of Chemistry, University of Turbat, Balochistan, Pakistan*

1. Introduction

In order to use chemists' abilities, expertise, and capabilities to diminish risks to both the environment and humans in all types of chemical processing, the concept of "green chemistry" (GC) first originated in the 1990s [1]. Paul Anastas recognized the important importance of the advancement of analytical methods and designated green analytical chemistry (GAC) as an emerging topic in GC 1 year before it began [2]. In Green Chemistry, analytical chemistry has two separate and conflicting functions. It helps preserve the environment by assessing the effect of chemical operations; however, because of the large amounts of harmful toxins utilized during an analytical method, as well as the energy needs, this could cause extra global pollution. GAC aims to rethink and reassess analytical processes by addressing solvent/reagent safety, hazardous laboratories waste production, worker safety, and clean energy. A chemical measurement technique consists of multiple phases, which are characterized as collecting, sample preparation, quantitative measuring, and data analysis. Samples are typically transformed to a form suitable for the equipment used for analysis during the sample preparation process, or they are cleaned of interfering matrix constituents. In certain cases, sample preparation comprises analyte enrichment to meet the sensitivities requirement of the analytical technique [3]. Earlier sample preparation techniques required a lot of work, time, and, most significantly, resources, which led to the creation of dangerous laboratory waste. Sample preparation was thus viewed as being essential to achieving the GC objectives since it was a significant contributor to the overall negative environmental effect of analytical methods. This is demonstrated in Paul Anastas's [2] initial paper articulating the purpose of creating analytical techniques in GC. Three case studies were presented, two of which demonstrated how traditional sample preparation procedures had a detrimental influence on the overall greenness of the analytical approach. Furthermore, the very similar study looked at advances achieved when current approaches were utilized instead of older ones. The focus of future investigations remained on the necessity of sample pretreatment for designing greener analytical procedures [4–6] till Gałuszka et al. [39] offered 12 criteria as main principles for developing green analytical procedures in 2013. The first concept of this strategy recommended avoiding sample preparation by using direct analytical methods. The same report also came to the conclusion that "green" steps undertaken throughout the sample processing phase (such as minimized energy use, operator safety, as well as the application of nonhazardous chemical or materials that come from sustainable sources) had a negative impact on the analytical process' accuracy, precision, selectivity, sensitivity, and detection range [7,8]. The first GAC concept is frequently misunderstood, leading individuals to believe that skipping the sample preparation process is a green strategy while completely ignoring the "green" technology advancements in the industry. Additionally, it completely ignores contexts in which conversion into an analysis-ready form is necessary since direct analysis is not an option and is required instead [9,10] and then immediately subjected to analysis, which reduces the method's sensitivities. Direct analysis's sensitivity and matrix-related issues can be partially resolved by the technical advancements in analytical apparatus that are currently available [11]. This method, however, typically demands the employment of costly equipment is indeed challenging to locate in several common labs. Due to the need for sample processing in several cases, the first criterion of GAC has not been fulfilled (i) aiming for selective analyte extraction from complicated matrix; (ii) preconcentrating the analyte(s) to attain the requisite sensitivity in techniques; (iii) converting the sample to a suitable condition appropriate for the measuring method; and (iv) the sample's cleanup. The "excluding" of sample preparation from GAC produced a gap; that rather than ignoring key steps, emphasis ought to have been spent to thoroughly describing in the framework of GC and the GAC methodology. And besides, GC was never concerned with what to do less of, but rather with what to do more of. Today's analytical scientists and practitioners encounter more complicated and interwoven difficulties both onsite and in the laboratory, as a result, sample preparation is usually necessary. Simultaneously, the present global environmental difficulties that society faces urge a shift toward green methods. In this context, it is more vital than ever to (re)define sample processing in the framework of GC and GAC to address environmental issues and advance the concept of sustainable sample processing (GSP).

Among the most labor and energy-intensive operations in analytical chemistry is sample preparation. To convert target analytes from their complicated state to a readily quantifiable form, a large number of organic solvents or compounds must be utilized throughout this procedure [12]. The primary goal of GAC would be to substitute dangerous chemicals with safer alternatives and to limit harmful reagent discharge into the environment. Analytical chemistry has been a breakthrough in terms of sample preparation and the development of new, long-lasting analytical procedures. In this context, a departure from traditional sample preparation approaches is required [13]. Anastas became acquainted with the subject of "green chemistry" for the very first time during 1998 [14]. GAC is a subset of GC that is based on the fundamental concepts of GC. These guidelines include easy operation and handling, minimal labor requirements, low waste output, and the elimination or reduction of the requirement for hazardous chemicals in favor of eco-friendly alternatives [15–19]. It seems to be critical to think about sample collection, sample preparation, chemicals, and equipment while developing a green analytical technique [20]. In terms of sample preparation, the discipline of analytical chemistry is rapidly developing and the creation of novel, efficient analytical methods. The sample preparation procedure presents a great potential for the use of GAC principles. This method mainly entails concentrating the target analyte, eliminating contaminants, and then applying it to the instrument analysis, each of which are usually difficult and essential to any analytical conclusion. Systematic inefficiencies may also be addressed by the approach because sample preparation often depends on extraction techniques that can be accomplished in a single or several processes. However, there are a number of limitations and disadvantages that must be highlighted, including the process' strong reliance on potentially dangerous organic solvents and the pollution it produces.

2. Green chemistry

In the 1990s, the concept of "green chemistry" evolved as a means of utilizing scientists' abilities, expertise, and capabilities to prevent hazards to both human and environmental health in all kinds of chemical processes [21]. Paul Anastas identified the critical importance of analytical technique development 1 year after his inception and designated GAC as an emerging topic in GC, important to both the scientific community and the business community [22]. Fig. 2.1 displays the progression of the publications on GC information generated from the Web of Science Core Collection database, taking into account the existence of the phrase "green chemistry," from 2010 to 2022. In GC, analytical chemistry serves two separate but compatible functions: It contributes to environmental management by measuring the consequences of chemical activity while also having the potential to exacerbate existing environmental issues because of the large amounts of hazardous substances that are consumed or produced during an analytical operation, as well as the high energy required. Sampling, sample preparation, analytical measurement, and

18 Green Analytical Chemistry

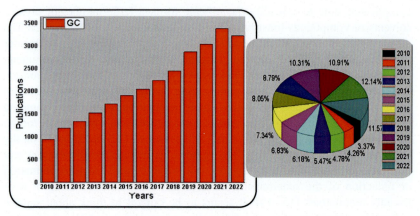

FIGURE 2.1 Displays the progression of the publications on green chemistry information generated from the Web of Science core collection database, taking into account the existence of the phrase "green chemistry," from 2010 to 2022.

data assessment are the processes that make up a chemical measurement technique. The sample processing stage involves either converting samples into a format that is suitable with the analytical tool being used or cleaning the samples of interfering analytes. In many other situations, sample preparation includes analyte enriching to satisfy the quantitative method's sensibility requirements [23]. Earlier sample preparation techniques required a lot of work, effort, and, more critically, resources, which led to the creation of dangerous laboratory waste. Until now, sample preparation was seen as essential to achieving the aims of GC since it was a significant contributor to the overall negative environmental effect of analytical techniques. For instance, Paul Anastas's [2] initial paper detailing the function of developing analytical methodologies in GC serves as an example of this. The significance of sample preparation in establishing greener analytical techniques remained the focus of following investigations [24–26] till 2013, after Gałuszka et al. [39] suggested 12 principles as broad recommendations for greening analytical procedures. The first premise of this strategy proposed avoiding sample preparation by using direct analytical methods [27]. This was an incorrect assumption, especially when taking into account the analytical capabilities of the mature sample preparation methods that were available at the time [28].

The urgent necessary sustainable development in all of our activities within a dynamic environment with the immense influence of human intervention has expedited the transition away from the chemical paragon and toward the environmental framework, wherein we need to really consider the ecological effects of the chemical activities [29]. Greening all parts of the analysis of all types of sample, not just for investigations in the environment, is the focus of intense study since GC [30–32] and GAC [33–37] emerged as the academic realm and entered the real world. We have no doubts whatsoever that GAC will be extremely helpful in the next years. The use of low-cost, quick, and ecologically safe methods for medical, industrial, and environmental analysis will raise living standards in poor nations [38]. Therefore, it is clear that GAC has been a crucial tool for moving analytical chemistry moving from chemical paragon and toward the environmental framework and producing sustainable instruments to fulfill the demands of the expanding need in analysis for a skillful integration of environmentally friendly and fairly cost approaches. Many green techniques have been created in recent years depending upon on 12 guiding principles of GAC [39], several academic publications recently released specialized issues on the use of GAC in research and practical laboratories, causing a surge that altered the concepts and methods of analysis and received positive feedback from the scientific communities. As a result, the development of analytical approaches' transition beyond qualitatively to quantitatively of the green feature has been significantly slower compared to the relevant scientific contribution. Life cycle assessment (LCA), a meticulous platform that considers most environmental exchanges (i.e., resources, energy, emission standards, and waste) that take place in all phases of the lifespan exercises, is helpful in this regard, particularly when tried to apply to products or services, which requires the life-cycle premise as well as its steps are firmly established [40]. The American Chemical Society's (ACS) Green Chemistry Institute (GCI) created a further semi-quantitative criterion. The National Environmental Methodologies Index (NEMI), a free online database of ecological method, was subjected to the criteria [41].

2.1 Green chemistry metrics

Many criteria were devised to evaluate the greenness of analytical techniques. Therefore, in the study, we want to offer a brief review of some typical GAC measures used to analyze the environmental effect of analytical techniques. Due to the abundance and complexity of sample matrix and analytical techniques, evaluating analytical approaches relating to GC is highly challenging. It is important to note that, in some circumstances, analytical methods cannot satisfy specific GC standards. In this case, efforts are required to enhance the procedure. Utilizing flow-injection systems to reduce the usage of chemicals that cannot be deleted or replaced is an excellent illustration of this strategy [42] or reducing the size of analytical systems [43]. The multistep nature of the analytical procedure presents another challenge for green analytical methods. Each stage should be taken into account in relation to the creation of more environmentally friendly analytical techniques. According to Tobiszewski et al. [44], sample preparation has been the lowest environmentally friendly step in the analysis process. The most effective technique uses directly analytical process to eliminate this stage. When designing new, greener analytical methods, also known as "Sustainable Analytical Procedures," four features can be recognized [45]: (1) method-oriented [42,46,47]; (2) analyte-oriented [24]; (3) process-oriented [48]; and (4) a combination involving three or four of the approaches mentioned above [49]. When looking for a green analytical approach, the main issue that might be posed is: "How to assess various processes in order to find the greenest one." There still are general guidelines for environmental practices, but they frequently include certain aspects of a particular strategy, (e.g., choosing of a more environmentally friendly solvent) [50,51]. A recent discussion focused on the use of GAC in an analytical process [47]. The vast majority of current methodologies for evaluating green technologies, unfortunately, have been predicated upon premises (i.e., not based on a scientific methodology, but on instinctual development the proposed system). Surprisingly, the first technique was determined to be greener in terms of life-cycle evaluation [47]. We require a much more qualitative methodology to select a green analytical procedure. Conventional GC metrics have limited use in GAC. Principals that are most notable [52−54]. This information concentrates on substances and procedures, which could be changed to reduce the amount of harmful pollution caused in experiments. Eco-Scale method to green organic synthesis assessment, assuming that a "perfect" reaction, with a score of 100 on the Eco-Scale, employs cheap molecules, is conducted at ambient temp with a 100% yield, and seems to be harmless for the both the operator and the environment. Demerit points are applied for any metric that departs from "the ideal value," decreasing the overall score. Higher cost-efficient and environment-effective organic preparation are indicated by a maximum possible score. The potential for evaluating green analytical techniques resides in this concept [55−59].

There are numerous ways of judging how green analytic methods are [60,61]. Several significant approaches for evaluating the environmental friendliness of analytical processes are the analytical atom economy, environmental factor, and energy intensity. To calculate the analytical procedure's greener scale, each of the following indicators considers various components of the process, as demonstrated in Fig. 2.2. Several measures for evaluating the greenness of analytical techniques have been established over time. Several are more general and useful for the significant number of analytical procedures, whereas some are designed for specific types of analytical processes [24,42,43,45,48,50−54,62−64].

3. Green analytical chemistry

It may be argued that because analytical chemistry uses comparatively tiny quantities of chemicals in comparison to synthetic and manufacturing chemical activities, the adverse consequences of analytical procedures are of little consequence. Furthermore, Paul Anastas recognized the significance of analytical measures within their works on GC, and the facts that analytical chemistry techniques are widely utilized in both academic and application

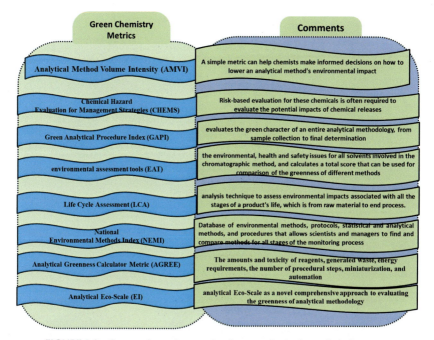

FIGURE 2.2 Commonly used green chemistry metrics in the analytical process.

laboratories made this topic particularly pertinent to daily life. Furthermore, considerable work has been done on theoretical GAC research, and this will help the current century's development of green approaches. Fig. 2.3 demonstrates the development of the literature on GAC the Web of Science Core Collection database considering the presence of the terms "green analytical chemistry." This subject has had a substantial influence on the analytical literature of the 21st century, particularly with the release in 2010 of the first book on GAC by Mihkel Koel and Mihkel Kaljurand, GAC, along with this progress in publications concerning GAC [61,65−68]. The significance of studies is on sample preparation using microextraction methods, as well as the high volume of citations they acquire each year [15,18,69−77].

3.1 Background of green analytical chemistry

Analytical chemistry is being always conscious of environmental compatibility, but during the middle of the 1990s, the green analytical technique approach have become more perceptible. The guidelines have been proposed in this regard to order to take more specific actions to reform and improve the analytical procedures [15,78]. Overview of the development process of GAC is shown in Fig. 2.4. A thorough approach to analysis preparation, pretreatment, and determination known as "GAC" uses a solvent, chemicals, and nontoxic,

22 Green Analytical Chemistry

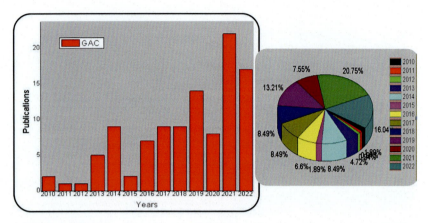

FIGURE 2.3 This figure shows the progression of the publications on green analytical chemistry information generated from the Web of Science core collection database, taking into account the existence of the phrase "green chemistry," from 2010 to 2022.

FIGURE 2.4 Background of the development process of green analytical chemistry.

recyclable, and ecologically friendly procedures that are nontoxic, biodegradable, and ecologically friendly. In addition, GAC emphasizes minimizing the size of the instruments and sensors as well as the amount of chemicals and materials utilized during the analysis [79]. An illustration of a GAC strategy is the ecologically safe and responsible fabrication of the electrode materials used in electrochemical sensor investigations [80]. A solvent must meet the following criteria in order to be considered green: it must be readily available, reasonably priced, of sufficient grade quality, easy and nontoxic to synthesize, nontoxic, biodegradable, effective, reusable, nonflammable, resilient thermal and chemical, and simple to carry and preserve [81]. Efficient and environmentally friendly solvents, certain fluorocarbon solvents, liquid polymeric materials, ionic liquids, and eutectic combinations like choline and glycolate are some examples of green solvents that have all or some of these characteristics. A green, sustainable, and safe solvent seems to be water. Supercritical liquids, such as supercritical carbon dioxide, as well as bio-based solvents made from sustainable material, like glycerol and glycerol-containing (not bio-based) [81]. Different extraction strategies are employed to simplify the study in complex mediums that enclose the target compounds at extremely minor levels. However, using those methods necessitates using a lot of organic solvents. The same scenario holds true for chromatographic techniques. For this reason, researchers are using several choices to provide a greener approach, such as modifying the methods for applying the approaches, the solvents employed or altogether changing the technique used [82—84]. Afterward 2010, the GAC progression progressed toward the development of green instruments, methodologies, and certifications that may be used on smartphones [78]. The cost can be decreased by switching to greener analytical techniques since there is a lesser waste to handle and less chemical use [78]. In recent times, the number of reviews and publishing articles is increased covering this topic due to the steady creation of the GAC method. In particular, da Silva et al. recently examined green extraction techniques and solvents for bioactive chemicals [85]. It is crucial to switch from hazardous organic solvents to ecologically sound solvents since solvents play crucial roles in analytical processes such as reagent preparation, extraction, washing of materials, etc. Research on the creation of solvent-free procedures has been done; however, they are challenging. Presently, selecting greener solvents has become the most popular alternative. Ionic liquids are important green solvents, with the exception of those that are costly and have dangerous qualities. In addition to being bio-derived eco-friendly substances, extraction techniques make use of deep eutectic solvents [85]. Del Valle examined numerous uses for green biological detectors [86]. This study placed a particular emphasis on mobile or intelligent sensing platforms. Integrating biosensors with remote testing, a greener strategy may be taken that employs the minimum solvents and chemicals and generates the minimum wastage. Contrarily, the majority of review papers assess GAC in general and outline the fundamental ideas and topic areas of that kind of technique [17,79,87]. It should also be mentioned that attaining an environmentally greener assessment without compromising the method's sensitivities and reliability is the most crucial objective [88].

The 12 principles of GAC along with GC have been given comparatively in Fig. 2.5. Anastas and Warner [1] introduced 12 GC principles in 1998. Only a few of these ideas could be directly applied to analytical chemistry because they were created to satisfy the demands of synthetic chemistry. The following principles are those that are applicable including both analytical and synthesis purposes: Additionally, efforts were undertaken to identify the analytical ramifications of the 12 GC principles [89]. However, among the 12 principles, principle number two, the maximizing of atom-economy, seems insufficient to analytical chemistry. Additionally, the 12 principles put out by Anastas and Warner [21] should not include several crucial GAC principles. Taking this into account, we propose that the 12 guiding principles of GC be revised to

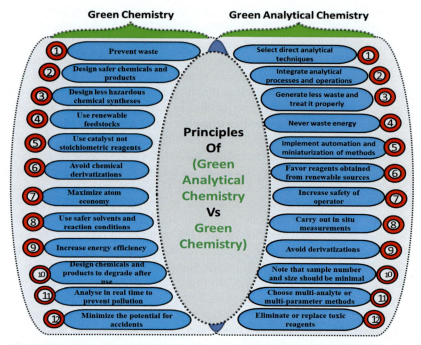

FIGURE 2.5 Comparison of the principles of green analytical chemistry versus green chemistry.

encompass analytical chemistry. Eight more principles that have significant implications for GAC were added to the four Anastas and Warner [1] principles that we included in our proposal.

4. Key steps used in sample reparation

Different important steps used in sample preparation techniques have been illustrated in Fig. 2.6. Sample collection is the initial stage in every analytical process. In some circumstances, analysis can be done immediately without having to bring the collected material to the device. The greenest method for collecting samples for this stage of the analytical process is known as in-line sampling [90]. The amount of time between collecting and receiving analytical data is significant for the environmentally friendly nature of the analytical process. In their study, Tobiszewski et al. [90] propose the removal of off-line sample collection and instead advocate for the use of on- and at-line sampling to be an additional sustainable alternative. Additional considerations, the energy consumption of equipment that requires a power supply is an important factor to take into account, collecting resources and promoting sustainability. The environmental impact of the process increases with the usage of more products. Preservation is necessary for maintaining the reliability of various types of samples, such as those related to the environment, biology, and food.

Principal concepts and guidelines for GAC **Chapter | 2** **25**

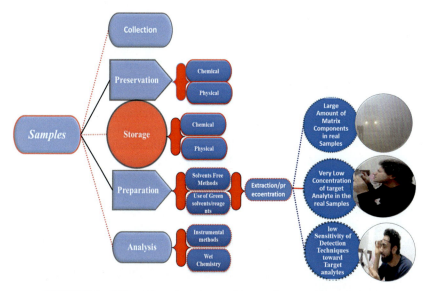

FIGURE 2.6 Different important steps used in sample preparation techniques.

The reason for this is that these samples may undergo various physical and chemical changes during the process of collection, transportation, storage, and preparation. Maintaining the credibility of the study as a whole is of utmost importance. One way to achieve this is by ensuring that the sample remains representative of the population. The appropriate methods of sample preservation depend on the type of sample being used and the specific analyte being tested [91]. Generally, these procedures can be divided into two groups: chemical methods (such as acidification, alkalization, and the addition of preservation agents) and physical methods (including cooling, freezing, and filtering). Maintaining samples during the analysis process is not environmentally friendly due to the energy and/or chemicals required. Ideally, the optimal approach for sample preservation in the green analytical method is to avoid it altogether. When beginning with an analytical method, it is crucial to consider the proper technique for sample movement. The procedure is not in line with the principles of GC because it consumes a significant amount of energy, especially in areas such as shipping and certain sample processing conditions like refrigeration. The reason for this is that GC was not developed until after the industrial revolution. If you use field analysis (also known as field screening), you have the option to skip this step, which offers several advantages. There are several benefits associated with this technology, including: (i) real-time analysis, (ii) minimal or no sample preparation required, (iii) portability of tools, (iv) reduced energy and chemical consumption, and (v) potential for profitability. It is recommended to minimize the distance between the collection site and the laboratory where the analysis will be conducted when transportation is needed. Sample storage is an optional step in an analytical method, which is usually determined based on the laboratory's analytical capacity. Samples can be stored at room temperature in a dark and well-ventilated area, or they can be stored under specific conditions such as freezing or cooling. Chemical preservation is essential for the long-term storage of samples. In terms of GAC, this option is the least favorable.

Sample preparation is an essential component in enhancing the environmental sustainability of an analytical procedure. The process involves multiple steps, some of which necessitate the use of environmentally harmful substances such as potent acids and organic solvents. There are analytical methods available that allow for optional or regular sample preparation. Dry ashing, digestion, extraction, postextraction cleanup, and derivatization are fundamental sample preparation methods commonly used in scientific research. The majority of analytical techniques used for measuring trace elements in solid samples necessitate the disintegration of the sample. Dry ashing and wet digestion are the two primary procedures employed for the disintegration of solid samples. Mineralization refers to the sequential procedure of dry ashing, followed by the subsequent dissolving of the resulting ash in acids. Dry ashing is a technique wherein a sample is subjected to elevated temperatures within a crucible positioned in a muffle furnace. After the process of ashing, the leftovers are dissolved in acids. Digestion is an alternative technique employed to decompose solid materials in order to ascertain the presence of trace elements. This process, similar to other approaches, necessitates the utilization of reagents, commonly mineral and oxidizing acids, as well as a substantial amount of energy. It is widely known that the most commonly used acids for the purpose of material degradation are hydrochloric, sulfuric, nitric, perchloric, and hydrofluoric (HF) acids. The idea is reinforced by the fact that each of these acids possesses a unique chemical structure [92,93]. They can be combined or used separately. Hot-plate or commercial sample-digestion machines are viable options for conducting acid digestion in open containers. Nevertheless, microwave digestion systems are commonly employed in laboratory settings to expedite and enhance the safety of the preparation process. Regrettably, the processes of dry ashing and digesting now lack viable means for enhancing their environmental sustainability. Moreover, it is recommended to employ HF acids for the analysis of various materials, including soil and geological samples. The most environmentally sustainable approach to mitigate the degradation of solid samples for trace element analysis is through the utilization of direct analytical procedures. Nevertheless, the applicability of these procedures may be limited in certain cases due to higher limits of detection (X-ray fluorescence) or another issue is the absence of matrix-matched standards and reference materials. One of the primary challenges encountered in the analysis of organic substances is the use of organic solvents, which can result in the generation of potentially hazardous waste materials. Derivatization, extraction, and cleansing procedures are three primary contributors to waste generation in the field of organic analysis. The avoidance of derivatization is crucial, as stated in the eighth principle of GC [94]. In certain cases, derivatization may present itself as a more favorable alternative to sample preservation [95]. Approximately 61% of the total time spent analyzing is dedicated to the preparation of organic component samples for assessment [31,96]. Additionally that consumes a significant quantity of chemicals. A wide array of options is accessible throughout this stage of the analytical procedure [97], each of the aforementioned factors ought to be employed in order to enhance the simplicity and accuracy of the study. The field of extraction has witnessed significant improvements throughout the realm of environmentally friendly analytical procedures. The optimal approach for sustainable sample preparation in the analysis of organic compounds involves the utilization of solventless procedures [98–101]. If the implementation of a solvent-free technique is not feasible, the environmentally friendly alternative would involve adopting a solvent-based reduction approach or substituting conventional organic solvents with environmentally sustainable solvents [50,102]. Ionic liquids, supercritical fluids, and subcritical water are widely acknowledged as "green solvents" due to their environmentally friendly properties. The main goal of conducting organic compound cleanup during the extraction process is to minimize the impact of interfering substances on the precise detection and analysis of the analyte. There are several different approaches that are based on various mechanisms [103]. The extraction of compounds is an additional step in an analytical framework that requires additional reagents and resources. As a result, this process generates more waste. In the field of GAC, it is crucial to carefully consider the possibility of removing this specific step, if it is determined to be possible.

It is important to remember that there is no one-size-fits-all sample preparation method that can be used in every situation. Instead, the process of sample preparation differs depending on factors such as the medium, constituents of the sample, and the technique used for measurement. Moreover, a technique that proves effective for one specific analyte may not be applicable for a broader chemical evaluation. In the last 10 years, there have been advancements in sample preparation methods. These methods have evolved from harsh techniques to more gentle approaches such as dissolving with ultrasound at room temperature or digesting with microwaves in a sealed container. These advancements have led to a simpler and safer procedure, especially when it comes to sample digestion and dissolution. The main objectives of analyte extraction are matrix isolation and analyte preconcentration. To achieve these goals, it is crucial to carefully select appropriate solvents and reagents, as well as effectively control the preconcentration process.

- Ensure that the analyte of interest is kept separated from the matrix;
- Enhance the concentration of the analyte of interest in the final solution for analysis, it is necessary to increase its amount; and
- Enhance the concentration of the analyte of interest in the final solution for analysis, it is necessary to increase its amount.

5. The principles of green sample preparation in the context of GAC

Sample preparation has served as the target of significant GAC research during the last 20 years because it represents the obstacle of analytical methods. It is important to note that the sample-preparation stage heavily influences the quality of the data acquired and serves as the principal cause of systematic mistakes and random inaccuracy in analytical procedures. In order to decrease possible impediments and matrix effects during the measurement step, the sample-treatment phase should ensure quantitative recovery of target contaminant while eliminating contaminants and enabling matrix separation as much as feasible [5]. Sample preparation is an important step in the analysis phase since it is necessary for the separating and enriching of target analytes, the removal or elimination of matrix interferences, and/or the compatibility of the testing instruments. As a result of the GAC highlighting sample preparation as being among the most important steps [104], mostly due to the customary high demands for solvents, sorbents, chemicals, acids, energy inputs, and other consumable materials or equipment. This is why the first GAC principle recommended eliminating sample preprocessing and adopting direct analysis techniques instead. However, it is not always possible to use direct analytical approaches in all situations [105–107], making the use of sample preparation techniques essential to overcoming analytical difficulties. Granted, the first GAC principle has contributed to the widespread misperception that skipping the sample preparation stage is a green strategy, completely ignoring the importance of this step and the advancements in technology in the field [88]. In this regard, a number of established and cutting-edge sample preparation techniques exist that do not harm the environment or human health and can result in more effective and metrologically better processes [108]. Until far, the 12 GAC principles have served as the foundation for metric instruments used to evaluate the sustainability of sample preparation techniques. However, the GAC approach's philosophical underpinnings make these metrics insufficient for delivering necessary levels of precision and specificity and therefore for measuring advancements in greening sample preparation. The necessity to construct a unique metric system for sample preparation is brought on by the vast range of factors that affect how environmentally friendly sample preparation is. The current effort intends to fill this gap by providing a robust yet user-friendly tool that will make it possible to evaluate the environmental effect of sample preparation, an important step in choosing green analytical procedures. The suggested metric tool emphasizes the importance of sample preparation and forecasts as well as finds areas that might be enhanced for greening the crucial phase of sample preparation.

Green sample preparation method offers comprehensive and systematic guidance for increasing the environmental friendliness of sample preparation techniques, then analysis techniques. The GSP and GAC techniques differ primarily in that the former has less emphasis on premeasurement preparation, while the latter has more emphasis on pretest preparation. All 10 GSP underlying principles are depicted in Fig. 2.7. They often focus primarily on sample processing and describe the purview of GSP, in contrasting to the GAC technique. They consolidate the necessary unique features to guarantee sustainability level in sample preparation and alleviate the adverse effects on the environment and public health by setting rules for solvents/reagents, substances, end up wasting, energy needs, frequency, miniaturization, processes rationalization, and operator security. The GSP approach additionally regards sample preparation to be a phase in the entire analytical chemical process, and it is linked to the sampling and measuring stages. GSP has aims that really are related to GAC's and moreover different in certain aspects. Comparing the GSP technique to the GAC method, the concept of sampling processing capacity called as the frequency during which samples get processed is included for the first time. This idea is linked to the amount of sampling produced in a given period (principle 6). Furthermore, the GSP approach acknowledges that solid

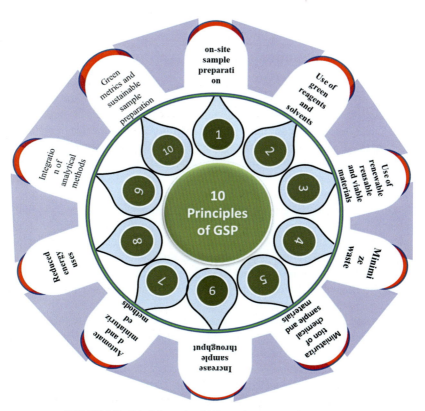

FIGURE 2.7 Principles and guidelines of green sample preparation.

residues (which including adsorbent in (micro) extraction methods) are commonly used during sample preparation and develops a basic concept (principle 3), which thus relates the technique's environmental quality to all these material properties seeming to be durable, obtained from renewable materials, and intended to be employed multiple times. According to mentioned in GSP guideline 2, the GAC method solely examines the environmental quality of substances (which including solvents, derivatization reagents, acidity and redox indications). The GAC method's description of in situ sample preparation varies from that of the GSP's principle 1. With the exception of sample preparation, sampling and measurement are related to in situ measurement in GAC. In contrast, GSP considers in situ sample preparation and connects it to in situ sampling and assessment. Furthermore, the GSP places a strong emphasis on sample preparation while also seeing it as an integral part of the overall analytical chemistry operation. In this context, especially in addition to in situ processes, GSP assesses the greenness of the postsample preparation setup for analysis (principle 9). Furthermore, it must be highlighted that in the GSP approach, these attributes seem obviously matched to the demands of sample preparation. The additional principles (principles 2, 4, 5, 7, 8, and 10) address features similar to those found in the GAC technique [88].

6. The main components of green analytical methods: Core principles of GAC

In order to identify a given analyte in distinct samples, modern analytical chemistry offers a variety of methods and tools. The main goals of "sustainable green" analysis techniques are: (1) eliminating or reducing the use of chemical substances; (2) reduction in energy use; (3) managing analytical discarded appropriately; and (4) promoting operator protection. The bulk of these obstacles call for cost-cutting solutions (such as sample size, reagents, energy, waste, danger, and hazard) that are described in Fig. 2.8.

One disadvantage of greening laboratory procedures is the necessity to find a balance among performance metrics and GAC needs. The majority of the analytical chemist suggestions suggested in our 12 principles may result in a loss in performance measures such as accuracy, precision, and sensitivity. When lowering sample size, using direct procedures, and miniaturizing devices, the trustworthiness of analytical methods can be easily questioned. However, rapid technology advancement and increased understanding of current difficulties would result in improvements in green analytical techniques. Usually, simple solutions may be used to fix these problems (for example, adjusting in-situ assessments by passing standard between samples to enhance standardization) [109]. Chemical evaluation is a difficult practice with several steps that could also utilize eco-friendly substitutes. This is crucial to evaluate how each analytical method's procedures and operations adhere to the principles of GC. Recently [12,50], Analytical Eco-Scale, a sustainable analytic evaluating

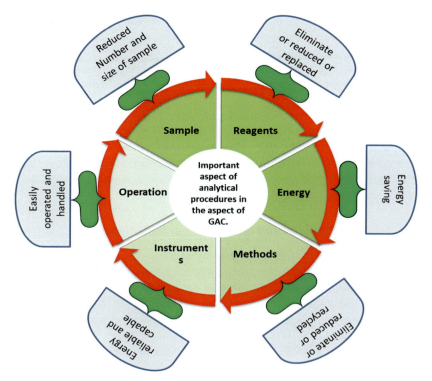

FIGURE 2.8 Important aspects of analytical procedures in the context of green analytical chemistry (GAC).

instrument, has been made available. for identifying and fixing the method's weakest link. Regardless of the analytical methodologies and analytes used, every analysis includes a sample and an operator. The majority of approaches generally produce analytical output and require the employment of at minimum a single chemical. This evaluation presents certain useful recommendations for the implementation of environmentally friendly practices utilizing a variety of implementations (environmental, food, medicinal, commercial, and biomedicine). Utmost newly established green analytical processes [39,110,111] are distinguished not only by their environmental friendliness but also by their comparatively high sensitivity, cheap analysis cost, decreased energy usage, simplicity, and time efficiency. As a result, it is extremely probable that certain conventional procedures will be phased out in favor of these new green ones.

6.1 Sample collection

Every analysis starts with sampling, excluding the use of direct analytical methods. The second principle of GAC is that samples should be small in both quantity and size. This approach is especially crucial for environmental sample analysis since it makes it simple to change the sampling technique. The number of samples can be decreased using one of two methods. The first entails the use of noninvasive techniques, such as various geophysical tools (e.g., portable XRF), or equipment (e.g., remote sensing, seismic, and magnetic surveys) for field screening and site selection for a thorough chemical investigation [112,113]. The second is to utilize statistics to pick sampling sites that will yield the most data for accurate results interpretation with the smallest possible sample size. Miniaturized analytical techniques [39] or other sample introduction technologies can be used to reduce sample size. Particularly when sampling heterogeneous materials, reductions in sample quantity and volume should be made with caution and consideration for the likelihood of sample representativeness loss [114,115].

6.2 Analytical process and quantification techniques

It is crucial for GAC to choose the greenest approach out of all the options that match the user's requirements (such as detection limits and precision). Due to this, the technique is mentioned in five of the 12 GAC principles. The greenest choice is ideally direct, automated, and miniaturized analytical approaches (principles number 1, 3, and 5). Integrating of analytical methods and activities (principle number 4) is particularly important since it allows all information to be obtained from a single analysis. Microdevices that execute different processes of chemical analysis (e.g., mixing, reaction, and separation) are the greatest examples of analytical system integration [116]. It was demonstrated that this rapid, easy method competes favorably with conventional analytical methods, takes little sample and reagent, and has good resolution and sensitivity. Additionally, it is frequently essential to identify many analytes or parameters in a single sample. In this situation, multiple-analyte or multiple-parameter approaches ought to be used (principle number 8). These techniques are particularly crucial in environmental monitoring when more than one analyte needs to be identified, such as when identifying potentially dangerous trace substances. A multielement approach will be more appropriate than time- and reagent-consuming procedures. Since derivatization and other extra stages in analysis require reagents and produce waste, it is better to eliminate them (principle number 6). In the event that derivatization is unavoidable, it may be combined with additional procedures (such as filtering and extraction) in a single system [117]. The greenest methods are frequently thought to involve the use of portable devices, tiny systems, and remote sensing [118–120]. Each analytical approach has unique criteria and issues that define it. As a result, it is crucial to assess, enhance, and concentrate on an analytical methodology or method's least green aspects. There are excellent illustrations of reviews on green spectroscopy, green electrochemistry, and green chromatography in the GAC literature [106,121–123]. Some of the GAC evaluations that have been published are focused to a particular analytical phase, particularly sample treatment [109]. The methodology and the tool are intertwined. In general, instrumental methods are favored for GAC, however certain traditional wet chemical techniques may be just as environmentally friendly as instrumental by titration, electrogravimetry, and AAS. The issue that has to be addressed in GAC is the significant ecological impact that is created during the development and upkeep of analytical instruments. This issue may be properly comprehended and resolved by examining the analytical instrument life cycle. Automated and miniature instruments promote operator safety (GAC principle number 12), decrease waste generation and reagent use, and support effective waste management (GAC concept 7) (GAC principle number 7) [124]. Miniaturized total analysis systems (lTAS) are the greatest example of miniaturization and integration of processes and operations (GAC principles four and 5), with sample treatment and measurement positioned extremely near to each other. When these systems were first developed in 1990 [125], they immediately gained popularity and were largely used in clinical/bioanalytical and environmental laboratories [126]. Another distinct benefit of lTAS is the use of small sample and reagent quantities (between pL and nL), which is advantageous for GAC (GAC principle number 2).

6.3 Reagents safety and secure operation

Eliminating or substitution of hazardous, prolonged chemicals with bioaccumulative qualities with less toxic alternatives is a significant trend in GAC (Principles no 11). Organic solvents are used to replace the bulk of reagents used in analytical processes. Alternative (green) solvents include ionic liquids and supercritical fluids. Numerous chromatographic and electromigration techniques, electrochemistry, and spectroscopic all make use of various ILs, notably those based on quaternary nitrogen cations [127,128]. ILs are alternative extraction solvents and can be used as stationary phases in gas chromatography [129,130]. Standard solvent extraction may be replaced by supercritical fluid extraction (SFE) and subcritical water extraction (SWE), both of which are environmentally friendly. SFE has grown into one of the foremost popular sample-preparation methods in chromatography due to the advantages of supercritical CO_2 as an alternative solvent [131]. SFC and packed column SFC have been created as a consequence of the utilization of SCF in chromatography [132]. The using chemicals sourced from renewable sources for assessment is among the most prominent GAC advancements (GAC principle number 10). The extract of many organisms, including plants, animals, and bacteria, may be used in biosensing, spectrometry, fluorimetry, acidity monitors, and redox indexes [133]. Although they have limitations in quantification, natural green chemicals have the potential to explain analytical chemistry [134]. Reducing the use of nonrenewable resources and increasing the use of renewables are priority for sustainable development. This should encourage the development of analytical techniques using natural reagents. Improved operator safety is facilitated by safer (less toxic, natural) reagents (GAC principle number 12). The employment of automated instrumental procedures, miniature systems, and online decontamination of analytical waste, however, also ensures lowering the danger for the operator.

6.4 Waste generation

Different analytical techniques yield different amounts of waste. The greenest techniques are those that produce little or no waste at all. In general, the amount of analytical waste increases with the number of stages in an analytical technique and the amount of reagents used. Therefore, reducing the usage of reagents using the methods covered above helps to reduce the amount of waste produced. Making sure analytical waste is treated properly is a crucial problem as well (GAC principle number 7). Recycling, degradation, and passivation of trash can lower the toxicity of waste; these processes are best carried out online [6].

7. Conclusion

GAC metrics offer the chance to evaluate various factors and analytical methods in order to identify the less environmentally friendly aspects; this might yet be developed to satisfy the needs of GAC. The GAC methodology covers all stages of the analytical process, including sample handling sample treatment, the choice and use of solvents and reagents, instrumentation, and waste generation. One advantage of GAC is its ability to evaluate the entire analytical approach and provide a comprehensive assessment of the environmental sustainability of different stages. The AGREE tool assesses analytical methodologies using the 12 principles set by the GAC. It presents a visual representation in the form of a greener. The establishment of a framework for GSP is essential to ensure that this process is not overlooked or disregarded. The objective of this study is to provide a structured framework for the gradual advancement of sample preparation techniques that are environmentally friendly. The research mentioned above explains the concepts of sample preparation in a clear and easy-to-understand way, in order to achieve this goal. The promotion of in situ over ex-situ sample preparation, the abolition, regeneration, or reduction of toxic solvents and reagents, the advancement of the usage of recyclable, ecological, and recyclable resources, and operator safety are just a few of the important factors considered when attempting to decrease the ecological effect of sample processing processes. Furthermore, the GSP aims to optimize resource utilization by generating reliable analytical data with the least amount of resources possible. It also strives to enhance sample velocity and reduce energy consumption and waste production. In addition to downsizing, the GSP strategy actively promotes the implementation of supporting strategies, including mobility, automation, and phase integration. Due to advancements in sample preparation, various techniques now offer the potential for developing more environmentally friendly procedures for sample preparation. These techniques also ensure that high extraction efficiencies are maintained. In recent years, significant advancements have been made in the field of sample preparation, providing analytical chemists with a wide range of diverse methods to choose from. These advancements have proven to be substantial and highly beneficial for researchers in this field. Choosing the appropriate instrument for a given task is one of the most crucial aspects of sample preparation and analytical techniques. The environmental friendliness of sample preparation processes is assessed using a set of 12 principles known as the GAC principles. Analysts have a social responsibility to engage in environmentally friendly sample preparation. These strategies promote environmental sustainability and help reduce pollution. It is important for all researchers, practitioners, and routine analyzers to take steps to minimize the environmental impact of sample preparation. This is crucial to safeguard both human health and the natural world. The act of preparing a sample is commonly known as "green sample preparation." The method of sample preparation is not entirely original but rather serves as a guiding principle for sustainability.

Acknowledgments

Naeemullah would like to express his sincere appreciation to TUBITAK for granting him the opportunity to participate in the "2221-Visiting Scientists Fellowship Program 2021/9" as a visiting scientist. Dr. M. Tuzen would like to express gratitude to the Turkish Academy of Sciences (TUBA) for their generous support.

References

[1] P.T. Anastas, J.C. Warner, Principles of green chemistry, Green Chemistry: Theory and Practice 29 (1998).
[2] P.T. Anastas, Green chemistry and the role of analytical methodology development, Critical Reviews in Analytical Chemistry 29 (3) (1999) 167—175.
[3] M. Stoeppler (Ed.), Sampling and Sample Preparation: Practical Guide for Analytical Chemists, Springer Science & Business Media, 2012.
[4] S. Armenta, S. Garrigues, De la Guardia, Analytical Chemistry 27 (2008) 497.
[5] S. Armenta, S. Garrigues, M. de la Guardia, The role of green extraction techniques in green analytical chemistry, TrAC, Trends in Analytical Chemistry 71 (2015) 2—8.
[6] S. Garrigues, S. Armenta, M. de la Guardia, Green strategies for decontamination of analytical wastes, TrAC, Trends in Analytical Chemistry 29 (7) (2010) 592—601.
[7] S. Armenta, S. Garrigues, M. de la Guardia, Green analytical chemistry, TrAC, Trends in Analytical Chemistry 27 (6) (2008) 497—511.
[8] H. Lord, J. Pawliszyn, Evolution of solid-phase microextraction technology, Journal of Chromatography A 885 (1—2) (2000) 153—193.
[9] M. Polet, W. Van Gansbeke, P. Van Eenoo, Development and validation of an open screening method for doping substances in urine by gas chromatography quadrupole time-of-flight mass spectrometry, Analytica Chimica Acta 1042 (2018) 52—59.

[10] B. Greer, O. Chevallier, B. Quinn, L.M. Botana, C.T. Elliott, Redefining dilute and shoot: the evolution of the technique and its application in the analysis of foods and biological matrices by liquid chromatography mass spectrometry, TrAC, Trends in Analytical Chemistry 141 (2021) 116284.

[11] M. Fresnais, W.E. Haefeli, J. Burhenne, R. Longuespée, Rapid drug detection in whole blood droplets using a desorption electrospray ionization static profiling approach—a proof-of-concept, Rapid Communications in Mass Spectrometry 34 (6) (2020) e8614.

[12] S. Armenta, F.A. Esteve-Turrillas, S. Garrigues, M. de la Guardia, Alternative green solvents in sample preparation, Green Analytical Chemistry 1 (2022) 100007.

[13] Y. Shen, B. Chen, T.A. van Beek, Alternative solvents can make preparative liquid chromatography greener, Green Chemistry 17 (7) (2015) 4073—4081.

[14] P.T. Anastas, L.B. Bartlett, M.M. Kirchhoff, T.C. Williamson, The role of catalysis in the design, development, and implementation of green chemistry, Catalysis Today 55 (1—2) (2000) 11—22.

[15] M. Tobiszewski, A. Mechlińska, J. Namieśnik, Green analytical chemistry—theory and practice, Chemical Society Reviews 39 (8) (2010) 2869—2878.

[16] J. Płotka-Wasylka, A. Gałuszka, J. Namieśnik, Green analytical chemistry: summary of existing knowledge and future trends, in: Green Analytical Chemistry, Springer, Singapore, 2019, pp. 431—449.

[17] R. Marcinkowska, J. Namieśnik, M. Tobiszewski, Green and equitable analytical chemistry, Current Opinion in Green and Sustainable Chemistry 19 (2019) 19—23.

[18] M. Tobiszewski, M. Marć, A. Gałuszka, J. Namieśnik, Green chemistry metrics with special reference to green analytical chemistry, Molecules 20 (6) (2015) 10928—10946.

[19] J. Płotka-Wasylka, A. Kurowska-Susdorf, M. Sajid, M. de la Guardia, J. Namieśnik, M. Tobiszewski, Green chemistry in higher education: state of the art, challenges, and future trends, ChemSusChem 11 (17) (2018) 2845—2858.

[20] J. Płotka-Wasylka, M. Fabjanowicz, K. Kalinowska, J. Namieśnik, History and milestones of green analytical chemistry, in: Green Analytical Chemistry, Springer, Singapore, 2019, pp. 1—17.

[21] P.T. Anastas, J.C. Warner, Green Chemistry: Theory and Practice, 30, Oxford University Press, New York, 1998.

[22] J. Płotka-Wasylka, H.M. Mohamed, A. Kurowska-Susdorf, R. Dewani, M.Y. Fares, V. Andruch, Green analytical chemistry as an integral part of sustainable education development, Current Opinion in Green and Sustainable Chemistry 31 (2021) 100508.

[23] S. Parimoo, B. Parimoo, Sample preparation in DNA analysis, Sample Preparation Techniques in Analytical Chemistry 162 (2003) 271—300.

[24] F. Pena-Pereira, W. Wojnowski, M. Tobiszewski, AGREE—analytical GREEnness metric approach and software, Analytical Chemistry 92 (14) (2020) 10076—10082.

[25] M. De la Guardia, S. Armenta, Green analytical chemistry (Cap. I. Origins of green analytical chemistry), Comprehensive Analytical Chemistry 57 (2011).

[26] S. Armenta, S. Garrigues, F.A. Esteve-Turrillas, M. de la Guardia, Green extraction techniques in green analytical chemistry, TrAC, Trends in Analytical Chemistry 116 (2019) 248—253.

[27] J. Liu, DNA-stabilized, fluorescent, metal nanoclusters for biosensor development, TrAC, Trends in Analytical Chemistry 58 (2014) 99—111.

[28] M. Sajid, M. Khaled Nazal, M. Rutkowska, N. Szczepańska, J. Namieśnik, J. Płotka-Wasylka, Solid phase microextraction: apparatus, sorbent materials, and application, Critical Reviews in Analytical Chemistry 49 (3) (2019) 271—288.

[29] H. Malissa, Changes of paradigms in analytical chemistry, in: E. Roth (Ed.), Reviews on Analytical Chemistry, Les Editions de Physique, Paris, 1988. Euroanalysis VI.
[30] L.H. Keith, L.U. Gron, J.L. Young, Green analytical methodologies, Chemical Reviews 107 (6) (2007) 2695−2708.
[31] P.T. Anastas, J.C. Warner, Green chemistry, Frontiers 640 (1998) 1998.
[32] L.A. Paquette, P. Anastas, T. Williamson, Green Chemistry: Frontiers in Benign Chemical Synthesis and Processing, 1998.
[33] M. Koel, Do we need green analytical chemistry? Green Chemistry 18 (4) (2016) 923−931.
[34] S. Garrigues, M. de la Guardia (Eds.), Challenges in Green Analytical Chemistry, 66, Royal Society of Chemistry, 2020.
[35] M. de la Guardia, S. Garrigues (Eds.), Handbook of Green Analytical Chemistry, John Wiley & Sons, 2012.
[36] M. Koel, M. Kaljurand, Green Analytical Chemistry, second ed., Royal Society of Chemistry, 2019.
[37] M. de la Guardia, S. Garrigues, Past, Present and Future of Green Analytical Chemistry, 2020.
[38] A.C. Duarte, Specialty grand challenges in environmental analytical methods, Frontiers in Environmental Chemistry 1 (2020) 4.
[39] A. Gałuszka, Z. Migaszewski, J. Namieśnik, The 12 principles of green analytical chemistry and the SIGNIFICANCE mnemonic of green analytical practices, TrAC, Trends in Analytical Chemistry 50 (2013) 78−84.
[40] D.L. Rocha, A.D. Batista, F.R. Rocha, G.L. Donati, J.A. Nóbrega, Greening sample preparation in inorganic analysis, TrAC, Trends in Analytical Chemistry 45 (2013) 79−92.
[41] US National Water Quality Monitoring Council, National Environmental Methods Index (NEMI), 2002. Available from:. (Accessed 14 December 2014).
[42] USEPA, Chemical Hazard Evaluation Management Strategies: A Method for Ranking and Scoring Chemicals by Potential Human Health and Environmental Impacts, BiblioGov, 2012.
[43] J. Płotka-Wasylka, A new tool for the evaluation of the analytical procedure: Green Analytical Procedure Index, Talanta 181 (2018) 204−209.
[44] M. Tobiszewski, W. Zabrocka, M. Bystrzanowska, Diethyl carbonate as a green extraction solvent for chlorophenol determination with dispersive liquid−liquid microextraction, Analytical Methods 11 (6) (2019) 844−850.
[45] Y. Gaber, U. Törnvall, M.A. Kumar, M.A. Amin, R. Hatti-Kaul, HPLC-EAT (environmental assessment tool): a tool for profiling safety, health and environmental impacts of liquid chromatography methods, Green Chemistry 13 (8) (2011) 2021−2025.
[46] I.V. Muralikrishna, V. Manickam, Life cycle assessment, in: Environmental Management, Elsevier, 2017, pp. 57−75.
[47] National Environmental Methods Index, 2023. https://www.nemi.gov/about/. (Accessed 28 July 2021).
[48] J.P. Brans, P. Vincke, Note—A preference ranking organisation method: (the PROMETHEE method for multiple criteria decision-making), Management Science 31 (6) (1985) 647−656.
[49] S. Nair, J.V. Gohel, A review on contemporary hole transport materials for perovskite solar cells, Nanotechnology for Energy and Environmental Engineering (2020) 145−168.

[50] A. Gałuszka, Z.M. Migaszewski, P. Konieczka, J. Namieśnik, Analytical eco-scale for assessing the greenness of analytical procedures, TrAC, Trends in Analytical Chemistry 37 (2012) 61−72.

[51] D. Gallart-Mateu, M.L. Cervera, S. Armenta, M. de la Guardia, The importance of incorporating a waste detoxification step in analytical methodologies, Analytical Methods 7 (13) (2015) 5702−5706.

[52] A. Ballester-Caudet, P. Campíns-Falcó, B. Pérez, R. Sancho, M. Lorente, G. Sastre, C. González, A new tool for evaluating and/or selecting analytical methods: summarizing the information in a hexagon, TrAC, Trends in Analytical Chemistry 118 (2019) 538−547.

[53] H. Al-Hazmi, J. Namiesnik, M. Tobiszewski, Application of TOPSIS for selection and assessment of analytical procedures for ibuprofen determination in wastewater, Current Analytical Chemistry 12 (4) (2016) 261−267.

[54] P.M. Nowak, P. Kościelniak, What color is your method? Adaptation of the RGB additive color model to analytical method evaluation, Analytical Chemistry 91 (16) (2019) 10343−10352.

[55] C. Jiménez-González, P. Poechlauer, Q.B. Broxterman, B.S. Yang, D. Am Ende, J. Baird, J. Manley, Key green engineering research areas for sustainable manufacturing: a perspective from pharmaceutical and fine chemicals manufacturers, Organic Process Research & Development 15 (4) (2011) 900−911.

[56] R.A. Sheldon, Fundamentals of green chemistry: efficiency in reaction design, Chemical Society Reviews 41 (4) (2012) 1437−1451.

[57] R.A. Sheldon, Organic synthesis-past, present and future, Chemistry and Industry (23) (1992) 903−906.

[58] D. Frey, C. Claeboe, J.,L. Brammer, Toward a 'reagent-free' synthesis, Green Chemistry 1 (2) (1999) 57−59.

[59] A.D. Curzons, D.J. Constable, D.N. Mortimer, V.L. Cunningham, So you think your process is green, how do you know?—Using principles of sustainability to determine what is green—a corporate perspective, Green Chemistry 3 (1) (2001) 1−6.

[60] M. Sajid, J. Płotka-Wasylka, Green analytical chemistry metrics: a review, Talanta 238 (2022) 123046.

[61] L.A. Currie, Nomenclature in evaluation of analytical methods including detection and quantification capabilities (IUPAC recommendations 1995), Pure and Applied Chemistry 67 (10) (1995) 1699−1723.

[62] I.M. Krishna, V. Manickam, A. Shah, N. Davergave, Environmental Management: Science and Engineering for Industry, Butterworth-Heinemann, 2017.

[63] J.R. Siche, F. Agostinho, E. Ortega, A. Romeiro, Sustainability of nations by indices: comparative study between environmental sustainability index, ecological footprint and the emergy performance indices, Ecological Economics 66 (4) (2008) 628−637.

[64] B. Uzun, A. Almasri, D. Uzun Ozsahin, Preference ranking organization method for enrichment evaluation (PROMETHEE), in: Application of Multi-Criteria Decision Analysis in Environmental and Civil Engineering, Springer, Cham, 2021, pp. 37−41.

[65] M. de La Guardia, J. Ruzicka, Guest editorial. Towards environmentally conscientious analytical chemistry through miniaturization, containment and reagent replacement, Analyst 120 (2) (1995), 17N-17N.

[66] R. Louw, Benign by Design, Alternative Synthetic Design for Pollution Prevention PT Anastas and CA Farris ACS, Washington DC, 1994, Series No. 577 xi+ 195 pp. $59.95 ISBN 0-8412-3053-6, 1996.

[67] W. Abdussalam-Mohammed, A.Q. Ali, A.O. Errayes, Green chemistry: principles, applications, and disadvantages, Chemical Methodologies 4 (2020) 408−423.
[68] E. Yilmaz, M. Soylak, Latest trends, green aspects, and innovations in liquid-phase−based microextraction techniques: a review, Turkish Journal of Chemistry 40 (6) (2016) 868−893.
[69] Y. Gao, Z. Shi, Z. Long, P. Wu, C. Zheng, X. Hou, Determination and speciation of mercury in environmental and biological samples by analytical atomic spectrometry, Microchemical Journal 103 (2012) 1−14.
[70] F. Pena-Pereira, I. Lavilla, C. Bendicho, Liquid-phase microextraction techniques within the framework of green chemistry, TrAC, Trends in Analytical Chemistry 29 (7) (2010) 617−628.
[71] M. Farré, S. Pérez, C. Gonçalves, M.F. Alpendurada, D. Barceló, Green analytical chemistry in the determination of organic pollutants in the aquatic environment, TrAC, Trends in Analytical Chemistry 29 (11) (2010) 1347−1362.
[72] C.J. Welch, N. Wu, M. Biba, R. Hartman, T. Brkovic, X. Gong, L. Zhou, Greening analytical chromatography, TrAC, Trends in Analytical Chemistry 29 (7) (2010) 667−680.
[73] V. Jalili, A. Barkhordari, A. Ghiasvand, A comprehensive look at solid-phase microextraction technique: a review of reviews, Microchemical Journal 152 (2020) 104319.
[74] J. Płotka-Wasylka, N. Szczepańska, M. de La Guardia, J. Namieśnik, Modern trends in solid phase extraction: new sorbent media, TrAC, Trends in Analytical Chemistry 77 (2016) 23−43.
[75] G.A. Price, D. Mallik, M.G. Organ, Process analytical tools for flow analysis: a perspective, Journal of Flow Chemistry 7 (3−4) (2017) 82−86.
[76] J. Moros, S. Garrigues, M. de la Guardia, Vibrational spectroscopy provides a green tool for multi-component analysis, TrAC, Trends in Analytical Chemistry 29 (7) (2010) 578−591.
[77] M. Herrero, M. Castro-Puyana, J.A. Mendiola, E. Ibañez, Compressed fluids for the extraction of bioactive compounds, TrAC, Trends in Analytical Chemistry 43 (2013) 67−83.
[78] J. Płotka-Wasylka, M. Rutkowska, K. Owczarek, M. Tobiszewski, J. Namieśnik, Extraction with environmentally friendly solvents, TrAC, Trends in Analytical Chemistry 91 (2017) 12−25.
[79] M. Koel, M. Kaljurand, Editorial overview: a closer look on green developments in analytical chemistry: green analytical chemistry is going mainstream, Current Opinion in Green and Sustainable Chemistry 31 (2021) 100541.
[80] P. Yáñez-Sedeño, S. Campuzano, J.M. Pingarrón, Electrochemical (bio) sensors: promising tools for green analytical chemistry, Current Opinion in Green and Sustainable Chemistry 19 (2019) 1−7.
[81] L. Montero, J.F. García-Reyes, B. Gilbert-López, Environmentally Friendly Solvents for Sample Preparation in Foodomics, 2021.
[82] C. Turner, Sustainable analytical chemistry—more than just being green, Pure and Applied Chemistry 85 (12) (2013) 2217−2229.
[83] M.A. Korany, H. Mahgoub, R.S. Haggag, M.A. Ragab, O.A. Elmallah, Green chemistry: analytical and chromatography, Journal of Liquid Chromatography & Related Technologies 40 (16) (2017) 839−852.
[84] A.A. Aly, T. Górecki, Green approaches to sample preparation based on extraction techniques, Molecules 25 (7) (2020) 1719.

[85] A. Gałuszka, Z. Migaszewski, J. Namieśnik, The 12 principles of green analytical chemistry and the SIGNIFICANCE mnemonic of green analytical practices, Trends in Analytical Chemistry (Reference Ed.) 50 (2013) 78–84, https://doi.org/10.1016/J.TRAC.2013.04.010.

[86] R. Hartman, R. Helmy, M. Al-Sayah, C.J. Welch, Analytical method volume intensity (AMVI): a green chemistry metric for HPLC methodology in the pharmaceutical industry, Green Chemistry 13 (2011) 934–939, https://doi.org/10.1039/C0GC00524J.

[87] P.I. Napolitano-Tabares, I. Negrín-Santamaría, A. Gutiérrez-Serpa, V. Pino, Recent efforts to increase greenness in chromatography, Current Opinion in Green and Sustainable Chemistry 32 (2021) 100536.

[88] Á.I. López-Lorente, F. Pena-Pereira, S. Pedersen-Bjergaard, V.G. Zuin, S.A. Ozkan, E. Psillakis, The ten principles of green sample preparation, TrAC, Trends in Analytical Chemistry (2022) 116530.

[89] M. de la Guardia, S. Garrigues, An ethical commitment and an economic opportunity, in: Challenges in Green Analytical Chemistry, Royal Society of Chemistry, 2011, pp. 1–12.

[90] M. Tobiszewski, A. Mechlińska, B. Zygmunt, J. Namieśnik, Green analytical chemistry in sample preparation for determination of trace organic pollutants, TrAC, Trends in Analytical Chemistry 28 (8) (2009) 943–951.

[91] S. Mitra, Sample Preparation Techniques in Analytical Chemistry, 2003.

[92] W. Wardencki, R.J. Katulski, J. Stefański, J. Namieśnik, The state of the art in the field of non-stationary instruments for the determination and monitoring of atmospheric pollutants, Critical Reviews in Analytical Chemistry 38 (4) (2008) 259–268.

[93] J.R. Dean, Methods for Environmental Trace Analysis 12, John Wiley and Sons, 2003.

[94] E.S. Beach, Z. Cui, P.T. Anastas, Green chemistry: a design framework for sustainability, Energy & Environmental Science 2 (10) (2009) 1038–1049.

[95] J. Namieśnik, P. Szefer, Preparing samples for analysis-the key to analytical success, Ecological Chemistry and Engineering S 15 (2) (2008) 167–244.

[96] W. Kamm, F. Dionisi, C. Hischenhuber, H.G. Schmarr, K.H. Engel, Rapid detection of vegetable oils in milk fat by on-line LC-GC analysis of β-sitosterol as marker, European Journal of Lipid Science and Technology 104 (11) (2002) 756–761.

[97] T. Hyötyläinen, Critical evaluation of sample pretreatment techniques, Analytical and Bioanalytical Chemistry 394 (3) (2009) 743–758.

[98] C. Nerín, J. Salafranca, M. Aznar, R. Batlle, Critical review on recent developments in solventless techniques for extraction of analytes, Analytical and Bioanalytical Chemistry 393 (3) (2009) 809–833.

[99] M. Urbanowicz, B. Zabiegała, J. Namieśnik, Solventless sample preparation techniques based on solid- and vapour-phase extraction, Analytical and Bioanalytical Chemistry 399 (1) (2011) 277–300.

[100] J. Curyło, W. Wardencki, J. Namieśnik, Green aspects of sample preparation–a need for solvent reduction, Polish Journal of Environmental Studies 16 (1) (2007).

[101] J. Namieśnik, W. Wardencki, Solventless sample preparation techniques in environmental analysis, Journal of High Resolution Chromatography 23 (4) (2000) 297–303.

[102] R. Majors, D. Raynie, The greening of the chromatography laboratory, LCGC North America 29 (2) (2011) 118–134.

[103] C. Zhang, Fundamentals of Environmental Sampling and Analysis, John Wiley & Sons, 2007.

[104] F. Chemat, M. Abert-Vian, A.S. Fabiano-Tixier, J. Strube, L. Uhlenbrock, V. Gunjevic, G. Cravotto, Green extraction of natural products. Origins, current status, and future challenges, TrAC, Trends in Analytical Chemistry 118 (2019) 248−263.

[105] M. Khanmohammadi, A.B. Garmarudi, Infrared spectroscopy provides a green analytical chemistry tool for direct diagnosis of cancer, TrAC, Trends in Analytical Chemistry 30 (6) (2011) 864−874.

[106] M. Tobiszewski, J. Namieśnik, Direct chromatographic methods in the context of green analytical chemistry, TrAC, Trends in Analytical Chemistry 35 (2012) 67−73.

[107] C. Bendicho, I. Lavilla, F. Pena-Pereira, V. Romero, Green chemistry in analytical atomic spectrometry: a review, Journal of Analytical Atomic Spectrometry 27 (11) (2012) 1831−1857.

[108] F. Pena-Pereira, I. Lavilla, C. Bendicho, Greening sample preparation: an overview of cutting-edge contributions, Current Opinion in Green and Sustainable Chemistry 30 (2021) 100481.

[109] C. Bendicho, I. De La Calle, F. Pena, M. Costas, N. Cabaleiro, I. Lavilla, Ultrasound-assisted pretreatment of solid samples in the context of green analytical chemistry, TrAC, Trends in Analytical Chemistry 31 (2012) 50−60.

[110] J.A. Rather, K. De Wael, Fullerene-C60 sensor for ultra-high sensitive detection of bisphenol-A and its treatment by green technology, Sensors and Actuators B: Chemical 176 (2013) 110−117.

[111] L. Mirmoghtadaie, A.A. Ensafi, M. Kadivar, P. Norouzi, Highly selective electrochemical biosensor for the determination of folic acid based on DNA modified-pencil graphite electrode using response surface methodology, Materials Science and Engineering: C 33 (3) (2013) 1753−1758.

[112] T. Missiaen, M. Söderström, I. Popescu, P. Vanninen, Evaluation of a chemical munition dumpsite in the Baltic Sea based on geophysical and chemical investigations, Science of the Total Environment 408 (17) (2010) 3536−3553.

[113] V. Balaram, Rare earth elements: a review of applications, occurrence, exploration, analysis, recycling, and environmental impact, Geoscience Frontiers 10 (4) (2019) 1285−1303.

[114] Z.M. Migaszewski, A. Gałuszka, P. Pasławski, The use of the barbell cluster ANOVA design for the assessment of environmental pollution: a case study, Wigierski National Park, NE Poland, Environmental Pollution 133 (2) (2005) 213−223.

[115] C.M. Brett, Novel sensor devices and monitoring strategies for green and sustainable chemistry processes, Pure and Applied Chemistry 79 (11) (2007) 1969−1980.

[116] O. Chailapakul, S. Korsrisakul, W. Siangproh, K. Grudpan, Fast and simultaneous detection of heavy metals using a simple and reliable microchip-electrochemistry route: an alternative approach to food analysis, Talanta 74 (4) (2008) 683−689.

[117] Y. Liu, C.D. Garcia, C.S. Henry, Recent progress in the development of μTAS for clinical analysis, Analyst 128 (8) (2003) 1002−1008.

[118] S. Garrigues, M. de la Guardia, Non-invasive analysis of solid samples, TrAC, Trends in Analytical Chemistry 43 (2013) 161−173.

[119] M. Kaljurand, M. Koel, Green bioanalytical chemistry, Bioanalysis 4 (11) (2012) 1271−1274.

[120] J. Wang, Real-time electrochemical monitoring: toward green analytical chemistry, Accounts of Chemical Research 35 (9) (2002) 811−816.

[121] Y. He, L. Tang, X. Wu, X. Hou, Y.I. Lee, Spectroscopy: the best way toward green analytical chemistry? Applied Spectroscopy Reviews 42 (2) (2007) 119−138.

[122] P. Sandra, G. Vanhoenacker, F. David, K. Sandra, A. Pereira, Green Chromatography (Part 1): Introduction and Liquid Chromatography, 2010.
[123] C. Brunelli, A. Pereira, M. Dunkle, F. David, P. Sandra, Green Chromatography (Part 2): The Role of GC and SFC, 2010.
[124] M. de la Guardia, S. Garrigues, The concept of green analytical chemistry, Handbook of Green Analytical Chemistry (2012) 1−16.
[125] A. Manz, N. Graber, H.Á. Widmer, Miniaturized total chemical analysis systems: a novel concept for chemical sensing, Sensors and Actuators B: Chemical 1 (1−6) (1990) 244−248.
[126] A. Ríos, A. Escarpa, M.C. González, A.G. Crevillén, Challenges of analytical microsystems, TrAC, Trends in Analytical Chemistry 25 (5) (2006) 467−479.
[127] M. Kaljurand, M. Koel, Recent advancements on greening analytical separation, Critical Reviews in Analytical Chemistry 41 (1) (2011) 2−20.
[128] J.F. Liu, G.B. Jiang, J.Å. Jönsson, Application of ionic liquids in analytical chemistry, TrAC, Trends in Analytical Chemistry 24 (1) (2005) 20−27.
[129] P. Sun, D.W. Armstrong, Ionic liquids in analytical chemistry, Analytica Chimica Acta 661 (1) (2010) 1−16.
[130] H. Zhao, S. Xia, P. Ma, Use of ionic liquids as 'green' solvents for extractions, Journal of Chemical Technology & Biotechnology: International Research in Process, Environmental & Clean Technology 80 (10) (2005) 1089−1096.
[131] J. Ding, T. Welton, D.W. Armstrong, Chiral ionic liquids as stationary phases in gas chromatography, Analytical Chemistry 76 (22) (2004) 6819−6822.
[132] Y. Hsieh, Supercritical fluids and green bioanalysis, Bioanalysis 2 (1) (2010) 1−4.
[133] I. Brondz, Yesterday, today and tomorrow of supercritical fluid extraction and chromatography, American Journal of Analytical Chemistry 3 (12A) (2012) 867.
[134] K. Grudpan, S.K. Hartwell, S. Lapanantnoppakhun, I. McKelvie, The case for the use of unrefined natural reagents in analytical chemistry—a green chemical perspective, Analytical Methods 2 (11) (2010) 1651−1661.

Further reading

[1] Y. Pico, Ultrasound-assisted extraction for food and environmental samples, TrAC, Trends in Analytical Chemistry 43 (2013) 84−99.
[2] M.M. Delgado-Povedano, M.L. de Castro, Ultrasound-assisted extraction and in situ derivatization, Journal of Chromatography A 1296 (2013) 226−234.
[3] B.K. Tiwari, Ultrasound: a clean, green extraction technology, TrAC, Trends in Analytical Chemistry 71 (2015) 100−109.
[4] K. Srogi, A review: application of microwave techniques for environmental analytical chemistry, Analytical Letters 39 (7) (2006) 1261−1288.
[5] A. Agazzi, C. Pirola, Fundamentals, methods and future trends of environmental microwave sample preparation, Microchemical Journal 67 (1−3) (2000) 337−341.
[6] F.E. Smith, E.A. Arsenault, Microwave-assisted sample preparation in analytical chemistry, Talanta 43 (8) (1996) 1207−1268.

Chapter 3

Conceptual density functional theory−based applications in extraction studies

Savaş Kaya
Department of Chemistry, Faculty of Science, Sivas Cumhuriyet University, Sivas, Turkey

1. Introduction

Density functional theory, based on the famous Hohenberg and Kohn theorems, which played an important role in the development of quantum chemistry, provides great convenience to theoretical and computational chemists in determining the structural and energetic properties of chemical species [1]. According to first Hohenberg-Kohn Theorem, it is the electron density, $\rho(r)$, that determines the external potential, $v(r)$, and the total number of electrons (N) is determined by normalizing the electron density as:

$$\int \rho(r)dr = N \tag{3.1}$$

Molecular Hamiltonian, H_{op} determining through N and $v(r)$, using Born-Oppenheimer approximation is calculated as:

$$H_{op} = -\sum_{i}^{N}\frac{1}{2}\nabla_i^2 - \sum_{A}^{n}\sum_{i}^{N}\frac{Z_A}{r_{iA}} + \sum_{i<j}^{N}\sum_{j}^{N}\frac{1}{r_{ij}} + \sum_{B<A}^{n}\sum_{A}^{n}\frac{Z_A Z_B}{R_{AB}} \tag{3.2}$$

In the given equation, summations over i and j run over electrons and summations over A and B run over nuclei. r_{ij}, r_{iA}, and R_{AB} represent the electron−electron, electron−nuclei, and internuclear distances, respectively. H_{op} calculated from Eq. (3.2) determines the energy of a system via Schrödinger's equation.

$$H_{op}\psi = E\psi \tag{3.3}$$

Here, ψ stands for the electronic wave function, and $\rho(r)$ determines the system's energy and all other ground state properties.

E is the functional of ρ and given as:

$$E = E_v[\rho] \tag{3.4}$$

Second Hohenberg-Kohn Theorem presents an ansatz to obtain ρ determining the $\rho(r)$ that minimizes the energy, E. It is important to note that energy does not change for optimal $\rho(r)$, provided that $\rho(r)$ always integrates to N as given in Eq. (3.1).

$$\delta(E - \mu\rho(r)) = 0 \tag{3.5}$$

Here, μ is the Lagrangian multiplier, and it is given as:

$$\mu = v(r) + \frac{\delta F_{HK}}{\delta \rho(r)} \tag{3.6}$$

In the equation, δF_{HK} is the Hohenberg-Kohn functional, and it contains the electronic kinetic energy functional, $T[\rho]$, and the electron−electron interaction functional, $V_{ee}[\rho]$. In parallel with the developments that form the basis of density functional theory, Parr and coworkers made a scientific advance regarding the mathematical definition of μ in Euler equation given by Eq. (3.6). Parr defined the chemical potential, μ, (Lagrangian multiplier) as the first derivative with respect to the number of the electrons (N) of total electronic energy (E) at a constant external potential [2].

$$\mu = \left(\frac{\partial E}{\partial N}\right)_{v(r)} \tag{3.7}$$

There is a parabolic relation between total electronic energy (E) and charge (q) can be given as [3]:

$$E = aq + bq^2 + cq^3 + dq^4 + \ldots \tag{3.8}$$

Here, $q = Z - N$ and Z represents the nuclear charge. According to Iczkowski-Margrave electronegativity definition [4], electronegativity is mathematically defined as:

$$\chi = -\left(\frac{\partial E}{dN}\right) \text{ or } \chi = \left(\frac{\partial E}{dq}\right) = a + 2bq + 3cq^2 + 4dq^3 \tag{3.9}$$

Because the cubic and quartic terms in Eq. (3.9) can be neglected, the relation with ionization energy and electron affinity of the electronegativity is obtained as [5]:

$$\chi = (I + A)/2 \tag{3.10}$$

Here, I and A are presented as:

$$I = E(N - 1) - E(N) \tag{3.11}$$

$$A = E(N) - E(N+1) \tag{3.12}$$

Chemical hardness concept introduced by Pearson [6] represents the resistance against the electron cloud polarization of the chemical systems. Parr and Pearson mathematically defined the chemical hardness as second derivatives with respect to the number of the electrons (N) of total electronic energy (E) at a constant external potential, $v(r)$ [7].

$$\eta = \left[\frac{\partial \mu}{\partial N}\right]_{v(r)} = \left[\frac{\partial^2 E}{\partial N^2}\right]_{v(r)} \tag{3.13}$$

In the light of the finite difference approach applied to Eq. (3.13), chemical hardness equation based on the ionization energy and electron affinity of the chemical systems is obtained as:

$$\eta = I - A \tag{3.14}$$

Softness (σ) is given as the inverse of the hardness.

$$\sigma = 1/\eta \tag{3.15}$$

Electrophilicity index (ω) is a reactivity descriptor combining the chemical hardness and electronegativity. Parr, Szentpaly, and Liu [8] introduced this descriptor with reference to a study performed by Maynard and coworkers [9] explaining ligand-binding phenomena in some biological systems. If an electrophilic ligand is immersed in an idealized sea of electrons with zero temperature and zero chemical potential, electron transfer continues until the chemical potentials of the sea and ligand become equal to each other. The energy change, ΔE, resulting from electron transfer, ΔN is mathematically presented as:

$$\Delta E = \mu \Delta N + \frac{1}{2} \eta \Delta N^2 \tag{3.16}$$

Here, μ and η represent the chemical potential and the chemical hardness of the ligand, respectively. If the electron sea provides sufficient electrons, the ligand become saturated with the electrons when $\Delta E/\Delta N = 0$, If so, one can write

$$\Delta E = -\frac{\mu^2}{2\eta} \quad \text{and} \quad \Delta N_{max} = -\frac{\mu}{\eta} \tag{3.17}$$

It is important to note that electron transfer process is favorable when $\eta > 0$ and $\Delta E < 0$. Parr, Szentpaly and Liu proposed that $\mu^2/2\eta$ ratio is a measure of the electrophilicity of the ligand. For that reason, first electrophilicity index (ω_1) is given as:

$$\omega_1 = \mu^2/2\eta = \chi^2/2\eta \tag{3.18}$$

Based on the ionization energy and electron affinity of the studied chemical systems, first electrophilicity index is given via the following equation.

$$\omega_1 = \frac{(I+A)^2}{8(I-A)} \tag{3.19}$$

First electrophilicity index has been derived through ground state parabola model. Second electrophilicity index (ω_2) derived by means of valence state parabola model is calculated from the following equation [10].

$$\omega_2 = \frac{I.A}{I-A} \tag{3.20}$$

In complexation reactions between ligands and metal ions, the electron donating and electron accepting capabilities of the ligands and metal ions are quite important to predict the strength of the bonds formed and to compare the complex stabilities. Gazquez and coworkers [11] derived the following equations to calculate two new reactivity descriptors named as electrodonating power (ω^-) and electroaccepting power (ω^+).

$$\omega^- = (3I+A)^2/(16(I-A)) \tag{3.21}$$

$$\omega^+ = (I+3A)^2/(16(I-A)) \tag{3.22}$$

The ionization energies and electron affinities of the molecules can be determined from the energies of cationic, anionic, and neutral forms of the system using Eqs. (3.11) and (3.12). Alternatively, researchers can use Koopmans Theorem [12] giving the following relations based on the frontier orbital energies for approximately prediction of ionization energy and electron affinities of the molecules.

$$I = -E_{HOMO} \tag{3.23}$$

$$A = -E_{LUMO} \tag{3.24}$$

where, E_{HOMO} and E_{LUMO} are the energies of HOMO and LUMO orbitals, respectively.

2. Hard and soft acid-base principle

The introducing of hard and soft acid-base (HSAB) Principle by Pearson in 1960s is among the remarkable developments in acid-base chemistry [13]. The prediction of the behaviors of the chemical systems in a generalized acid-base reaction is very important. As an example, cations such as Sc^{3+} and Al^{3+} prefer to react with F^- rather than I^-, and cations such as Hg^{2+} and Pt^{2+} prefer to react with I^- rather than F^-. The first clues to the validity of the HSAB principle were reported very early by Berzelius. He reported that some metals like Mg and Ca are found in nature in the form of oxides and carbonates, and some metals like Hg and Cd are found in the form of sulfides. HSAB Principle offers logical explanations for all these situations. Before Pearson introduced the chemical hardness concept and HSAB Principle, in independent researches, Chatt and Schwarzenbach [14] divided metal ions into two groups, taking into account their affinity for various electron-donating atoms in the ligands as class (a) and class (b). This classification is given in Table 3.1.

As < P < Se < S ~ I ~ C < Br < Cl < N < O < F is the electronegativity order of the donor atoms. Class (a) cations prefer to react with the donor atoms on the right in the order given, and class (b) cations prefer to react with the donor atoms on the left in the order given. After these developments, Pearson classified all Lewis acids and bases as hard, soft, and borderline. With the explanation "hard acids prefer to coordinate to hard bases and soft acids prefer to coordinate to soft bases." Hard and soft acid-base Principle was imparted to the literature. In short, the compatibility of acids and bases in terms of chemical hardness makes the interaction between them strong. Class (a) and class (b) acids appearing in Table 3.1, are also hard and soft acids, respectively. Hard, soft and boderline bases are given in Table 3.2.

By examining the chemical systems in Tables 3.1 and 3.2, the properties of hard and soft acids and bases can be given as follows.

TABLE 3.1 Class (a) and class (b) classification of Chatt and Schwarzenbach.

Class (a)/hard	Class (b)/soft
H^+, Li^+, Na^+, K^+	Cu^+, Ag^+, $Au+$, $T1^+$, Hg^+, Cs^+
Be^{2+}, Mg^{2+}, $Ca,^{2+}$, Sr^{2+}, Sn^{2+},	Pd^{2+}, Cd^{2+}, Pt^{2+}, Hg^{2+}
Al^{3+}, Se^{3+}, Ga^{3+}, In^{3+}, La^{3+},	CH_3Hg^+
Cr^{3+}, Co^{3+}, Fe^{3+}, As^{3+}, Ir^{3+},	Tl^{3+}, $T1(CH_3)_3$, RH_3
Si^{4+}, Ti^{4+}, Zr^{4+}, Th^{4+}, Pu^{4+}, VO_2+	RS^+, RSe^+, RTe^+
UO_2^+, $(CH_3)_2Sn^{2+}$	I^+, Br^+, HO^+, RO^+
BF_3, $BC1_3$, $B(OR)_3$	I_2, Br_2
$A1(CH_3)_3$, $Ga(CH_3)_3$, $In(CH_3)_3$	Heavy metal ions
RPO^+, $ROPO^+$.	
RSO^+, $ROSO^+$, SO_3	
I^{7+}, I^{5+}, Cl^{7+}	
R_3C^+, RCO^+, CO_2, NC^+	

TABLE 3.2 Classification as hard, soft, and borderline of Lewis bases.

Hard bases	Soft bases
H_2O, OH^-, F^-	R_2S, RSH, RS^-
$CH_3CO_2^-$, PO_4^{3-}, SO_4^{2-}	I^-, SCN^-, $S_2O_3^{2-}$
Cl^-, CO_3^{2-}, ClO_4^-, NO_3^-	R_3P, R_3As, $(RO)_3P$
ROH, RO^-, R_2O	CN^-, RCN, CO
NH_3, RNH_2, N_2H_4	C_2H_4, C_6H_6, H^-, R^-
Bonderline bases	
$C_6H_5NH_2$, C_5H_5N, N_3^-	
Br^-, NO_2^-, SO_3^-, N	

3. Hard acids

- Acidic (acceptor) atom has small ionic radius.
- They tend to interact electrostatically.
- The acidic (acceptor) atom has a high positive charge.
- They do not contain electron pairs in their valence shells.
- They have low electron affinity.
- They do not contain external electrons that can be easily excited.
- Their polarizability is very low.

4. Soft acids

- Acidic (acceptor) atoms have large radii.
- Acidic (acceptor) atom has a low positive charge.
- They do not contain electron pairs in their valence shells.
- They interact covalently.
- They contain outer electrons that can be easily excited.
- They have high polarizability.

5. Hard bases

- Donor atoms are small atoms with high electronegativity.
- They tend to interact electrostatically.
- They have a high negative charge.
- They contain electron pairs in their valence shells.
- They have high electron affinity.
- They have low polarizability.

6. Soft bases

- Donor atoms are large atoms with low electronegativity.
- They have low electron affinity.
- They contain electron pairs in their valence shells.
- They interact covalently.
- They can be oxidized easily.

Now let us give some examples of the strength of hard acid hard base and soft acid and soft base interactions. To show the power of soft acid-soft base interactions, we can present an explanation for why heavy metal ions are toxic. Most heavy metal ions such as Hg^{2+} and Pb^{2+} are highly toxic. The reason for this can be explained by the HSAB principle. Heavy metal ions are soft acids and have a high tendency to react with the S^{2-} ion, which is a soft base. S is also found in the structure of methionine and cysteine amino acids, and these amino acids are included in the structure of proteins and enzymes that have important vital activities. Heavy metal ions taken into the body bind to the sulfur in the structure of amino acids. Thus, the structure of the protein is disrupted and deactivated. Long-term exposure to heavy metal ions causes death.

To explain the power of the interaction between hard acid and hard bases, we can give example some papers including complex formation. Z.A. Begum and coworkers [15] investigated the stability of the complexes formed with Fe^{3+} and Cr^{3+} ions of some chelating ligands such as DL-2-(2-carboxymethyl) nitrilotriacetic acid (GLDA) and 3-hydroxy-2,2'-iminodisuccinic acid (HIDS) that their complexes with some divalent ions (Cu^{2+}, Ni^{2+}, Pb^{2+}, Zn^{2+}, Cd^{2+}) had been studied in a previous study performed by same research group [16]. In Fig. 3.1, chemical structures of GLDA and HIDS chelating ligands are given. As can be seen from the structures that both ligands include hard electron donor atoms such as oxygen and nitrogen. Combining the data obtained in both studied performed by Begum and coworkers, the comparative stability order of the metal ion complexes of GLDA and HIDS ligands was presented as: $\log K_{FeL} > \log K_{CrL} > \log K_{CuL} > \log K_{NiL} > \log K_{PbL} > \log K_{ZnL} > \log K_{CdL}$ for both GLDA (15.27 > 13.77 > 13.03 > 12.74 > 11.60 > 11.52 > 10.31) and HIDS (14.96 > 12.67 > 12.63 > 11.30 > 10.21 > 9.76 > 7.58).

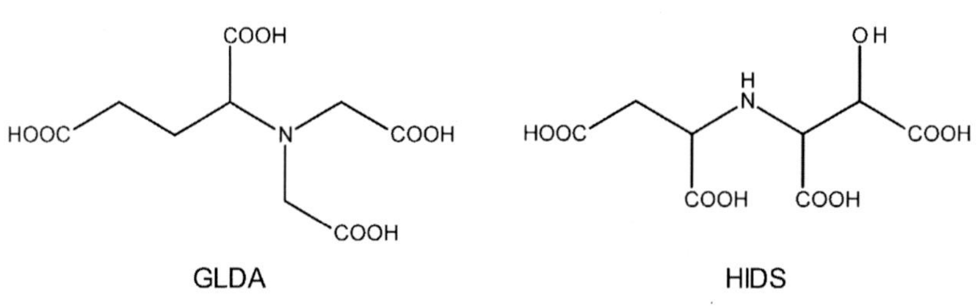

FIGURE 3.1 Chemical structures of DL-2-(2-carboxymethyl)nitrilotriacetic acid (GLDA) and 3-hydroxy-2,2'-iminodisuccinic acid (HIDS) ligands.

The obtained results showed that GLDA and HIDs ligands interact more powerful with trivalent ions like Fe^{3+} and Cr^{3+}. On the other hand, the stabilities of the complexes formed with divalent ions are relatively lower. These observations are in good agreement with the predictions made via HSAB Principle. As can be seen from Table 3.1 that Fe^{3+} and Cr^{3+} ions are among hard acids, while mentioned divalent ions are generally among soft acids. For that reason, trivalent ions interacted more strongly and formed more stable complexes with GLDA and HIDS ligands including hard electron donor atoms.

7. Maximization and minimization in stable states of hardness, polarizability, electrophilicity, and magnetizability

Some of the chemical reactivity descriptors such as hardness, polarizability, electrophilicity and magnetizability are maximized in steady states, some ones are minimized. Shortly, thanks to calculated values of these descriptors, global chemical reactivities of atomic and molecular systems can be easily compared. The relations with chemical reactivity/stability of the reactivity descriptors are given through the electronic structure principles or rules. One of the most popular electronic structure principles in the literature is Maximum Hardness Principle (MHP) [17,18]. According to MHP, "there seems to be a rule of nature that molecules arrange themselves so as to be as hard as possible." Within the framework of this information, it can be said that chemical hardness is a measure of the stability and hard molecules are more stable than soft ones. In a Density Functional Theory (DFT) study performed by Kaviani and coworkers [19], complex formations between desferrithiocin molecule and some metal ions (Mg^{2+}, Al^{3+}, Ca^{2+}, Mn^{2+}, Fe^{3+}, Co^{2+}, Ni^{2+}, Cu^{2+}, Zn^{2+}) were investigated. Calculated chemical hardness (η) and stability constant values of the desferrithiocin complexes of various metal ions in this study are given in Table 3.3. Authors made all calculations via CAM-B3LYP/6-31G(d) level of the theory in the water phase. As a result, it was shown that harder metal ion complexes exhibit high stability as compatible with Maximum Hardness Principle.

Hard molecules are the chemical systems with low polarizability. Namely, there is an inverse relation between hardness and polarizability. In some papers penned by Ghanty and Ghosh [20], it was noted that softness (the multiplicative inverse of the hardness) is proportional to the cube root of the polarizability. Considering the inverse relation between hardness and polarizability, Chattaraj and Sengupta [21] introduced the Minimum Polarizability Principle. Minimum Polarizability Principle states that polarizability is minimized in stable states. The molecules with low polarizability exhibit high stability. Electrophilicity index is among widely used chemical reactivity descriptors in CDFT-based analyses. The electronic structure principle regarding to this descriptor is Minimum Electrophilicity Principle introduced by Chamorro, Chattaraj, and Fuentealba [22]. These authors proposed the minimization of

TABLE 3.3 Calculated chemical hardness (η) and stability constant values of the desferrithiocin complexes of various metal ions (chemical hardnesses are given in a.u unit).

Metal ion	η	log β$_2$
Mg^{2+}	0.04881	9.15
Al^{3+}	0.09300	22.07
Ca^{2+}	0.05186	11.35
Mn^{2+}	0.06403	15.28
Fe^{3+}	0.09514	23.88
Co^{2+}	0.06915	17.02
Ni^{2+}	0.08164	18.69
Cu^{2+}	0.09225	22.05
Zn^{2+}	0.08870	20.19

electrophilicity index in stable states with an explanation "in an exothermic reaction, the sum of the electrophilicity indexes of the products should be smaller than that of the reactants." When considered in terms of stability, it can be said that electrophilicity is minimized in stable states, just like polarizability. In a paper, Noorizadeh and Shakerzadeh [23] investigated the validity of Minimum Electrophilicity Principle in some Lewis acid-base complexes. In the study, the complexes of some boron trihalides; BX_3 (X = F, Cl and Br) with strong (NH_3, H_2O, $N(CH_3)_3$ and $O(CH_3)_2$) and weak (CO, CH_3F, HCN and CH_3CN) bases were considered. Authors noted that Minimum Electrophilicity Principle is successful in the explaining of the power of the interactions between the mentioned base groups and boron trihalides. In addition, the obtained stability order via Minimum Electrophilicity Principle of the studied complexes was compatible with experiments. Here, in addition to the studies showing the Minimum Electrophilicity Principle as very useful chemical reactivity analysis tool, we should also briefly mention the limitations of this principle. In the paper entitled "Why and when is electrophilicity minimized? New theorems and guiding rules." penned by Szentpaly and Kaya [24], authors investigated validity and limitations of Minimum Electrophilicity Principle. In the study, while it was reported in the study that the second electrophilicity index provides more compatible results with Minimum Electrophilicity Principle compared to the first electrophilicity index, it was seen that this principle is not very successful in predicting the exothermic or endothermic nature of solid state double exchange reactions. Minimum Magnetizability Principle [25] presents the relation with chemical stability of magnetizability. According to this principle, "A stable configuration/conformation of a molecule or a favorable chemical process is associated with a minimum value of the magnetizability." From here, it is clear that in stable states, magnetizability is also minimized. The magnetizability is a closely related parameter to the softness and polarizability. The relation between softness and magnetizability was illuminated by Chattaraj [26]. In a recent paper, Kaya [27] introduced a new method called as "Property Prediction from Structural Differences" to compute the molar diamagnetic susceptibilities of organic chemical systems. With this method, for small and large molecular systems, magnetizability values can be calculated and compared.

8. Conceptual DFT-based applications in extraction studies

In this part of the chapter, some applications of CDFT in extraction studies are given with conspicuous examples in the literature. It is well-known that in chemistry extraction is the separation process of a matter from a matrix. This is based on exactly how the analyte moves from the initial solvent into the extracting solvent. It is not difficult to predict that for an effective extraction procedure, chemical species that strongly interact with the substances to be analyzed should be taken into account and used.

One of the oldest studies regarding to the using of HSAB Principle in extraction studies was published by Kawamoto and coworkers [28]. Here authors determined the extraction constants of 2-thenoyltrifluoroacetonate (TTA) chelates formed with some metal ions. For the estimation of the extraction constants (K_{ex}) of the metal chelates of divalent metal ions (M^{z+}, $z = 2$), the authors proposed the following relation.

$$\log K_{ex} + zpK_a = 0.385 S_A + 0.080 \sigma_A \quad (3.25)$$

In the given relation, pK_a, S_A, and σ_A stand for the acid dissociation exponent of the chelating agent, strength factor and soft factor of metal ion, respectively. For a metal ion, S_A and σ_A are calculated from the following relations.

$$S_A = -\frac{2\Delta S_h^0}{2.30 R} \quad (3.26)$$

$$\sigma_A = \frac{\Delta H_f^0}{2.30 RT} \quad (3.27)$$

In the given equations, ΔS_h^0, ΔH_f^0, R and T are he hydration entropy of a metal ion, the heat of formation of an aqueous metal ion, the gas constant and the absolute temperature, respectively.

Kudo and coworkers [29] studied the extraction of CdI_2 with 18-Crown-6 Ether into 10 different diluents. In the study author determined extraction constants. In the experiments, the ion-pair formations in the different diluents were evaluated in the light of Hard and Soft Acid-Base Principle. Ramirez-Silva and coworkers [30] investigated complexation constants of the complexes formed as a result of the interactions with the thymol blue (TB) molecule of Hg^{2+}, Pb^{2+} and Cd^{2+} ions. Authors calculated the chemical hardness of TB molecule as 1.48 eV. Experimentally determined chemical hardness values for Hg^{2+}, Pb^{2+} and Cd^{2+} ions are 10.29, 8.46 and 7.7 eV, respectively. Fig. 3.2 presents diagram showing the comparison of the chemical hardness of TB and Hg^{2+}, Pb^{2+}, and Cd^{2+} ions. It can be seen from this diagram that chemical hardness of TB is more close to that of Hg^{2+} ion. So, HSAB Principle implies that TB molecule will interact more powerful with Hg^{2+} ion and a higher complexation constant for HgTB complex will be obtained. For complexation constants regarding to the molecular systems in this study, HSAB Principle proposes the order $K_{HgTB} > K_{PbTB} > K_{CdTB}$. In the experimental part of the study, log K values determined for HgTB, PbTB, and CdTB complexes were 16.04, 5.59, and 5.09, respectively, and experimental data supported the predictions made via HSAB Principle.

Sasaki and coworkers [31] developed four new extractants, TODGA, DOODA, MIDOA, and NTAamide. The chemical structures of these molecules are given in Fig. 3.3. As a result of the analyses checking the extraction behaviors of these molecules for many metal ions including both hard and soft Lewis acids, the authors noted that TODGA and DOODA, which include hard oxygen donors in their central frames, prefer to form

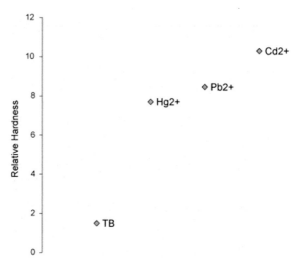

FIGURE 3.2 A diagram showing the comparison of the chemical hardness of TB and Hg^{2+}, Pb^{2+}, and Cd^{2+} ions.

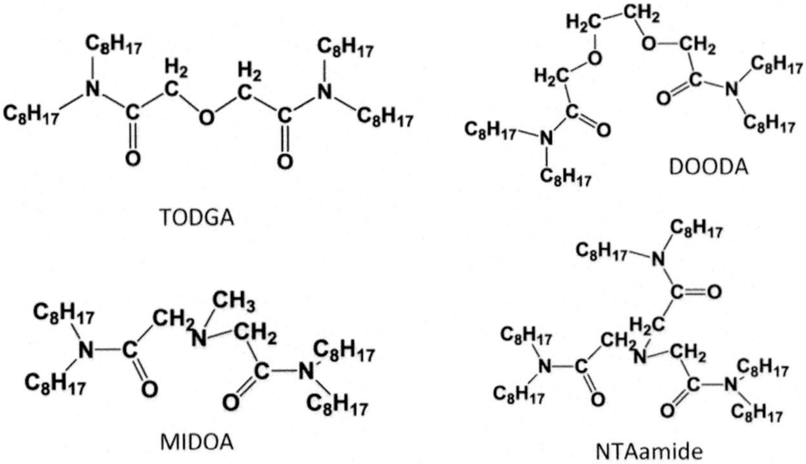

FIGURE 3.3 Chemical structures of ODGA, DOODA, MIDOA, and NTAamide [31].

metal complexes with hard acids, while MIDOA and NTAamide, which have relatively softer N donors in the center of their backbones prefer the binding to soft acids.

Nemati and coworkers [32] investigated the triazole pesticides (triticonazole, tebuconazole, diniconazole, hexaconazole, and difenoconazole) extraction by Ti_2C nanosheets as a sorbent in dispersive solid phase extraction. The interactions of the triazole pesticides with Ti_2C were investigated via DFT- and CDFT-based computations. Table 3.4 includes the calculated characteristics of the complex systems obtained with the interactions between Ti_2C nanosheets and

TABLE 3.4 Calculated characteristics of complex systems obtained with the interactions between Ti_2C nanosheets and studied pesticides triticonazole, tebuconazole, diniconazole, hexaconazole, and difenoconazole [32].

Complex system (C)	E_{HOMO} (eV)	E_{LUMO} (eV)	η (eV)	E_b (eV)	Dipole moment, (Debye)
Triticonazole/Ti_6C_4	−3.896	−2.269	1.627	3.693	5.922
Tebuconazole/Ti_6C_4	−3.793	−2.125	1.668	1.973	5.304
Diniconazole/Ti_6C_4	−3.586	−1.916	1.670	4.517	6.541
Hexaconazole/Ti_6C_4	−3.943	−2.318	1.625	3.666	7.676
Difenoconazole/Ti_6C_4	−3.404	−2.033	1.371	1.782	2.148

studied pesticides. As mentioned in the previous part of the chapter, chemical hardness is a measure of the hardness. The link between hardness and stability is given through Maximum Hardness Principle. It is seen from Table 3.4 that the hardness system is diniconazole/Ti$_6$C$_4$ and the binding energy (E_b) calculated for this system is greater than the binding energy values calculated for other systems. The obtained results can be considered the proofs of the usefulness in extraction studies of Maximum Hardness Principle. Chemical systems will form stable complexes with the analyte under study can be predicted through calculations taking into account the Maximum Hardness Principle.

In the paper published by Altunay and coworkers [33], authors presented an ultrasonic-assisted microextraction technique for determination of melamine residues in dairy products. In the experimental part of the article, authors checked the effects of some divalent ions such as Fe^{2+}, Co^{2+}, Cu^{2+}, and Zn^{2+} on extraction process, and they noted that the best results were obtained in experiments where the Fe ion was taken into account. In the theoretical part of the study, the stabilities of the melamine complexes of the mentioned divalent ions were checked and discussed via CDFT-based calculations. Fig. 3.4 visually presents HOMO, LUMO, and optimized structures of the melamine complexes of Fe^{2+}, Co^{2+}, Cu^{2+}, and Zn^{2+} ions. Using calculated chemical hardness and electrophilicity index values and considering Maximum Hardness and Minimum Electrophilicity Principles, authors reported that the most stable complex among the studied metal–ion complexes is Fe^{2+}–melamine complex. This theoretical prediction made via some CDFT-based electronic structure principles supported the experimental observations.

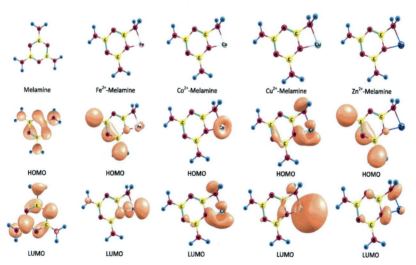

FIGURE 3.4 HOMO, LUMO, and optimized structures of the melamine complexes of Fe^{2+}, Co^{2+}, Cu^{2+}, and Zn^{2+} ions.

9. Conclusion

Conceptual Density Functional Theory known the chemical reactivity related branch of DFT has and will continue to have important applications in extraction studies. This book chapter introduces in detail the equations, principles, and approaches brought to science within the scope of Conceptual Density Functional Theory and explains the applications of CDFT in extraction studies with some examples. It has been observed that the concept of chemical hardness has found important applications in extraction studies. Since the mechanism and strength of the interactions between the species to be analyzed in extraction studies are important, popular electronic structure principles of CDFT (HSAB, maximum hardness, minimum polarizability and minimum electrophilicity principles) should be considered while searching for answers to the questions of which chemical species interacts more strongly with which chemical species or forms more stable complexes.

References

[1] P. Hohenberg, W.J.P.R. Kohn, Density functional theory (DFT), Physical Review 136 (1964) (1964) B864.
[2] R.G. Parr, W. Yang, Density-functional theory of the electronic structure of molecules, Annual Review of Physical Chemistry 46 (1) (1995) 701–728.
[3] S. Kaya, C. Kaya, A new method for calculation of molecular hardness: a theoretical study, Computational and Theoretical Chemistry 1060 (2015) 66–70.
[4] R.P. Iczkowski, J.L. Margrave, Electronegativity, Journal of the American Chemical Society 83 (17) (1961) 3547–3551.
[5] R.G. Parr, R.G. Pearson, Absolute hardness: companion parameter to absolute electronegativity, Journal of the American Chemical Society 105 (26) (1983) 7512–7516.
[6] R.G. Pearson, Chemical hardness and density functional theory, Journal of Chemical Sciences 117 (2005) 369–377.
[7] S. Kaya, A. Robles-Navarro, E. Mejía, T. Gómez, C. Cardenas, On the prediction of lattice energy with the fukui potential: some supports on hardness maximization in inorganic solids, The Journal of Physical Chemistry A 126 (27) (2022) 4507–4516.
[8] R.G. Parr, L.V. Szentpály, S. Liu, Electrophilicity index, Journal of the American Chemical Society 121 (9) (1999) 1922–1924.
[9] A.T. Maynard, M. Huang, W.G. Rice, D.G. Covell, Reactivity of the HIV-1 nucleocapsid protein p7 zinc finger domains from the perspective of density-functional theory, Proceedings of the National Academy of Sciences 95 (20) (1998) 11578–11583.
[10] L. von Szentpály, Ruling out any electrophilicity equalization principle, The Journal of Physical Chemistry A 115 (30) (2011) 8528–8531.
[11] J.L. Gázquez, A. Cedillo, A. Vela, Electrodonating and electroaccepting powers, The Journal of Physical Chemistry A 111 (10) (2007) 1966–1970.
[12] T. Koopmans, Über die Zuordnung von Wellenfunktionen und Eigenwerten zu den einzelnen Elektronen eines Atoms, Physica 1 (1–6) (1934) 104–113.
[13] R.G. Pearson, Hard and soft acids and bases, Journal of the American Chemical Society 85 (22) (1963) 3533–3539.
[14] R.G. Pearson, Chemical Hardness, vol 10, Wiley-VCH, Weinheim, 1997, 3527606173.

[15] Z.A. Begum, I.M. Rahman, H. Sawai, Y. Tate, T. Maki, H. Hasegawa, Stability constants of Fe (III) and Cr (III) complexes with dl-2-(2-carboxymethyl) nitrilotriacetic acid (GLDA) and 3-hydroxy-2, 2′-iminodisuccinic acid (HIDS) in aqueous solution, Journal of Chemical & Engineering Data 57 (10) (2012) 2723−2732.

[16] Z.A. Begum, I.M. Rahman, Y. Tate, Y. Egawa, T. Maki, H. Hasegawa, Formation and stability of binary complexes of divalent ecotoxic ions (Ni, Cu, Zn, Cd, Pb) with biodegradable aminopolycarboxylate chelants (dl-2-(2-carboxymethyl) nitrilotriacetic acid, GLDA, and 3-hydroxy-2, 2′-iminodisuccinic acid, HIDS) in aqueous solutions, Journal of Solution Chemistry 41 (2012) 1713−1728.

[17] R.G. Pearson, The principle of maximum hardness, Accounts of Chemical Research 26 (5) (1993) 250−255.

[18] S. Kaya, C. Kaya, A simple method for the calculation of lattice energies of inorganic ionic crystals based on the chemical hardness, Inorganic Chemistry 54 (17) (2015) 8207−8213.

[19] S. Kaviani, M. Izadyar, M.R. Housaindokht, A DFT study on the complex formation between desferrithiocin and metal ions (Mg2+, Al3+, Ca2+, Mn2+, Fe3+, Co2+, Ni2+, Cu2+, Zn2+), Computational Biology and Chemistry 67 (2017) 114−121.

[20] T.K. Ghanty, S.K. Ghosh, Correlation between hardness, polarizability, and size of atoms, molecules, and clusters, The Journal of Physical Chemistry 97 (19) (1993) 4951−4953.

[21] P.K. Chattaraj, S. Sengupta, Popular electronic structure principles in a dynamical context, The Journal of Physical Chemistry 100 (40) (1996) 16126−16130.

[22] E. Chamorro, P.K. Chattaraj, P. Fuentealba, Variation of the electrophilicity index along the reaction path, The Journal of Physical Chemistry A 107 (36) (2003) 7068−7072.

[23] S. Noorizadeh, E. Shakerzadeh, Minimum electrophilicity principle in Lewis acid−base complexes of boron trihalides, Journal of Molecular Structure: THEOCHEM 868 (1−3) (2008) 22−26.

[24] L. von Szentpaly, S. Kaya, N. Karakus, Why and when is electrophilicity minimized? New theorems and guiding rules, The Journal of Physical Chemistry A 124 (51) (2020) 10897−10908.

[25] A. Tanwar, S. Pal, D. Ranjan Roy, P. Kumar Chattaraj, Minimum magnetizability principle, The Journal of Chemical Physics 125 (5) (2006).

[26] P.K. Chattaraj, T.A. Murthy, S. Giri, D.R. Roy, A connection between softness and magnetizability, Journal of Molecular Structure: THEOCHEM 813 (1−3) (2007) 63−65.

[27] S. Kaya, Property prediction from structural differences: I. Molar diamagnetic susceptibilities of organic chemical systems, Chemical Physics Letters 836 (2024) 141046.

[28] H. Kawamoto, H. Itabashi, A. Mitsuyama, Critical evaluation of the extraction constants for 2-thenoyltrifluoroacetonato chelates on the basis of the HSAB principle, Analytical Sciences 10 (4) (1994) 675−677.

[29] Y. Kudo, Y. Ishikawa, H. Ichikawa, CdI2 extraction with 18-crown-6 ether into various diluents: classification of extracted Cd (II) complex ions based on the HSAB principle, American Journal of Analytical Chemistry 9 (11) (2018) 560−579.

[30] P. Balderas-Hernández, A. Rojas-Hernández, M. Galván, M.R. Romo, M. Palomar-Pardavé, M.T. Ramírez-Silva, Determination of the complexation constants of Pb (II) and Cd (II) with thymol blue using spectrophotometry, SQUAD and the HSAB principle, Spectrochimica Acta Part A: Molecular and Biomolecular Spectroscopy 66 (1) (2007) 68−73.

[31] Y. Sasaki, Y. Tsubata, Y. Kitatsuji, Y. Sugo, N. Shirasu, Y. Morita, T. Kimura, Extraction behavior of metal ions by TODGA, DOODA, MIDOA, and NTAamide extractants from HNO3 to n-dodecane, Solvent Extraction and Ion Exchange 31 (4) (2013) 401−415.

[32] E.M. Khosrowshahi, M.R.A. Mogaddam, Y. Javadzadeh, N. Altunay, M. Tuzen, S. Kaya, M. Nemati, Experimental and density functional theoretical modeling of triazole pesticides extraction by Ti2C nanosheets as a sorbent in dispersive solid phase extraction method before HPLC-MS/MS analysis, Microchemical Journal 178 (2022) 107331.
[33] N. Altunay, A. Elik, S. Kaya, A simple and quick ionic liquid-based ultrasonic-assisted microextraction for determination of melamine residues in dairy products: theoretical and experimental approaches, Food Chemistry 326 (2020) 126988.

Chapter 4

Sorbent-based extraction procedures

Mohammad Reza Afshar Mogaddam[1,2], Sarina Beiramzadeh[1,3], Mohammad Nazari Koloujeh[4], Aysan Changizi Kecheklou[1,5], Mir Mahdi Daghi[1,3], Mir Ali Farajzadeh[5,6] and Mustafa Tuzen[7]

[1]*Food and Drug Safety Research Center, Tabriz University of Medical Sciences, Tabriz, Iran;*
[2]*Pharmaceutical Analysis Research Center, Tabriz University of Medical Science, Tabriz, Iran;*
[3]*Faculty of Chemistry, Tabriz Branch, Islamic Azad University, Tabriz, Iran;* [4]*Department of Chemistry, Faculty of Science, University of Maragheh, Maragheh, Iran;* [5]*Analytical Chemistry Department, Faculty of Chemistry, University of Tabriz, Tabriz, Iran;* [6]*Engineering Faculty, Near East University, Nicosia, North Cyprus, Turkey;* [7]*Tokat Gaziosmanpaşa University, Faculty of Science and Arts, Chemistry Department, Tokat, Turkey*

1. Introduction

Implementing proper sample preparation protocols before instrumental analysis like chromatography or spectroscopy is a crucial yet often overlooked necessity. Real-world samples from environmental, biological, food, pharmaceutical, and forensic sources contain extremely complex matrices rife with interfering components like proteins, lipids, pigments, salts, and countless other substances alongside the actual compounds of interest [1]. Additionally, target analytes embedded inside these "complex" matrices frequently occur at the concentrations below from detector sensitivity threshold. Hence, there is the necessity of preliminary extraction steps that serve to isolate analytes from high-interference sample environments and boost trace analysis [2]. Simply put, direct analysis of raw samples containing anything without extraction is a formula for deleterious outcomes spanning inaccurate quantitation, poor analytical sensitivity, low column lifetime, and general assay failure [3]. Some appropriate sample preparation serves to "clean up" messy samples through steps like liquid−liquid extraction (LLE), solid phase extraction (SPE), protein precipitation, and quick-easy-cheap-effective-rugged-safe (QuEChERS) to remove confounding matrix constituents [4]. It serves as the gateway to obtaining contamination-free concentrates of analytes that lead to accurate, precise instrumental analytical and robust data. Hence, understanding the pivotal role of sample preparation constitutes the first step toward designing fit-for-purpose bioanalytical methods and assays [5]. The most ancient method of sample preparation belongs to LLE in which it is difficult to determine the trace concentration of the analyte [6]. Some sources mention the use of LLE in the early 1910s and 1920s to separate organic acids and bases [7]. Additionally, LLE was commonly used in the petroleum industry in the 1930s—1940s to extract aromatic hydrocarbons. However, traditional solid-phase extraction dates back to commercial development in the late 1970s and early 1980s [8]. Before 2000, LLE was very popular, but it was limited when green analytical chemistry (GAC) was firstly proposed in 2000 as a solution to lessen or completely eradicate the harmful impacts that analytical procedures have on operators and environment. GAC is largely concerned with improving methods and equipment, reducing the negative environmental effects of chemical analyses, and encouraging the energy and economic sustainability of analytical laboratories [9]. So, SPE is more friendly to the environment than the previous method due to the positive aspects compared to the previous method, such as low solvent consumption or even no solvent consumption. It can be said that it was more interested in analytical chemists and has made significant improvements in recent years. As a result, GAC now has a framework, thanks to the adoption of green chemistry rules and principles [10]. Furthermore, the components of green analysis are well-known and have been included in numerous critical evaluations. That is why, in analytical chemistry, the entire analytical procedure often consists of multiple steps [11] that are shown in Fig. 4.1.

To improve the subsequent separation and detection stages, sample preparation techniques such as SPE are frequently utilized for the isolation and concentration of the compounds of interest from complicated sample matrices. All of these analytical stages are necessary to provide reliable results, but it is often acknowledged that the sample preparation stage takes up more than 80%

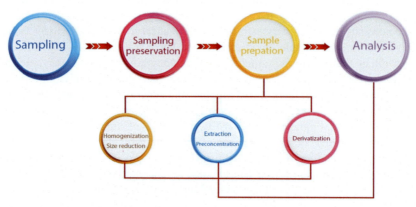

FIGURE 4.1 An analytical procedure step.

of the analysis time [12]. As a result, any curriculum designed to teach analytical chemistry should cover sample preparation, which is a crucial part of the analytical process. These actions are taken in order and issues with any one of them could jeopardize the analysis as a whole. For instance, an improperly handled sample could render the entire experiment incorrect [13]. Instrument maintenance expenses are further decreased by using cleaned samples obtained through methods such as SPE. There are two approaches to SPE: batch (Fig. 4.2) and column (Fig. 4.3). The solid phase in the column technique, which is one of the continuous and static methods, is quiet and may be connected straight to the analytical equipment, such as gas and liquid chromatography [14]. Cartridges are the most popular way to use the adsorbent phase in the column technique. They are made of plastic or glass tanks with an open bottom that is loaded with adsorbent phase particles. The syringe holding the sample may be connected to the hole in the cartridge's entry section. To preserve the cartridge's accumulation, a 20 μm-thick polyethylene coating containing netted glass powder is applied to both ends. Cartridges' inability to wash analytes quickly enough and with adequate power and resistance is one of the main drawbacks [15].

One of the discontinuous and dynamic approaches is the batch method. The adsorbent is added straight to the analyte's solution when using batch techniques. Stirring and raising the temperature both shorten the time it takes to achieve extraction equilibrium. Compared to the column approach, it has higher extraction speed and less quantity of adsorbent needed [16]. Traditional

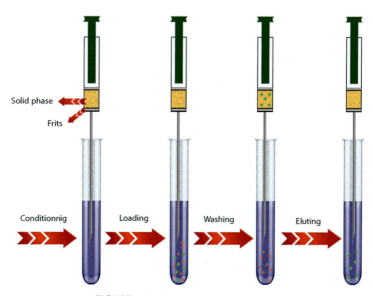

FIGURE 4.2 The procedure of batch SPE.

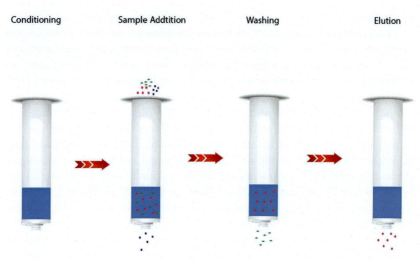

FIGURE 4.3 Column SPE.

LLE is still used in many analytical methods for sample preparation, despite advancements in chromatographic separation and detection techniques. Long extraction times and large consumption of dangerous chemical solvents are two major drawbacks of LLE. Solvent waste results from this, which may have detrimental effects on the environment [17].

The SPE approach that described above can be compared with LLE considering:

I. Selectivity
   - Depending on the sorbent selected, SPE can be more selective from LLE. Multiretention mechanisms are combined in mixed-mode sorbents.
   - LLE lacks specific selectivity and tends to isolate a wider variety of chemicals.
II. Consumption of organic solvents
   - SPE exclusively uses organic solvents for the eluting of retained analytes.
   - For LLE, comparatively larger volumes of organic solvents are required.
III. Automation capability
   - Column layouts and auto-samplers make SPE easily automated.
   - LLE automation is more intricate.
IV. Enrichment factor (EF)
   - Due to the limited space in which analytes are eluted, SPE offers greater concentration factors.

- The ratios of the organic to aqueous phases restrict LLE concentration factors.
V. Environmental impact
- SPE produces far less waste containing harmful chemicals.
- There is a lot of solvent waste produced by LLE.

In conclusion, SPE outperforms standard LLE in terms of selectivity, automation, economically, and environmental friendliness [18]. Notably, methods such as solidified floating organic drop microextraction (SFODME) can also lower solvent quantities, but they need a lot of work and time. On the other hand, SPE offers a straightforward, quick, and automated method for separating analytes from sample matrices while using low amount of solvent possible [19]. SPE has several benefits, including broad applicability to a variety of analytes and sample types, high recoveries and enrichment factors, enhanced chromatographic behavior from interference-free extracts, and inexpensive per-sample costs. The use of SPE in forensic, therapeutic, food safety, pharmaceutical, and environmental applications has increased dramatically in recent years. Traditional SPE involves passing aqueous samples through cartridges filled with solid sorbent, which hold the analytes while letting interferences pass through. A tiny amount of organic solvent is subsequently used to recover the retained analytes. Thus, in this instance, SPE has several advantages over LLE, including lower solvent consumption, greater concentration factors, quicker extraction, and easier automation [20]. Although, this method has a lot of advantages but it still has some disadvantages that must be modified.

The following are a few drawbacks of SPE [21]:

I. The apparent challenge of learning how to use this process for method development.
II. It can be difficult to completely understand the procedure due to the large variety of compounds and interactions, which provide a multitude of alternatives for adjusting solvent and pH conditions.
III. It is common to need to take more than one step, which adds to the time needed.

Novel approaches to microextraction have been presented to overcome these constraints. For example, stir-bar sorptive extraction (SBSE), solid-phase microextraction (SPME), magnetic SPE (MSPE), etc. are some of them. The advantage of these approaches over the multistep procedures involved in SPE is that they require less time and work. Additionally, they make it possible to combine several processes including extraction, enrichment of analytes above the limit of detection (LOD) of a method, sampling, and the separation of analytes from sample matrices that cannot be directly injected into measuring devices. This chapter presents a thorough analysis of advances in adsorbent-based SPE. We go into great details on SPE techniques that have been downsized, referencing the most recent, representative, and pertinent scientific works. As a more environmentally friendly option to conventional sample preparation techniques, it may be concluded that SPE will keep growing [22].

1.1 Solid phase extraction

1.1.1 A brief history of SPE

SPE was originally used in the 1940s and rapidly expanded to various applications in the 1970s due to its many benefits over other conventional approaches. The first adsorbent in the SPE for extracting colors from reaction mixtures was most likely animal charcoal. SPE started to gain acceptance as a legitimate scientific method in the 1970s. Three stages of sample preparation have evolved since the field's finding in 1968: 1968—77, 1977—89, and 1989—present [23]. Sorbents have become much more popular throughout these periods. Simultaneously, sorbents kinds and shapes have evolved in response to technological advancements. The first adsorbents of SPE were some synthetic polymers, such as styrene-divinylbenzene resins. Beginning in the 1950s, SPE was initially used experimentally to analyze organic residues in water samples [24]. Over the last several years, hundreds of publications presenting SPE as a water analytical technique and method for measuring organic molecules have been published in professional journals. The process was made more convenient and entered a new stage of development in 1977 with the introduction of prefilled cartridges and columns containing silica sorbents. This invention was featured on a laboratory equipment cover in 1978. Furthermore, the bonding process was published for the first time using SPE on silica and explained how to remove histamine from wine using the Sep PakTM Cig [25]. Applications in the fields of clinical, pharmaceutical, and environmental sciences have been made possible by the development of stable and covalently bonded adsorbents, particularly those having a reverse phase. C_{18} may be employed in SPE or high-performance liquid chromatography (HPLC) procedures that do not call for polar interactions due to its carbonaceous composition. SPE sorbents are usually bead-shaped and are often packaged into cartridges or columns [26]. Adding sorbent particles to disk format is another area of active research and use. The disk format has shown the ability to provide high flow rates and excellent extraction recoveries because of its compact porous particles and quick mass transfer [27]. SPE development entered a new phase in 1989 with the introduction of SPE disks or membranes. An absorbent substance is sandwiched inside the matrix or between Teflon or fiber glass pads in these disks. A very short and accurate SPE cartridge is made with this design. In the 1990s, stiff polymeric monoliths were widely used as an alternate stationary phase for liquid chromatography. Monolithic polymer stationary phases entered the SPE domain shortly thereafter. Poly(styrene-codivinylbenzene) (PS-DVB) was the first monolithic polymer used in an SPE device [28]. Huck et al. used both PS-BVD

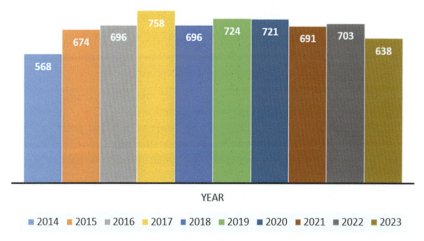

FIGURE 4.4 Status of the published paper from 2014 based on Scopus data.

and octadecyl silane (ODS) phases to study the recovery of 13 pesticides. When it came to all chemicals, PS-DVB was found to have an average recovery of 77%, whereas ODS only achieved 69%. Because PS-DVD copolymers are naturally hydrophobic, they have shown to be very successful at recovering nonpolar chemicals [29]. Additionally, great materials for high-throughput SPE tests were monolithic sorbents. Monolithic SPE sorbent's highly linked pores and exceptional permeability allow it to provide better mass transfer characteristics. However, compared to nonpolar analytes, polar analytes are generally less efficiently retained by PS-DVB copolymer. As a result, multifunctional polymer-based sorbents have been developed to improve hydrophilic interaction—based polar analyte retention [30]. According to the chart below (Fig. 4.4), extracted from Scopus data, which is Elsevier official and cited blog, since 2014, the usage of the preparation method of SPE has remained relatively constant, but it can be said that by modification of sorbent and optimization, different parameters that are involved in this method are likely to be increased.

1.1.2 Principle of SPE

The basic tenets of SPE are the selective adsorption of analytes on solid sorbents. The sorption of analytes on a solid sorbent inside a polypropylene cartridge forms the basis of the SPE principle. Material that is jam-packed into a well plate, disk, or cartridge but still permits interferences to flow through. Analytes are simultaneously extracted and enriched when the initial aqueous sample solution is run down the SPE column, where they are trapped on the sorbent and subsequently eluted with a suitable volume of an appropriate organic solvent [31]. The general procedures for carrying out SPE are as follows [32]:

I. Pretreat the sample (pH correction, dilution, etc.).
II. Run solvent or water through the cartridge to condition it.
III. Loading sample.

Spotlight: Various mechanisms may be occurred:

- Interactions between charged analytes and oppositely charged sorbent sites, known as ionic or electrostatic interactions.
- Hydrogen bonding: This occurs between sorbent sites and polar analytes.
- π–π interactions: Certain sorbents and aromatic rings.
- Affinity binding: Using enzymes, molecular imprinted polymers, and immobilized antibodies.
- Hydrophobic interactions: Applicable to partially polar or nonpolar analytes.

IV. Washing the interferences (elimination of interferences from absorbent).
V. Eluting (recovering analyte selectively in the desired solution).

This approach has drawn a lot of interest in the past and present because, in contrast to the methods like LLE, which need extremely high volumes of dangerous organic solvents, it offers quick, easy, and preconcentration of analytes from sample matrices before analysis. Principal benefits consist of [33]:

- Minimal waste production and solvent consumption
- High enrichment factors derived from adsorbed analyte concentrations
- Quick and easy extraction procedures
- Higher selectivity when utilizing specially designed sorbents
- Broad application to various matrices and analytes

Various analytical parameters, such as EF and analyte extraction recovery (ER), are indicated in SPE. These parameters highlight the effectiveness and efficiency of the method. The analytes' increased consent ratio following elution from the SPE sorbent in comparison to LLE is reflected in the EF. The EF is defined as the ratio of the concentration of the analyte in the final organic solvent (C_{org}) or the concentration of the analyte in the final eluent (C_{eluent}) to the initial sample concentration ($C_{initial}$) before loading onto the SPE sorbent or the concentration of the analyte in the initial aqueous solvent (C_{aq}), taking into account the following equation:

$$EF = \frac{C\text{ eluent}}{C\text{ initialration}} \tag{4.1}$$

or

$$EF = \frac{C(org)}{C(aq)} \tag{4.2}$$

The following equation is used to calculate ER%, which is defined as the proportion of the initial loaded analyte amount that is recovered in the eluent after passing through the SPE sorbent. This indicates the effectiveness of adsorption on the sorbent bed and subsequent elution.

Calculated as ER (%) = (mass of analyte in eluent/initial mass of analyte loaded) × 100,

Also, it can be shown by:

$$ER\% = \frac{v(\text{org})}{v(\text{aq})} \times EF \times 100 \tag{4.3}$$

V_{aq} is the volume of the aqueous phase, while V_{org} displays the final volume of the organic phase [34].

By changing parameters such as sorbent type, sample pH, elution solvent, etc., it is simple to optimize both of the aforementioned factors. SPE's dependability and simplicity have resulted in exponential growth. As sorbent technology and online/automated formats continue to progress, SPE is becoming a crucial sample preparation procedure for environmentally friendly analytical techniques.

1.1.3 SPE efficiency-related parameters

Optimizing various parameters is crucial for achieving a high EF and ER since they have an impact on extraction efficiency. A variety of strategies have been used to choose appropriate variables in SPE. Due to its ease of usage, one-factor-at-a-time optimization is a popular approach. A few additional chemometrics techniques, such as central composite design, have also been used in the past [35].

1.1.3.1 Type of sorbent

The kind of sorbent used in SPE is a crucial factor in figuring out how effective the extraction procedure is overall. Sorbents are available with various functional groups, surface areas, pores size, and polarities. Making the right sorbent choice that has a high affinity for the target analytes is essential to optimizing the extraction process. Increased selectivity in sorbents may optimize the extraction of analytes while reducing the number of interferences that are removed from the sample matrix concurrently. Choosing the right sorbent that is compatible with the particular physicochemical characteristics of the sample and the analytes is ultimately one of the most crucial factors influencing SPE effectiveness [36].

1.1.3.2 Amount of sorbent

Extraction efficiency is strongly impacted by the amount of sorbent utilized in SPE. Overloading the sorbent bed with too much analytes lowers analyte separation and increases the chance of breakthroughs into the collected fractions. Poorer analyte recoveries and decreased extraction efficiency result from overloading caused by insufficient sorbent amounts. The characteristics of the analytes and sample matrix, together with the sorbent's capacity, determine the ideal sorbent quantities. By choosing the right quantity, the analytes are captured effectively and as well as possible while using the least amount required [37].

1.1.3.3 Type of extraction

The total extraction efficiency is significantly impacted by the kind of SPE used. Normal phase, reversed phase, ion exchanger, mixed mode, and other specialized extraction mechanisms are among the several modes of SPE. The compounds of interest should be paired with the appropriate extraction technique according to their unique chemical and physical characteristics. For instance, reversed-phase SPE works best for adsorption and eluting nonpolar or hydrophobic analytes. Analytes that are charged, however, need mixed-mode SPE or ion exchanger. Analytes are adsorbed by the SPE sorbent by complementary chemical interactions, which are obtained by choosing the best extraction mechanisms. To optimize efficiency, it permits the analytes from the sample matrix to be concentrated under the right circumstances. The SPE technique with the incorrect extraction mechanisms will have extremely low extraction efficiency, coextraction of interferences, and poor analyte adsorption. Achieving high efficiency therefore requires careful selection of the SPE extraction mode according to the distinct features of the analyte [38].

1.1.3.4 Type and volume of elution solvent

An essential parameter of efficiency is the kind and amount of the solvent utilized to elute adsorbed analytes from SPE sorbent bed. Several key factors must be evaluated when selecting the optimal solvent, including the solvent's viscosity and its ability to effectively disrupt the analyte—adsorbent interactions quantitatively and have the ability for injection into analytical devices. If the elution solvent is not suitable, the analyte—sorbent interactions will not be sufficiently disrupted, which might result in permanent binding and significant losses. The capacity of the elution solvent to alter the sorption equilibrium and to have a high recovery of the analytes must be taken into consideration. Furthermore, inadequate elution solvent volume, even when used appropriately, might lead to ineffective retention analyte removal. Overly short eluent volumes result in low contact time and readsorption. Conversely, large quantities raise the concentration step needlessly. To ensure efficient desorption and prevent readsorption in the most concentrated form feasible, each analyte and sorbent must have its ideal elution solvent and minimum volume [39].

1.1.3.5 Ionic strength

The ionic strength of solutions used during SPE procedures significantly influences the extraction efficiency. Changes in ionic strength can dramatically impact the sorption equilibrium between the sorbent and analytes. High ionic strength buffers can compete with analytes for binding sites on ion exchanger sorbents, while lower ionic strengths may strengthen sorbent—analyte interactions. Additionally, adjustments of ionic strength using salt addition are often used to facilitate the elution of the adsorbed analytes. Optimizing ionic strength favors both maximizing adsorption from the sample matrix and recovery in the elution step. Appropriate ionic strength also prevents nonspecific binding of interferences which reduces efficiency. Determining and controlling the ionic strength using salts or other modifiers at each step of SPE method is essential for obtaining high extraction efficiencies. The ionic strength parameter must be balanced based on the unique properties of the analytes, sample matrix, and sorbent to achieve an optimal SPE procedure [40].

1.1.3.6 pH

An important factor impacting extraction performance is the pH of the solutions used in SPE processes. The ionization state of analytes and functional groups on the sorbent surface are affected by pH. To minimize undesired charge-based binding or repulsions and increase desirable interactions, pH control is required. pH must be adjusted for neutralization and hydrophobic adsorption of acidic or basic analytes, whereas silica sorbents' ionizable silanol groups are affected by pH. Moreover, pH may influence the solubility of the analytes and viscosity of sample solution, both of which affect interactions. Inadequate pH management leads to reduced analyte adsorption, capacity and overload limitations, or interferences extraction. All of which may cause poor SPE performance. To maximize selectivity and achieve high extraction efficiency, the ideal pH must be determined and maintained [40].

1.1.3.7 Temperature

Extraction effectiveness is also highly dependent on the temperature employed in SPE processes. The kinetic of contact between analyte and sorbent surface are governed by temperature. Because sorption is often an exothermic process, thermodynamic factors at higher temperatures might decrease adsorption. At high temperatures, binding capacity and matrix removal may sometimes be enhanced by viscosity and solubility effects. Low temperatures may amplify the interactions between sorbent and analyte, while higher temperatures enhance the disruptive effects of organic solvents to enhance elution. Temperature also affects diffusion rates, which regulate the mass transport along the sorbent bed. Unsuitable circumstances, such as irreversible binding, breakthrough, or readsorption problems during elution, result from improper temperature regulation. Determining the ideal conditions for optimizing SPE efficiency within a process therefore requires balancing thermodynamic, kinetic, and mass transfer effects via testing various temperatures [41].

1.2 Solid phase microextraction

SPE and SPME are among the most important sample clean-up and pretreatment methods, which have been utilized in the preconcentration and extraction of a wide range of analytes in biological, pharmaceutical, environmental, industrial, and food samples. SPME as a substitute for cartridge-based SPE was first presented in 1990. In the past, carbon-based materials, polymer-based sorbents, and chemically bonded silica-based with various functional groups were the most often used traditional sorbents in SPE process [42]. However, a significant effort has been made recently to introduce new solid sorbents for SPE to facilitate the extraction process in complex matrices. This is because conventional sorbents have certain constraints, including restricted capacity, limited reusability, and limited physical and chemical stability at low or high pHs. Because of this, SPME has established itself as one of the most widely used environmentally friendly method for pretreating and enriching a wide range of analytes from complex samples based on novel nanomaterial adsorbents from the biological media separation, concentration, and analysis of various analytes. Recently, both in vivo and ex vivo investigations have made extensive use of SPME [43]. Ex vivo SPME extraction involves the extraction of analytes from samples taken from living species, whereas in vivo sampling involves the direct use of SPME in both the headspace (HS) and direct immersion (DI) modes on living organisms.

1.2.1 Principle of SPME

SPME is a solvent-free sample preparation technique that was introduced in the early 1990s for chemical analysis. It integrates sampling, extraction, concentration, and sample introduction into a single solvent-free or solventless step. The basic principle involves using a solid phase as a sorbent that can be in different formats. To extract analytes, the sorbent is extended out and exposed to a sample (DI) or its HS. Analyte's partition extracts them from the sample matrix onto the desired sorbent. After extraction, the sorbent is washed with a suitable solvent for desorption (like methanol, acetonitrile, and ethanol), and then the solvent is injected into an analytical instrument (typically gas chromatograph or liquid chromatograph) for analysis. SPME is an equilibrium extraction technique. The amount of analyte extracted depends on the partition coefficient of the analyte between the sample matrix and the solid phase.

Factors like exposure time, temperature, salt concentration, stirring rate, and sample pH can influence extraction efficiency. Analyte preconcentration, extraction, and sampling all be able to be completed in a single step using SPME, ideally with little to no solvent usage and little invasiveness [44]. The sorbent used in SPME plays a critical role in extraction efficiency and selectivity. Some key effects of the SPME sorbent include:

Firstly, extraction efficiency is affected by sorbent, the chemical nature of the sorbent material determines its affinity for different types of analytes and therefore greatly affects extraction efficiencies. For nonpolar compounds, nonpolar sorbents and for more polar compounds, polar sorbents are used. Mixed sorbents can extract a wider range of analytes. Extraction selectivity is the second factor that is affected by sorbent. Sorbents can be tailored to extract specific classes of analytes from complex matrices. Using specific sorbents allows selective extraction of the target analytes from sample background interferences. So that extraction kinetics is influenced by sorbent. The amount of the sorbent affects the quantity of extraction. So, an appropriate amount is used based on the required sensitivity and analysis time. Finally, thermal stability is the last factor that is affected. The upper-temperature limit of the SPME sorbent determines the range of sampling conditions like boiling points of solvents or high-temperature HS sampling. Some sorbents like poly dimethyl silane (PDMS) have high thermal stability, while many polar polymer coatings have lower limits. In summary, optimization of the type and amount of sorbent allows efficient, selective, and controlled extraction of analytes from complex samples for accurate qualitative and quantitative analysis after chromatographic separation [45].

1.2.2 Advantages of SPME

I. Less solvent consumption or even solvent-free

Does not need the preparation, purchasing, or disposal of solvents, ecologically favorable.

II. Easy

No external analytical sample preparation is required; sampling, extraction, concentration, and introduction are all integrated into a single operation.

III. Being simultaneous

Able to extract several analytes from various complicated matrices.

IV. Quick

Because the sample and coating come into close touch, most extractions take a few minutes.

V. Low cost

Reusable, inexpensive fibers that are inexpensive. Generally, less expensive than alternative techniques for sample preparation.

VI. Portability

Direct immersion sampling and field analysis are made possible by fibers.

VII. Low sample volume: small sample sizes are necessary, which is beneficial for priceless samples.

VIII. Automatable

Auto-samplers make it simple to automate extraction.

1.2.3 Drawbacks of SPME (based on fiber)

1. Limited capacity

The low phase ratio of fiber coating to the sample restricts the amount extracted.

2. Carryover

The analytes remained in coating causing cross-contamination between samples.

3. Fragility

Fibers are delicate and can be damaged with improper handling.

4. Matrix effects

Complex sample composition can negatively impact extraction efficiency.

5. Qualitative

Best suited for qualitative identification rather than precise quantification.

6. Limited coating chemistry

Restricts the range of analytes that can be extracted [46].

1.2.4 Applications of SPME

1.2.4.1 Environmental analysis

SPME is commonly used to analyze pollutants in water samples like river, lake, groundwater, and wastewater sample. It allows rapid on-site sampling and extraction of contaminants without requiring solvents. Analytes such as pesticides, polycyclic aromatic hydrocarbons (PAHs), polychlorinated biphenyls (PCBs), and volatile organic compounds (VOCs) are frequently analyzed by SPME sampling of the HS above aqueous samples.

1.2.4.2 Food and flavor analysis

SPME can extract volatile and semivolatile components from various food products and beverages for aroma profiling. It is a quick and simple preparation method for qualitative analysis of flavors, off-flavors, fragrances, essential oils, etc. in items like milk, meat, fruit, wine, beer, etc.

1.2.4.3 Forensic toxicology

SPME is used in forensic labs to extract and quantify toxic drugs and metabolites in biological fluids like blood and urine. The small sample volumes needed make it useful for analyzing valuable evidentiary specimens.

1.2.4.4 Air sampling

Air sampling devices allow rapid SPME extraction of VOCs and semivolatile organic compounds (SVOCs) from air in both indoor and outdoor environments. Portable field devices are also available. It is a solventless method useful for the analysis of air pollutants.

1.2.4.5 Pharmaceutical and clinical analysis

SPME has been used for the measurement of drug molecules from formulations as well as biological samples such as plasma [47].

1.2.5 Sorbents in SPME

Sorbents of SPME are widely used in several fields and some of these sorbents include:

PDMS, polyacrylate (PA), divinyl benzene (DVB), and lab-made fibers based on the use of biocompatible coatings like C18, polydopamine (PDA), molecularly imprinted polymers (MIPs), mixed C_{18} and cation exchange particles, polyacrylonitrile (PAN), hydrophilic−lipophilic balanced particles (HLBs) have been proposed with the final aim of improving both method sensitivity and selectivity [48].

1.2.6 Future trends of SPME

In essence, HS-SPME is a variant SPME that makes use of more sensitive extraction forms. SPME designed in two ways for headspace extraction or DI extraction, respectively, by applying the basic SPME extraction process [49].

1.2.7 Classification of SPME

1.2.7.1 Headspace SPME

Two forms of SPME techniques that are widely used for sample preparation in analytical chemistry are HS-SPE and (DI-SPME). One of the oldest microextraction methods is HS-SPME, which invented by Janusz Pawliszyn's group at the University of Waterloo in Canada in 1990. The idea was based on SPME; however, instead of directly immersing into the sample, it employed the idea of a coated fused silica fiber into the headspace above the sample. Based on taking advantage of analyte partitioning into the gas phase, HS microextraction provides a straightforward yet effective method for concentrating VOCs and SVOCs into detectable amounts [50].

Some benefits of the headspace SPME (HS-SPME) approach are listed below:

I. Enhanced sensitivity

Because HS microextraction enables analytes to be preconcentrated in the HS above the sample before extraction, it can lead to increased sensitivity for volatile chemicals. When working with complicated matrices or doing trace-level analysis, this might be helpful.

II. Reduced matrix effects

By concentrating the volatile molecules found in the HS and reducing interference from nonvolatile components in the sample matrix, HS microextraction can aid in the reduction of matrix effects. This might raise the analysis's precision and dependability.

III. Simplicity and ease of use

Since HS microextraction eliminates the need for direct contact between the sample and the extraction phase, it is frequently seen as a simpler and easier method than SPME. This can improve accessibility and usability for analysts and researchers.

IV. Wide applicability

HS microextraction is a flexible method with a broad variety of applications in different areas including environmental, food, beverage, and pharmaceutical analyses. It may be utilized for a wide range of sample types and analytes.

Overall, HS-SPME and DI-SPME both have advantages and disadvantages [51]. However, HS microextraction may be preferred in some circumstances because of its wide range of applications, enhanced sensitivity, decreased matrix effects, sample compatibility, and ease of use. Furthermore, ongoing developments in microextraction are lowering detection limits through new interfaces like needled microtraps for in vivo and air monitoring applications [52].

1.2.7.1.1 Principles of HS-SPME In analytical chemistry, volatile and semivolatile chemicals may be extracted and concentrated from the headspace of a sample matrix using a process called HS-SPME. A sample container with an HS capacity, a SPE cartridge or sorbent, and a sampling tool like a syringe are the main components of HS-SPME. According to Fig. 4.5, to enable volatile substances to equilibrate between the sample matrix and the HS, the sample is first put in a sealed container. After extraction, analytes are typically desorbed, which are then transferred to an analytical instrument for detection and quantification [53]. This technique offers advantages such as simplicity, speed, and high sensitivity, making it valuable in various applications. Another important issue is the extraction parameters, which include temperature, pressure, static extraction time, elution solvent type, and volume should be optimized for the analyte/matrix combination to achieve high extraction efficiency while minimizing coextraction of interferences.

Since the analyte's total mass must stay constant and equal to the initial quantity, it is feasible to say that: If the sampling duration is sufficient to attain equilibrium, then

$$C_0 V_s = C^{\infty}_e V_e + C^{\infty}_h V_h + C^{\infty}_s V_s \quad (4.4)$$

That C_0 highlights the analyte concentration in the sample before the extraction stage; C^{∞}_e, C^{∞}_h, and C^{∞}_s illustrate the analyte concentration which once equilibrium has been gained in the extractant, the HS and the sample,

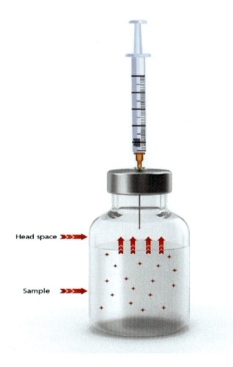

FIGURE 4.5 HS-SPME procedure.

respectively. Also, V_s, V_e, and V_h show the sample, extractant, and HS volumes, respectively.

In general, there are a few fundamental steps in HS-SPME approach:

1. Analytes must move from the bulk sample to the contact between the sample and gas.
2. They must endure evaporation from the surface of the sample to the gas phase next to the surface.
3. There must be transport via the bulk gas phase to the gas-extractant phase contact.
4. The extractant phase must undergo sorption of the analytes.
5. There must be transport to the majority of the extractant phase.

It is feasible to identify among those stages related to mass transfer that are restricting the extraction process's kinetics.

1.2.7.1.2 Advantage and disadvantages of HS-SPME As a sample preparation method for the analysis of volatile and semivolatile chemicals, HS-SPME has several significant benefits. By removing direct contact with the sample matrix, HS-SPME may reduce matrix effects that might compromise the precision and efficiency of extraction. It is an ecologically friendly, green procedure since it does not use much solvents. Because of the analyte preconcentration on the fiber coating, HS-SPME may be very sensitive. Since the balance between HS and fiber coating may be attained rapidly, it is comparatively fast. When parameters are adequately tuned and controlled, the procedure exhibits high repeatability. The ability to interface with GC equipment makes it automated-friendly, enabling high-throughput analysis with less need for manual sample processing.

While HS-SPME provides several benefits as a sample preparation technique for the analysis of volatile chemicals, it has some drawbacks that must be considered when selecting an appropriate extraction method. HS-SPME is restricted to volatile analytes in the HS, and may not extract semivolatiles or nonvolatiles in the sample matrix itself, restricting the range of compounds amenable to this technique. Additionally, the equilibration of analytes from the sample into HS can result in losses through diffusion and adsorption before extraction, requiring careful optimization. Though HS-SPME can reduce matrix effects compared to DI-SPME, complex sample matrix components may still impact efficiency and selectivity. The extraction capacity of the SPME fiber also limits the quantity of analytes extracted before saturation; samples with high concentrations may not be suitable. Variables like temperature, equilibration time, sample composition, and fiber properties can substantially influence the repeatability and efficiency of HS-SPME, necessitating extensive optimization to obtain reliable results. Finally, specialized equipment, multistep sample processing, and significant expertise are required to develop and carry out HS-SPME methods successfully [54].

1.2.7.2 Direct solid-phase microextraction

One other technique in SPME is DI-SPME. A frequent version of SPME technique is called DI-SPME, in which the sample is submerged immediately into the sample solution without the need for any further extraction or sample preparation activities. It appears from recent studies that no clear record exists that identifies DI-SPME as the invention that came first. However, in 2010, published studies began to use DI-SPME devices and terminology. Different analytes may be extracted from complicated sample matrices quickly and easily using the straightforward extraction technique known as DI-SPME. The main advantage of this method is that it can be used to quickly and effectively extract analytes from liquid samples, including water, drinks, biological fluids (similar to the previous method), and environmental samples. It also does away with the need for laborious sample preparation steps. High extraction efficiencies and enhanced sensitivity for analyte identification in analytical procedures may be obtained using DI-SPME. The analytes are selectively adsorbed and concentrated in DI-SPME by submerging fiber into sample solution [55].

1.2.7.2.1 Principles of DI-SPME In order to selectively extract desired analytes from a sample matrix, a solid sorbent is directly submerged into the sample matrix using DI-SPME method. In DI-SPE, as opposed to standard SPE, where the sample passes through a cartridge that contains the sorbent, the sorbent is submerged in a vial or other container that contains the sample, like Fig. 4.6. The sorbent's functional groups engage noncovalently with the analytes to remove them from the sample matrix and concentrate them onto its surface. Ideally, interfering matrix elements are remained in the sample. After DI, the extracted analyte-containing sorbent is usually taken out and relocated for a washing step to get rid of any leftover interferences. Then, using the proper elution solvent, the analytes are eluted from the sorbent into a concentrated, purified extract that is ready for instrumental analysis [56].

1.2.7.3 Comparison of HS-SPME and DI-SPEM

The following displays a comparison of two methods:
The target sample's preparation:

- HS-SPME method is used to extract the analytes from liquid sample above. For extraction, the sample is typically heated to release volatile analytes into the HS.
- The fiber is immediately immersed into the sample solution; no sample preparation steps are necessary. The analytes are extracted directly out of the matrix of the sample.
    1. Usage
        - HS-SPME is an often-used technique for extracting VOCs or SVOCs that may be partitioned into the HS above a liquid sample. Its common use is the measuring of VOCs in materials including food, soil, and water.

FIGURE 4.6 DI-SPME procedure.

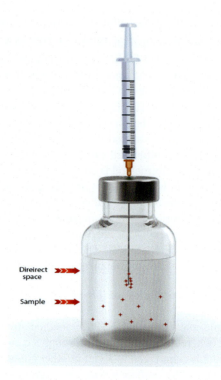

- DI-SPME may be used for a wide range of analytes, including both volatile and nonvolatile compounds. Without necessitating several intricate sample preparation steps, it is very beneficial for extracting analytes from liquid samples.
2. Sensitivity and selectivity
   - Since the HS-SPME enriches the analytes, it may provide high-sensitivity results for volatile compounds. Selectivity stems from the choice of sorbent material.
   - DI-SPME may give excellent sensitivity and selectivity over a larger range of analytes since it comes into direct contact with the sample matrix. The sorbent material choice may also affect selectivity.
3. Quickness and user-friendliness
   - In HS-SPME, heating the sample is often required to liberate volatile analytes into the HS, potentially extending the extraction duration. It can need more steps in comparison to DI-SPME.
   - There is no need for additional sample preparation steps with DI-SPME, making it a rapid and easy extraction procedure. It might be a faster and more useful option for certain applications.

Finally, both HS-SPME and DI-SPME techniques have advantages and perform well with different types of analytes and sample matrices [57].

1.2.7.4 Applications of HS-SPME and DI-SPME

HS-SPME and DI-SPME effectively extract and analyze volatile chemicals, so that using a broad range of porous solid phase sorbents is possible depending on the analyte's types. Molecular sieves, silica gel, activated carbon, alumina, florisil, and ion exchange resins are examples of common materials. For the extraction of polar semivolatiles, silica gel with having large surface area and adjustable polarity is advantageous.

The following are some typical uses for HS-SPME and DI-SPME:

1. Environmental analysis: VOCs in environmental samples such as air, water, soil, and sediment are analyzed using HS-SPME. It may be applied to environmental quality assessment, pollutant monitoring, and source identification.
2. Food and beverage analysis: The food and beverage sector frequently use HS-SPME to analyze pollutants, taste compounds, and off-flavors in food items. It is frequently used for authenticity testing and quality control of different food and beverage goods.
3. Pharmaceutical analysis: To extract and quantify volatile chemicals from medication formulations, packaging materials, and biological samples, HS-SPME is used. Drug release kinetics analysis, impurity profiling, and stability investigations can all benefit from its use.
4. Forensic analysis: In forensic science, volatile chemicals in crime scene samples, arson investigations, drug analyses, and toxicology studies are detected and analyzed using HS-SPME. It can be useful in locating traces of evidence, examining flammable residues, and connecting suspects to the locations of crimes.
5. Flavor and fragrance analysis: In order to extract and identify volatile aroma components from natural goods, essential oils, perfumes, and cosmetics, HS-SPME is used. It makes it possible to assess the quality of the product and characterize intricate odor characteristics.
6. Biomedical research: To analyze volatile biomarkers in biological fluids, breath samples, and tissues, biomedical researchers utilize HS-SPME. It may be applied to analyte physiological changes, tracking metabolic processes, diagnosing diseases, and evaluating the effectiveness of treatments.
7. Petrochemical analysis: The petrochemical industries use HS-SPME to analyze fuels, lubricants, gases, and petroleum products for the presence of sulfur compounds, volatile hydrocarbons, and other pollutants. In the petrochemical industry, it supports environmental monitoring, process optimization, and quality control. HS-SPME is a flexible method with a wide range of applications in various fields of study and industry where volatile compounds extraction and analysis are critical for process comprehension, contamination detection, maintaining product quality, and assisting scientific inquiries [58].

1.3 Fiber solid phase extraction

SPME, the concept of fiber SPE originated. In SPME, the analytes are extracted from samples using a coated thin fused silica fiber. Throughout the 2000s, different fiber coatings and formats were utilized to create variations of SPME. As these alternative methods began to deviate from the traditional SPME, they eventually became known as fiber SPE or fiber SPME. Fiber SPME is a sample preparation technique that involves extracting and concentrating analytes from liquid or solid samples onto a solid fiber coated with an extracting phase. The analytes to be targeted are adsorbed onto the fiber coating and then thermally desorbed into a mass spectrometer or chromatograph, for instance. The principle behind fiber SPME is based on the partitioning of analytes between the sample matrix and the sorbent coating on the fiber. The selectivity and affinity of the coating material allow the compounds of interest to be adsorbed on the fiber, while matrix interferences are washed away. After extraction, the adsorbed analytes are typically desorbed from the fiber into a compatible solvent.

The desorption process concentrates the extracted compounds into a purified and minimal volume of solvent that is compatible with analytical instrumentation. Precision, accuracy, and recovery can exceed traditional techniques [59]. Fiber-SPME works on a similar fundamental premise as standard SPE procedures, with interfering chemicals eliminated and the analytes kept onto the sorbent material according to their physicochemical qualities. Following the extraction process, the analytes are concentrated and purified in preparation for further analysis by being eluted from the sorbent using an appropriate solvent.

Additional principles that underpin the effectiveness and versatility of fiber SPME include achievable enrichment factors over 1000×, high reusability that minimizes cost and waste, and ease of automation for high-throughput analysis. The mating of selective extraction phases with a fiber format makes SPME amenable to in situ or remote sampling and miniaturized analytical workflows [60].

1.3.1 Sorbent type

The type of sorbent is a crucial factor in this method which should be optimized. To extract analytes from samples, SPME uses fused silica fibers covered with different liquid polymers, sol-gel, or carbon-based sorbent coatings. PDMS is the most often utilized coating; it removes organic molecules that are nonpolar or moderately polar via adsorption onto the coated layer. To extract more polar analytes, polar coatings such as polyacrylate are used and also carbowax-divinylbenzene and other mixed-phase coatings enhance the extraction of analytes with a broader [61].

Comparing fiber SPME to alternative sample preparation methods reveals a number of benefits, such as:

1. High ER

Fiber-based sorbents have high surface area-to-volume ratio makes it possible to extract and preconcentrate analytes from complicated sample matrices with high efficiency.

2. Rapid extraction

Because of the fast mass transfer rates connected to the tiny sorbent particles, fiber SPE may be completed rapidly, resulting in a shorter sample processing time.

3. Minimal sample concentration and volume

Fiber SPME works well with trace analytes or in situations where sample volume is limited since it only requires tiny sample quantities.

4. Versatility

By choosing the right sorbent materials and extraction conditions, fiber SPME may be readily adjusted to various analyte classes and sample types.

5. Automation

Robotic systems or autosamplers can be used to automate fiber SPME, enabling high-throughput sample processing and minimizing manual labor.

1.4 In-tube solid phase microextraction

In-tube SPME is another sample preparation technique in which the solid extracting phase is placed inside a capillary column rather than on an external device. In 1996, the earliest research paper was published proposing an "in-column extraction" method by Supelco's Vigh team, later called In-tube SPME. In in-tube SPME, a section of the capillary column is coated or packed with an extracting solid phase material. A sample solution is then passed through the coated section where analytes are extracted by sorption onto the sorbent coating. After extraction, the sorbent section containing the concentrated analytes is connected in line to a separation column for further analysis. By integrating sorbent extraction into a capillary format, In-tube SPME provides sample clean-up, preconcentration, and injection all in one fully automated set-up compatible with liquid and gas chromatography systems. Advantages include simplicity, precision, and minimal need for organic solvents compared to traditional methods [62].

1.4.1 Principles of in-tube SPME

Generally, in-tube SPME combines the principles of SPME with the convenience of a miniaturized microformat, allowing for efficient and rapid sample processing. The basic concept of in tube-SPME involves introducing the liquid sample containing the compounds of interest into the tube packed with the solid phase sorbent. The analytes are adsorbed on the sorbent based on their physicochemical properties, while interfering compounds are washed away and are not adsorbed onto the desired sorbent. After a desorption step, the analytes are eluted from the sorbent using a suitable solvent, leading to their concentration and purification [63].

1.4.2 Comparison between conventional SPME and in-tube SPME

In-tube-SPME offers several advantages over conventional SPE techniques, including [64]:

1. **Miniaturization:** In-tube-SPME requires smaller sample volumes and sorbent amounts compared to traditional SPME, making it suitable for applications requiring limited sample volume or when dealing with trace analytes.
2. **Automation:** In-tube-SPME can be easily automated using autosamplers, allowing for high-throughput sample processing and reducing manual labor.
3. **Rapid analysis:** In-tube-SPME enables fast extraction and preconcentration of analytes, leading to reduced analysis time and improved sample throughput.
4. **Enhanced sensitivity:** Concentration of analytes on a small sorbent bed in in-tube-SPME can lead to higher sensitivity in analytical measurements.
5. **Reduced matrix effects:** In-tube-SPME can effectively remove interfering matrix components, enhancing the selectivity and accuracy of analytical methods.

Normal-phase silica sorbents and C_{18} or C_8 reverse-phase sorbents are used in tube SPME, enabling extracting nonpolar to moderately polar analytes from polar matrices like biological fluids. However, these sorbents show low selectivity, especially in complex samples leading to poor analyte recovery or purification. This droves research into more selective in tube SPME sorbents including specialized chemically bonded silica, immunoaffinity sorbents, ion exchange materials MIPs, restricted access media, activated carbon, and polymeric resins. Each sorbent offers unique properties for adsorption of specific classes of analytes more selectively from within complex sample matrices. For example, immunoaffinity SPME sorbents uses antibodies to provide very high selectivity toward the analyte through antigen-antibody binding. MIPs can also be generated to mimic the binding sites of the target molecule leading to high selectivity. Ion exchangers additionally enable the adsorption of charged analytes which may not be adsorbed well on other media. Using multiple mechanisms, mixed-mode sorbents have also been introduced to integrate the capabilities of different sorptive compounds into the in-tube SPME method [65].

1.4.3 Differences between fiber and in-tube methods

Generally, target analytes are extracted and preconcentrated from liquid samples using SPME sorbent material in two different sample preparation techniques: in-tube SPME and fiber-SPME. The fundamental idea of extraction is the same in both to separate target analytes from liquid samples. Sorbents are used in both fiber SPME and in-tube SPME. However, the forms are very different. The sorbent is placed inside a tube or column housing in in-tube SPME. The chemicals in the solid phase are retained when the sample passes through this densely packed bed. On the other hand, sorbent coatings mounted onto solid fiber supports and immediately submerged into solutions in fiber SPME. The open structure of fiber SPME offers benefits over constricted tube housing, including quick mass transfer and small space requirements. Another area where the two methods differ is in how samples are analyzed. While fibers may be used for portable direct sampling, in-tube SPME is used in automated systems pump materials. However, both methods use selected sorbent phases to concentrate trace analytes, indicating the adaptability of SPME in various configurations. In-tube SPME is frequently used for batch processing, allowing for the simultaneous processing many samples utilizing a column-based configuration. Nonetheless, fiber-SPME is frequently chosen for applications that need the quick and effective extraction of analytes from intricate matrices or tiny sample quantities. The decision between the two procedures relies on several parameters, including sample volume, analyte characteristics, necessary extraction efficiency, and convenience of use. Both in-tube SPME and fiber-SPME have benefits and drawbacks. Based on these factors, researchers frequently choose the best approach to obtain the best outcomes in their sample preparation procedures [66].

1.5 Matrix-dispersed solid phase extraction

A common sample preparation technique for solid, semisolid, or viscous samples—such as animal tissues and meals—is matrix dispersed solid-phase extraction (MDSPE). The procedure includes mixing the sample with a solid phase, allowing the matrix cell to break, and then using an appropriate elution solvent to extract the analytes. Because of the process's viability and adaptability, MDSPE employment and innovations have continued to rise since its introduction in 1989. In contrast to traditional extraction techniques, which are costly and time-consuming due to the need for several samples, sorbents, organic solvents, and clean-up stages, MDSPE is quicker, less labor-intensive, and more environmentally friendly. Depending on the analytes and the equipment used for their detection, further sample clean-up may be necessary after extraction. The MDSPE protocol is seen as a viable substitute for Soxhlet, microwave-assisted extraction (MAE), supercritical fluid extraction (SFE), and pressurized liquid extraction (PLE) for all of the aforementioned reasons [67]. The adsorption characteristics of MDSPE appear to be a combination of ion exchange, partition, and adsorption. The composition of the sorbent materials and the elution solvent used are the only factors that precisely determine MDSPE selectivity, regardless of the precise processes involved. Lipophilic sorbent materials, such as C18-bonded silica and less commonly C8-bonded silica, are used in the majority of MDSPE applications. According to Barker et al., the lipophilic component coupled to solid support plays a similar effect to the surfactants and detergents used in traditional techniques in dispersing tissues and rupturing cell membranes. A preliminary classification of MDSPE can be made based on the dispersing sorbent characteristics of the various dispersing materials (underivatized silica, sand, synthetic polymers, florisil, graphitic fibers, etc.) that were tested in the subsequent developments and applications. The type of solid material has a direct bearing on the solvents chosen for elution. Solid, semisolid, viscous, and liquid sample typologies have all been used with MDSPE. Stronger analyte—matrix interactions and, generally speaking, more interferences in solid samples than in liquid ones make up the primary distinctions. The robust analyte—matrix interactions of solid samples also suggest that proper spiking with target chemicals for recovery tests may be more challenging. The ratio of the sample to the dispersing material is another crucial factor, in addition to the sample's make up. By using modifiers during the blending process, such as acids, bases, salts, chelators (like EDTA), and so on, that can affect the disruption, distribution, and elution profile, and the yield of the MDSPE process. Furthermore, it is crucial to remember that porous materials, such as porcelain, have been demonstrated to cause analyte and sample loss. Instead, glass or agate should be used as the mortar and pestle [68] (Fig. 4.7).

In the upcoming years, MDSPE is probably going to continue growing and developing because of a number of important elements. Reproducibility, efficiency, and broader adoption across analytical processes may be enhanced by integrating robots and intelligent optimization algorithms to facilitate the shift from manual to automated forms. Miniaturization of parts and procedures using microextraction techniques might enable field analysis and make biological applications with small sample sizes easier. Rapid, environmentally friendly analysis and smooth extraction may be made possible by expanding the online connection of MDSPE to analytical tools like LC—MS. Furthermore, the creation of multifunctional sorbents that are selective and include nanoparticles or MIPs shows promise for managing increasingly complex matrices in a variety of sectors, including forensics and food testing. Overall, MDSPE distinct matrix-compatibility advantages and cost-effectiveness, bolstered by these developing technologies, position it for significant future expansion across a wide variety of high-impact analytical applications.

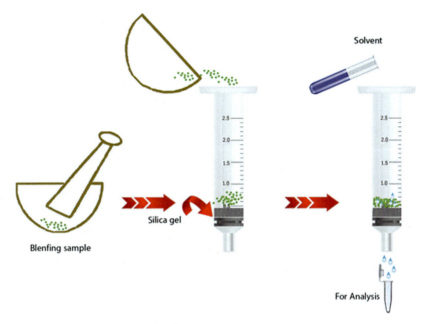

FIGURE 4.7 MDSPE procedure.

1.6 Sorbent-based dynamic extraction

Generally, SPE which was established the groundwork for contemporary sorbent-based extraction methods, is the origin of sorbent-based dynamic extraction. Without a doubt, it can be said that it is innovative procedures for sample preparation known as "sorbent-based dynamic extraction" make use of a variety of sorptive materials to effectively extract target analytes from complex sample matrices. These methods improve the kinetics and capacity of extraction by relying on the sorbent material's constant mobility and modification. They have considered as environmentally friendly and solvent-free substitutes for traditional extraction techniques used to extract analytes from liquid and gas samples.

Sorbent-based dynamic extraction basic idea is to continually expose the sample to new sorbent binding sites. This is accomplished by using a variety of sorbent dynamic methods, including flow through the sample, vibration, mixing, and rotation. Dynamic stir bar sorptive extraction (DSBSE) and rotating disk sorptive extraction (RDSE) are among those methods with RDSE; the sample is rotated by a disk coated with sorbent material, which renews the surface. A stir bar with immobilized sorbent is rotated in DSBSE. The mass transfer process is accelerated by the constant cycling of new sorbent sites, allowing for the thorough and quick extraction of target analytes even in small sample quantities while most of the sorbent's capacity is used.

Sorbent-based dynamic extraction is extremely effective because of its kinetic advantage and selectivity. Finally, the extracted analytes are recovered using the proper desorption technique in order to be determined [69].

In order to the wide benefits of this method, it can be used in various fields. Applications for sorbent-based dynamic extraction may be found in food analysis, pharmaceutical analysis, environmental monitoring, and forensic science. Typical uses for them are in environmental fields like organic pollutants, pesticides, and medicines that are extracted and preconcentrated from environmental samples, such as water, soil, and air. In food analysis field, they are used to extract and concentrate pollutants, pesticides, mycotoxins, and other residues, from food samples. In pharmaceutical analysis pharmaceuticals, metabolites, and drug residues in biological fluids, pharmaceutical formulations, and environmental samples are analyzed using sorbent-based dynamic extraction. Also, forensic science, drugs, poisons, and other substances are extracted and purified from biological samples, such as blood, urine, and tissues, using this method. There are some basic differences between dynamic methods and traditional SPE. The flow dynamics and analyte adsorption mechanism of sorbent-based dynamic extraction and conventional SPE are the primary differences. Dynamic extraction allows for constant contact between the analytes and the sorbent material since the sample passes through the sorbent bed under carefully regulated circumstances. The mass transfer kinetics are improved by this dynamic flow, which also makes it possible to extract analytes from the sample matrix effectively. In contrast, stationary phase contact is used in classic SPE, in which the sample is put onto a solid-phase sorbent bed, given a period of time to interact with the sorbent, and then eluted using a solvent. The equilibrium partitioning between the sample solution and the sorbent material is the basis for analyte retention in SPE. In comparison to conventional SPE techniques, sorbent-based dynamic extraction has several benefits, including quicker extraction kinetics, increased sample throughput, enhanced selectivity, and lower solvent consumption. For a variety of analytical applications, it offers a more effective and adaptable method for sample preparation and analyte extraction [70].

Sorbent-based dynamic methods have been improved and used in order of their wide advantages. One of the main benefits of sorbent-based dynamic extraction is its capacity to decrease interferences from complex sample matrices while selectively targeting the compounds of interest with customized sorbent materials. The method also allows for analysis down to ultra-trace levels and improves detection sensitivity by concentrating trace analytes on the renewable sorptive surface. These techniques are also very adaptable since a variety of sorbent types may extract a broad spectrum of chemicals. High throughput and repeatable extractions with little effort are made possible by automated systems. This method reduces the amount of solvent used compared to traditional preparations, leading to environmentally friendly, economically viable procedures that meet contemporary analytical requirements. In conclusion, sorbent-based dynamic extraction is flexible, efficient, and environmentally conscious, which increases its promise as an ideal sample preparation platform for a variety of applications. It is worth noting that selective extraction, increased sensitivity, versatility across analyte and sample types, process automation capabilities, and favorable sustainability profiles round out its advantages. Although sorbent-based dynamic extraction techniques offer significant benefits, certain associated drawbacks need to be taken into account. Sample matrices may bias quantitation accuracy in different ways or interfere with analyte site binding. Additionally, for suitably concentrated or initial samples, limiting sorbent capacities may lead to analyte breakthroughs. In general, multistep processes require more time to prepare than simpler methods. Importantly, the greater costs associated with acquiring sorbents and specialized dynamic extraction equipment may discourage adoption. The knowledge required to choose sorbents correctly and adjust important experimental parameters to produce effective and selective extraction presents additional challenges. Therefore, while determining whether the distinct advantages of sorptive dynamic extraction outweigh the drawbacks for their particular application against selecting standard sample processing options, analysts must carefully consider objectives, tolerated costs, and in-house skills. Although promising, developing practical methods is essential to effectively address the current obstacles. To make a long story short in the upcoming years, sorbent-based dynamic extraction is expected to see a sharp increase in both acceptance and technological improvement. The creation of novel selective sorbents with optimal morphologies and the integration of many sorptive processes into a single medium to increase application across various analyte classes are two important areas of progress. Computational modeling will allow more specialized, application-specific designs through comprehensive sorbent screening and extraction optimization. Further downsizing of instrumentation footprints can yield point-of-care and field-deployable technologies for on-site analysis and instantaneous decision-making. Furthermore, new developments in immobilized and inline formats will make it easier to couple dynamic extraction with analytical detectors for smooth operations. Given its distinct kinetic benefits, sorptive dynamic extraction is poised to become a widely used sample preparation paradigm, aided by these rapidly developing and merging technologies. When it is used more widely, it will be seen as a flexible, effective, and sustainable alternative to traditional extraction in high-growth fields including drug discovery, clinical testing, omics research, and point-of-use monitoring [15].

1.7 Stir bar sorptive extraction

The first method of the subset of dynamic methods described here is SBSE. It should be mentioned one of the solvent-free techniques for sample preparation is SBSE which was first presented in 1999 by Baltussen et al. It is said this method is related to the dynamic category initially that is used to extract and concentrate organic molecules from aqueous matrices. The process is based on sorptive extraction, in which the analytes are drawn out by an adsorption process and placed onto a polymer covering that is stirred magnetically. Then they are released from absorbent by washing with a suitable solvent, and finally the achieved solvent is injected into analytical devices for analysis. Furthermore, it is worth noting that the analyte partitioning coefficient between the polymer coating and sample matrix as well as the phase ratio between the polymer coating and sample volume are important in the extraction process [71].

1.7.1 Principles of stir bar sorptive extraction

In aqueous samples, the extraction of analytes from the aqueous phase into an extraction medium is governed by the analytes partitioning coefficient between the extractive phase and the aqueous phase. Sorptive extraction is an equilibrium procedure. Research has shown a correlation between this partitioning coefficient and octanol−water distribution coefficient ($k_{o/w}$). $k_{o/w}$, while not entirely accurate, provides a useful indication of how efficiently a particular analyte may be extracted using SBSE or SPME. It is also critical to understand that the sorption equilibrium depends on the phase ratio and, consequently, on the quantity of sorbent (for example, PDMS) used. This connection is illustrated by Eq. (4.5):

$$K_{o/w} \approx K_{PDMS/w} = (C_{PDMS})/(C_w) = (M_{PDMS})/(M_w)\ (V_w)/(V_{PDMS})$$
$$= (M_{PDMS})/(M_w)\ \beta \qquad (4.5)$$

The ratio of an analyte concentration in the polydimethylsiloxane phase (C_{PDMS}) to its concentration in the water (C_w) at equilibrium is known as the distribution coefficient between PDMS and water ($K_{PDMS/w}$). This ratio is equivalent to the mass ratio of the analyte in the aqueous phase (M_w) times the phase ratio β (where $\beta = V_w/V_{PDMS}$) multiplied by the mass of the analyte in the PDMS phase (M_{PDMS}).

The recovery may be defined as the ratio of the extracted analyte (M_{PDMS}) to the initial analyte in the water ($M_0 = M_{PDMS} + M_w$). This ratio is derived from the phase ratio β and the distribution coefficient $K_{PDMS/w}$, as delineated in Eq. (4.2):

$$(M_{PDMS})/(M_0) = ((K_{PDMS/w})/\beta)/(1 + (K_{PDMS/w})/\beta) \qquad (4.6)$$

This formula may be used to determine the theoretical recovery for an analyte using given phase ratio and known partition coefficient. However, the resulting value is only attained in the entire equilibrium state (see below). Eq. (4.2) makes it evident that as $K_{PDMS/w}$ grows, so does extraction efficiency. Since $K_{PDMS/w}$ and $K_{o/w}$, are comparable, extraction efficiency on PDMS generally falls as polarity increases. The phase ratio (β) (volume of sample/volume of PDMS) is also significant in addition to the $K_{PDMS/w}$ factor. The extraction efficiency increases and the β value decreases as the quantity of PDMS increases [72].

This method has some advantages and disadvantages which are noted below:

When compared to conventional sample preparation techniques, SBSE has a number of noteworthy benefits. High-throughput applications find it appealing due to its ease of use, speed, and capacity to handle big sample quantities. Very good extraction capabilities are provided by the thick polymer coating the stir bar, which adsorbs a high concentration of analytes. Cost-effectiveness, environmentally friendliness, and trash minimization are other benefits of not requiring complicated solvent handling procedures. Furthermore, there is a high degree of automation available for SBSE stages including stir bar rinsing and analyte extraction. Potential issues with extremely volatile or polar analytes that do not partition well into the polymer coating are among the drawbacks. If insufficient desorption or ineffective thermal cleaning procedures are applied between analyses, carryover may also become a problem. Certain liquid samples may be too viscous to mix sufficiently to allow for representative sampling. In order to minimize the existing constraints of SBSE and utilize its unique benefits, further process advancements to enhance extraction phase chemistry, stirring parameters, and thermal desorption integration are being made [73].

1.7.2 Future trends of stir bar sorptive extraction

More attention should be paid in the future to the creation of innovative SBSE coating materials that facilitate thermal desorption in GC. Only a small number of SBSE publications have focused on the structural alteration of stirrer bar devices, such as substrates without a coated, coated shaped like dumbbells, and membrane-protected stirrer bar, whereas the majority of SBSE reports have addressed the structural modification of coated substrates. Additionally, the automation of SBSE technology has only been the subject of a small number of papers to date, even though it is a promising avenue for future SPE research.

1.8 Rotating disk solid phase extraction

Berijani et al. reported rotating disk solid extraction (RDSPE) for the first time in 2010. RDSPE was introduced as a unique sorptive extraction method in this research, and it is one of the dynamic methods. A little earlier in 2009, rotation was used for the first time to convenient an extraction device's surface in order to speed up the extraction. Kabir et al. called this as dynamic fabric phase sorptive extraction (FPSE). An innovation in sample preparation RDSPE offers a very effective preconcentration and extraction approach for a variety of analytes in intricate sample matrices. It functions as a substitute for conventional solid-phase extraction methods. RDSPE extracts compounds by partly immersing a rotating disk into the sample solution, as opposed to packing a solid phase into a cartridge or on a disk [74].

1.8.1 Principles of RDSPE

The underlying idea of RDSPE is the extractive phase renewal, which is made possible by the disk rotation. A layer of extraction phase, consisting of a polymer, ionic liquid, or any other substance with affinity for the analytes, is present on the revolving disk. The analytes are adsorbed onto the extractive phase while the disk rotates inside the sample solution. In order to maximize extraction efficiency, the rotation causes new sorptive material to be continuously exposed to the sample solution. Next, in a different step, the extracted analytes are desorbed using a tiny volume of a suitable extraction solvent. It can be said that absorbents are typically the same as the former method [75].

The rotation speed, extraction duration, type of extraction phase, and physicochemical interactions between the analytes and extractive phase are the primary factors influencing the extraction efficiency. Longer extraction periods permit more interaction, but faster rotation renews the surface more quickly. As a result, mass transfer to the surface and analyte sorption kinetics may be controlled by RDSPE and tailored for specific uses. Overall, extraction rates, capacity, and selectivity are improved by the rotating motion's constant renewal of the extractive phase. These characteristics provide RDSPE with a viable, eco-friendly, and effective method of sample preparation [75].

1.8.2 Applications of RDSPE

The extraction, preconcentration, and detection of trace analytes in complicated samples are common uses for RDSPE, as a flexible sample preparation method. High extraction efficiency, quick extraction kinetics, low solvent consumption, less sample preparation, and compatibility with a variety of analytes and sample matrices are just a few benefits that come with RDSE overall. Because of its versatility, RDSPE is a useful tool for analysts and researchers in a variety of sectors who need to prepare samples in a sensitive and targeted manner [76].

1.8.3 Advantages and disadvantages of RDSPE

The technology of RDSPE is appealing due to its several significant benefits. Analytes may be extracted from the sample solution with maximum extraction efficiency and quick kinetics, thanks to the constant renewal of the sorbent coating caused by the disk rotation. This is one of the main advantages. Furthermore, by choosing the right sorbent materials, RDSE is a flexible method that may be used with a wide variety of analytes polarity. For high-throughput analysis, it can also be automated with integrated rotating systems and auto-samplers. Furthermore, because the desorption stage requires small volumes of solvent, RDSPE is compliant with green chemistry principles. In conclusion, RDSPE provides superior sustainability credentials through solvent reduction, excellent extraction efficiency and enrichment factors, flexibility in sorbent selection and analyte compatibility, automation possibilities for expedited analysis, and more. These complementing performance characteristics highlight RDSPE potential as a useful substitute for traditional sample preparation techniques [77]. Although RDSPE has a number of noteworthy benefits, there are also certain drawbacks to the method to take into account. One major drawback is that the rotating limited surface area and finite sorbent capacity limit the usefulness of the device to samples with extremely high analyte concentrations or unclean matrices where the capacity is prone to overloading. Problems with sample loss may also result for analytes in the sorbent coating not fully being adsorbing or desorbing. Additionally, samples with multiple matrix interferences may exhibit reduced selectivity, which might result in nonspecific binding and accuracy issues. Analytes with undesirable polarity, volatility, or affinity characteristics that restrict sorptive interactions with the selected sorbent may also not be suitable for RDSPE. Practically speaking, RDSPE calls for specialized apparatus that involves large initial investments and continuous maintenance concerns for disk rotation, sample handling, and solvent delivery. Therefore, even though they offer obvious advantages, analysts should consider performance requirements in addition to RDSPE constraints when choosing the best sample preparation techniques for a particular application [77].

1.8.4 Future trends of RDSPE

Considering its high extraction efficiency, green analytical capabilities, and automation amenability, RDSPE is expected to grow and be adopted rapidly in the future. Beyond the present uses, which are mostly concentrated on food, beverage and environmental testing, growth areas are probably going to include the pharmaceutical, semiconductor, and biomedical sectors. Furthermore, developing sensor technologies will make it easier to incorporate detection, in situ extraction, and RDSPE into tiny "lab on a disk" devices for inexpensive, portable field analysis. Significant prospects exist for broadening the spectrum of extractable analytes by the creation of innovative, selective multicomponent sorbent coatings having microscale topologies that are tuned using computational modeling techniques. Advanced robotics automation may reduce labor and solvent use while increasing throughput, repeatability, and user-friendliness. Adoption rates are expected to rise in the upcoming years as RDSPE hardware/software and multi-functional sorptive microextraction phases coevolve. This is because RDSPE offers a number of advantages over established sample preparation methods, such as SPE or SPME, across a variety of applicable industries.

1.9 Magnetic solid phase extraction

Because of the benefits of this format over traditional packed SPE, the usage of magnetizable materials has grown in importance and popularity among the many SPE techniques. Šafaříková and Šafařík invented the technique known as magnetic SPE (MSPE) in 1999. It involves dispersing a magnetic substance into a sample solution and recovering it quickly and easily using a magnetic field. Common difficulties with traditional packed SPE, such as excessive back pressure, and sorbent packing concerns, are avoided using MSPE. Analytes are separated from sample solutions using MSPE, a sample preparation method that employs magnetic particles coated with a functionalized sorbent substance. Moreover, magnetic sorbents offer flexibility in extraction schemes by facilitating dispersive-MSPE, in situ MSPE, and online MSPE in microfluidic lab-on-a-chip platforms as well as direct coupling with analytical systems. These features make MSPE well-suited for high-throughput analysis, while greatly minimizing solvent/labor costs. The recent years have therefore witnessed tremendous progress in expanding the applicability of MSPE across diverse fields of analysis. MSPE methods have become highly reliable alternatives to conventional SPE procedures [78].

1.9.1 Principle of MSPE

MSPE makes the use of magnetic particles that have been functionalized with specific functionalized groups, such as coated films or surface-immobilized ligands. These particles are usually micro- or nanoparticles made of iron oxides or other magnetically sensitive materials. To selectively extract the analytes, these magnetic sorbents are simply added to sample solutions containing the analytes. After sorption, the magnetic particles migrate in the direction of the magnet due to the application of an external magnetic field. This effectively purifies and concentrates the targets in a single step by allowing the separation of the magnetic sorbent particles from the original sample solution and any analytes recovered from them. Before the final elution of the concentrated target analytes, the sorbent materials bound to the magnet can be cleaned using tiny quantities of suitable organic solvents or solutions to eliminate loosely adherent interferences. Analytes can be recovered for sensitive identification and quantification using chromatography or mass spectrometry systems thanks to magnetic sorbent which integrate and streamline previously multistep sample preparation methods based on the magnetic field-enhanced separation principle. These methods allow for rapid, selective, yet gentle extraction and preconcentration [79] (Fig. 4.8).

1.9.2 Applications of MSPE

Because of their simplicity, ease of surface modification, and most importantly, their adaptability, magnetic nanoparticles (MNPs) are utilized extensively in a variety of fields, including analytical chemistry, biotechnology, and

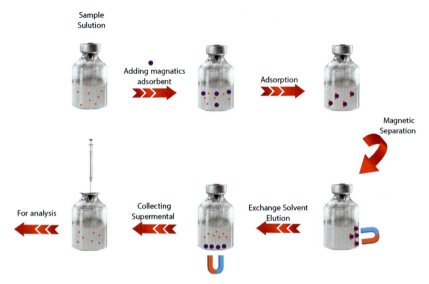

FIGURE 4.8 MSPE procedure.

medicine. These materials are utilized in MSPE to isolate and enrich analytes, frequently from complex matrix samples including biological, environmental, and dietary samples. The selection of an acceptable approach to detect and quantify analytes should be taken into consideration in addition to the appropriate modification of the magnetic core coating of MNPs, as analytes in various samples have variable physicochemical characteristics [80].

1.9.2.1 Biological samples

Chemical stability, degree of dispersion, and biocompatibility are the main constraints on the bioanalytical use of MNPs. MNPs' appropriate size, shape, surface chemistry, and composition must all be precisely specified. Because of its physicochemical characteristics, silica is the material most frequently utilized to cover the magnetic core for bioanalytical applications.

1.9.2.2 Environmental samples

There are various chemical contaminants and each environmental contaminant is a distinct chemical. Because of the nature of the matrices, the location of collection, and the variety of sample components that might act as interfering agents in later stages of the analytical process, environmental samples are extremely complicated study materials. For this reason, the process of getting samples ready for analysis is required. The majority of the time, analytes are hydrophobic, while the matrix of environmental samples is hydrophilic. The majority of sorbents concentrate based on their analyte affinity toward analytes. Nevertheless, their compatibility with the sample matrix is not given enough thought, which might lead to some operational difficulties or even negatively impact how well the sample is prepared for analysis in the future. It has been documented that MNPs coated with carbon may be used to remove PAHs from aqueous samples.

1.9.2.3 Foods

Consumers and organizations that are widely described as public health are particularly interested in food safety and quality. Chemical pollutants are one of the factors used to evaluate a product's safety before it is intended for human consumption, which is why foodborne chemical contamination has been closely monitored for many years. Nonetheless, food samples include an extremely complex matrix, and chemical contaminants in food have distinct chemical and physical characteristics. Much studies have been done on the use of MNPs in SPE to separate impurities from food samples. Nevertheless, a great deal of scientific and technological work has to be done before they can be used in food analysis. A careful evaluation of MNPs' possible health concerns to people is necessary.

1.9.3 Comparison of MSPE with other methods

Compared to conventional SPE techniques and other comparable methods, MSPE provides a number of benefits.

1. Short extraction time: Because the MNPs utilized in MSPE can effectively and quickly are collected in the presence of magnetic fields, MSPE can greatly shorten the extraction time compared to standard SPE procedures.
2. Greater extraction efficiency: Compared to conventional SPE sorbents, the MNPs used in MSPE have a better extraction efficiency because of their large surface area, which can offer more active sites for adsorption.
3. Less solvent consumption: Generally speaking, MSPE uses less solvent than conventional SPE techniques, which makes it a greener choice.
4. Ease of separation: Unlike traditional SPE procedures, which need time-consuming and arduous filtering or centrifugation stages, MSPE uses magnetic nanoparticles that are quickly separated from the sample using an external magnetic field.
5. Reusability: MSPE is a financially advantageous solution over time since the MNPs utilized in it are simply regenerable and may be used repeatedly without losing their extraction effectiveness.

All things considered, MSPE provides a quicker, more effective, and more ecologically friendly substitute for conventional SPE techniques for sample preparation and extraction [81].

The use of MNPs as sorbents endows unique advantages to MSPE including rapid and efficient phase separation via external magnets, improved extraction kinetics owing to the high surface area to volume ratio of the magnetic micro/nanoparticles, higher extraction capacity and recovery of analytes, excellent compatibility with different complex matrices as well as solvents, and amenability to automation. Additionally, the magnetic separation enables selective and thorough purification of analytes from intricate sample matrices, since the interference from sample particulates as well as other matrix constituents is greatly reduced prior to analysis [82].

1.9.4 Future trends of MSPE

MNPs' exceptional qualities provide a very bright future for their use in this area. Therefore, optimizing current MNPs synthesis techniques to achieve monodispersed and regulated MNP form is crucial. Isolating tiny quantities of analytes from complex matrices, particularly biological samples, is the main problem. More focus should be placed on creating novel coatings that will enhance dispersity in aqueous solution, guarantee the magnetic core's effective protection, and produce multifunctional MNPs.

1.10 Thin film microextraction

Thin film microextraction (TFME) is a novel approach to sample preparation that offers a rapid, easy, and effective substitute for more traditional techniques like SPE and LLE. Analytes from liquid or gas samples are extracted and concentrated on a thin polymer film that is positioned on a solid substrate. The use of a bigger volume of the extractive phase with a larger surface area is the primary way in which this innovative technology differs from traditional SPME. In doing so, the ultimate objective of greater sensitivity is achieved without shortening the sampling period. By submerging a thin coated strip in a sample solution or subjecting it to a gaseous sample, analytes are adsorbed by the coated film. Usually ranging from 10 to 200 μm in thickness, the film may be made using various polymers, offering selectivity for distinct categories of analytes. PDMS, PA, metal−organic framework, covalent organic framework, and styrene polymers are frequently used materials as coating. Following extraction, the strip is taken off, and the concentrated analytes are either thermally desorbed in an analytical device for further analysis or eluted with an appropriate solvent [83].

1.10.1 Principles of thin film microextraction

Regarding their suitability for various instruments, TFME may be categorized into two primary groups:

(1) Thermally desorbed TFME that works with devices that let analytes to be introduced into gas phase.
(2) Devices for solvent-desorbed TFME that may be utilized in conjunction with any type of analytical instrument capable of handling liquid samples.

There are several extractive materials and thin film preparation techniques available for their synthesis. With this approach, the device's speed of withdrawal from the slurry mostly controls the thickness of the extractive phase. By applying several homogeneous layers (i.e., dipping the same device many times), the coating thickness may be reproducibly raised using a certain extractive phase and withdrawal speed. Because dip coating is so practical and easy to use, it is widely employed in the manufacturing biocompatible TFME devices in which extractive particles are immobilized in PAN. In recent years, there has been a growing need for ultra-thin extractive phases for very sensitive applications like mass spectrometry direct coupling, where rapid desorption kinetics are crucial. One of the best coating techniques for creating smooth surfaces with coating thicknesses as thin as a few micrometers in these kinds of applications is dip coating. The majority of dip coating techniques produces a coating by allowing the extractive phase to physically accumulate on the surface without forming a chemical connection. However, a chemical bonding step can be required depending on the coating type. To encourage the creation of functional moieties that easily interact with the extractive phase in this situation, the supporting surface has to be activated or prefunctionalized. This is not the same as the physical deposition method described above since, in most cases, the initial immersion occupies the majority of the surface-active groups on the support, leading to a thinner layer of extractive phase. Furthermore, designing and optimizing the extraction phase is more time-consuming and difficult because coating deposition with surface activation or prefunctionalization necessitates the creation of chemical bonds to the surface. Examples of TFME created by chemical modification include aptamer immobilized paper spray devices and fabric phase TFME, which was made via a sol-gel technique. An alternative coating technique is called spin coating, in which a disk is coated with a liquid extractive phase, which is then spun at a high speed to disseminate the phase as a thin layer across the disc's surface. If the extractive phase (such as PDMS) is sufficiently self-supportive, the spin coating can be employed to manufacture TFME on a supporting surface or it can be utilized without support. This process yields coatings with various sorbents in each layer and allows for the deposition of numerous layers of extractive phases. A primary constraint of this technique is its potential to impact interdevice repeatability because of differences in thickness between the produced bulk extractive phase's outside and inner sections. The bar coating technique may be used to create TFME devices with or without a supporting core, much like spin coating [84].

1.10.1.1 1-TFME with thermal desorption

Heat-stable extractive material thin films can be immediately thermally desorbed into GC system via a temperature-programmable volume thermal desorption unit or a Curie point injector, or they can be desorbed in a headspace sampler unit before being injected into the GC. There are not many thermally stable extractive phases and not every laboratory has access to large-volume thermal desorption equipment, hence there are not many advancements and uses in this field. An intralaboratory validation research, for example, employed PDMS/DVB and PDMS/DVB-carbon mesh-supported membranes for the extraction of pesticides from aqueous samples. While both TFME devices had higher sensitivity, a comparison of the output from two different membrane types and the output from a reference technique (LLE) revealed similarities in accuracy. Therefore, for laboratory samples that have been collected, TFME devices not only offer improved quantitation sensitivity at ultra-trace levels, but they also provide a process that is less labor-intensive and significantly reduces the use of organic solvents. The TFME device's strong construction and ease of use allow for direct on-site deployment. The sol-gel method was first introduced by Alcudia-Leon et al. via chemically bonded poly(dimethyl diphenyl siloxane) (PDMDPS) to silica fiber glass fabric. The coated cloth was then put in a makeshift holder that was attached to a pump and utilized as an active on-site air sampler in the field, primarily focusing on insect sexual pheromones. The analytes that were desorbed from the sampler were promptly transferred into GC–MS for further separation and detection when the TFME field sampling was completed. The holder was then put straight into a headspace desorption unit. Many additional on-site applications might benefit from the usage of such a handy field sampler, especially as an addition to current sample preparation techniques that can be more laborious and unfeasible. MIP thin films were created by Hijazi and Bottaro in a different investigation to enable the selective extraction of semivolatile thiophenes from seawater. The authors of this study also came to the conclusion that in order to achieve full desorption from such a coating, a small amount of acetonitrile which serves as a phase transfer mediator and reduces response variability had to be applied. For several analytes in seawater samples, subppb detection limits and low matrix effects were obtained, confirming that the approach would be a useful analytical tool for the measurement of thiophenes in developing environmental conditions, such as hazard analysis following an oil spill. It is important to note that solvent desorption applications can also be employed with thermally stable TFME devices. They therefore do not limit themselves to GC-based applications [85].

1.10.1.2 TFME with solvent desorption

Solvent desorption from TFME is more difficult than heat desorption. The absence of standard interfaces for the one-line connection of TFME devices to the analytical instrument workflow is one of the primary challenges. This would make it easier to insert the extracted analytes straight into the analytical device. Additionally, during the desorption process, a greater amount of eluate is needed due to the comparatively big TFME device. This does not fully utilize the large extractive phase because only a small portion of the solution is introduced to the instrument after solvent desorption. Unless additional evaporation and reconstitution steps are used to generate a smaller, more concentrated sample volume to account for the extra dilution; this does not fully utilize the large extractive phase. Furthermore, the liquid phase diffusion kinetics are slower than those of the gas phase, leading to comparatively longer desorption periods [86].

1.10.1.2.1 TFME materials devices PAN is often employed as a binding glue to fixate the extractive phase on the supporting surface during the fabrication of biocompatible TFME devices. When PAN is utilized as an overcoating, it not only ensures that the device is biocompatible (i.e., nontoxic), but it also functions as a molecular cut-off filter, further shielding the extractive phase from macromolecular bio-fouling. A recent study provides a thorough summary of the uses of HLB, C_{18}, C_{18} with benzene sulfonyl acid moieties (i.e., mixed-mode), and other extractive phases utilized with PAN-based TFME devices in food, environmental, and clinical studies [87].

These kinds of devices may be repeatedly reused because of their inherent feature. However, PAN-based devices cannot be reused because they are not stable enough for direct desorption in GC. Even single-use devices might provide a high background because of significant breakdown products. They have not yet been reported for direct thermal desorption in GC due to these reasons. However, when utilized in conjunction with a DART source, where the extractive phase is exposed to high temperatures for brief periods, PAN-based devices can withstand such temperatures without the binder decomposing. Notwithstanding these drawbacks, the field's prevailing benefits such as the simplicity of automating the sample process and the abundance of accessible or potentially useful extraction phases and supports continue to spur new developments.

An effective technique for creating extractive films with ultrafine nano-to-micro-sized particles and nonwoven fibrous structures is electrospinning, which is based on pulling charged threads of polymeric solution with electric force. This process offers a cutting-edge platform for the creation of novel TFME devices. Compared to traditional coating techniques, the thin films generated by the electrospinning method provide a number of benefits, including large coating-specific surface areas and glue-free immobilization of the polymeric extractive phase, which guarantee extremely quick extraction kinetics. Similarly, it is also feasible to fine-tune the parameters of the extraction phase by adding different dopants. A number of recent researches have been carried out using this methodology [88].

1.10.2 Comparison and applications of TFME

It has been demonstrated that, in comparison to traditional SPME, TFME offers better sensitivity and extraction kinetics using the mathematically developed equations below. By using SPME under equilibrium circumstances and the basic mass balance, the TFME technique's increased sensitivity may be achieved. The inherent link between the quantity of analyte (number of moles) extracted into the extractive phase at equilibrium (n_{eq}) and its initial concentration (C_s^0) in the sample is illustrated by Eq. (4.7), which is obtained from the mass balance under equilibrium circumstances. The distribution constant of the analyte for the extractive phase and sample matrix is represented by K_{es} in this equation, where V_s and V_e stand for the sample and extractive phase volumes, respectively. In the end, this formula proves mathematically that greater extractive phase volume will lead to higher analytical sensitivity.

$$n_{eq} = K_{es}V_sV_e/(K_{es}V_e + V_s) + C_s^0 \quad (4.7)$$

In the abovementioned equation, the equilibrium extraction time (t_e), defined as the time required to extract 95% of the equilibrium amount, can be calculated by Eq. (4.2); where, D_s, b, K_{es}, and d refer to the diffusion coefficient of analyte in the sample, the thickness of the extractive phase, the equilibrium constant relating to the extraction into the thin film, and the thickness of the boundary layer, respectively. This equation illustrates that for a given amount and type of extractive phase, a faster extraction equilibrium can be achieved when a thin layer of extractive phase is used [89].

$$t_{e}^{1/4} = 3\,\delta K_{es}(b)/D_s \quad (4.8)$$

1.10.3 Future trends of TFME

Researchers anticipate that when TFME finds new uses and advances the technology, it will continue to expand and change in the future. The following are a few possible future developments:

1. A rise in the application of environmental monitoring
2. Combination with other analytical methods
3. Automation and miniaturization
4. Creation of novel materials

The creation of innovative materials that improve this technique's performance might be a future trend. All things considered, TFME future trends are probably going to be more technological innovation and advancement, which will expand its applications and boost its use across a range of industries.

1.11 Dispersive solid phase extraction

An improved version of the SPE method known as dispersive SPE (DSPE) has gained a lot of attention since Anastasiadis et al. published it in 2003 as a useful cleanup step for extracting pesticides from food. This extremely simple technique uses centrifugation and agitation alone to apply the solid sorbent directly into the volume of a liquid sample solution. This straightforward method guarantees a large surface area of contact between the sorbent and the sample, facilitating a speedy attainment of the extraction equilibrium. Because extremely little sorbent and solvent are used, the resultant process is both quick and effective. Compared to normal SPE, it is also more environmentally friendly.

The most appealing aspect of DSPE, in addition to its simplicity, versatility, and ease of handling as compared to conventional approaches, is its capacity to reduce sample treatment times, allowing for the analysis of more samples in less time. Because of its adaptability, durability, and selectivity, DSPE has been used as a clean-up approach since its development [90].

1.11.1 Principles of DSPE

DSPE is often used to clean up the matrix, which implies that any potential matrix interferences are left on the sorbents after the sorbents have been dispersed in the solution or matrix containing the analytes. The target analytes are then gathered in the supernatant after the sorbents are centrifuged out of the desired solution. On the other hand, the analytes can be trapped in the sorbents to carry out D-m-SPE. Centrifugation or filtering is used to separate the sorbent that contains the target analytes after extraction. After that, the analytes can be desorbed or eluted using the proper desorption solvent. Additionally, DμSPE provides the following benefits over DμSPE: less time commitment, less solvent use, and easier operation. Target analytes can be separated from matrix solution using D-m-SPE by adsorption on a solid sorbent. Analyte adsorption is dependent on a variety of processes, including pep interactions, dipole-dipole interactions, and hydrogen bonding. For DμSPE. To yield an accurate, sensitive, and selective analyte detection, the physicochemical characteristics of the solid sorbent are crucial [91]. Enhancing agitation improves the mass-transfer coefficient (b) of analytes from an aqueous phase onto a solid sorbent by reducing the thickness of the Nernst diffusion film (d). These occurrences result in an increase in the efficiency of the extraction procedure in a short period. DμSPE, similar to SPE, is a method that relies on surface properties since its rate of adsorption is directly influenced by the area of contact between the substances being analyzed and the solid sorbent. DμSPE facilitates the separation of desired substances from a solution containing various components by capturing them on a solid sorbent through adsorption. The adsorption of analytes is determined by several processes, including hydrogen bonding, dipole–dipole interactions, and $\pi-\pi$ interactions. The composition and physicochemical features of the solid sorbent play a crucial role in achieving precise, sensitive, and selective detection of target analytes in DμSPE. Practically, a reliable sorbent must meet the following key criteria: (a) rapid and complete adsorption and desorption, (b) a large surface area and high capacity, and (c) excellent dispersibility in liquid samples [91].

1.11.2 Advantages and disadvantages of DSPE

DμSPE offers a number of significant benefits over conventional SPE techniques. First, it employs significantly smaller amounts of sorbent (<1 mg), which lowers the cost of materials. Furthermore, quick extraction without the need for column-passing stages is made possible by the direct injection of sorbent into samples. It also makes very large throughputs possible. Furthermore, DμSPE reduces waste and solvent consumption for better sustainability without the need for specialized SPE apparatus. Many applications find DμSPE attractive because of its speed, cost-effectiveness, and environmental friendliness. Even with the many benefits, there are several restrictions. Phase interactions are complicated, making method development difficult. Inadequate control over particles mixing and sample modifications in between repetitions can also negatively impact reproducibility. Furthermore, in certain DμSPE techniques, sorbent recovery may provide challenges that restrict sorbent reusability. Concerns with back pressure have also been reported when integrating with capillary electrophoresis (CE) or LC systems. Addressing these shortcomings is being aided by ongoing innovation [92].

1.11.3 Future trends of DSPE

It is widely anticipated that as many sciences advance, there will be a greater number of synthetic sorbents available in the future, and additional DμSPE applications will be considered. Despite its benefits, DμSPE applicability is regrettably restricted to the analytical scale. We believe that collaboration between industry and researchers is necessary for the wider implementation of DμSPE. It is well known that the characteristics of the solid sorbent play a crucial role in achieving precise, sensitive, and selective target analyte detection using DμSPE. Therefore, new approaches for the large-scale, cost-effective manufacturing of effective sorbents must be presented by researchers in order to realize the industrial success of DμSPE. Green and environmentally friendly sorbent manufacture requires more work.

1.12 Packed sorbent microextraction

Since the packed sorbent microextraction (PSME) introduction in 2004, it has gained recognition as a desirable choice and effective sample preparation method fit for addressing analytical and bioanalytical tasks. This new method preconcentrates and extracts analytes from various matrices using a well-known sample preparation method. With a few key distinctions, PSME is a scaled-down version of the SPE method for sample purification and extraction. In contrast to SPE, PSME integrates the packing right into the syringe rather than having it in a separate column. As a result, unlike in SPE, no additional device is required to apply the sample to the solid phase. For plasma or urine samples, MEPS can even be used more than 100 times, whereas a traditional

SPE column is only used once. PSME is capable of handling both small and large sample volumes (10 µL to milliliter levels) and may be linked online to capillary electrochromatography (CEC), LC coupled to mass spectrometry (LC—MS), and GC—MS. Furthermore, PSME is a versatile method for preparing samples in ion-exchange, reversed phase, normal phases, and mixed mode versions. Using the same syringe, PSME may be used as an online sampling device that is completely automated, encompassing the extraction, injection, and sample processing stages [93].

1.12.1 Principles of PSME

Generally speaking, the bases of this method are not very different from traditional SPE, and it is based on analyte adsorption/desorption process. To the best of our knowledge in PSME, ~2 mg of the solid sorbent is packed as a plug or as a cartridge between the barrel and the needle in a syringe (100—250 µL). On the packed sorbent, sample extraction and enrichment may be completed. The PSME syringe and the PSME cartridge also referred to as the BIN, are the two components of this apparatus that integrate sample extraction, preconcentration, and clean-up. Built within the syringe needle, the packed PSME bed (BIN) provides a sturdy support that holds onto the target analytes as the sample goes through it. When using the BIN, a gas-tight PSME syringe (100-µL or 250-µL) is employed, allowing for fluid handling at standard SPE pressures. It is simple to swap out the BIN by unscrewing the locking nut, taking the BIN out, and then replacing it when it runs out or another phase is needed. There are several ways to control the entire equipment, including manually and online. The PSME method of sample preparation works well with ion-exchange chemistries, mixed modes, reversed phases, and normal phases. Generally, as it said at the beginning MEPS is an SPE adaption that packs all the desired features into a smaller device with an average retention volume of less than 10 µL. This, together with the fact that it works with autosampler syringes, makes the MEPS format perfect for an analytical method [94].

1.12.2 Differences between PSME and SPE

The solution flow is one way (up to down) in SPE, whereas it is two ways (up and down) in PSME which is the difference between the two approaches. An introduction to the analytical instrument as well as sections on conditioning, loading, washing, and elution are usually included in a PSME operating stage. This method can be expedited in a short amount of time by the PSME completely automated sample preparation apparatus. A one-way check valve and a PSME syringe allowed for the realization of automated or semiautomatic PSME. Automation was used for sample loading, washing, and elution. Sample preparation is rapid since MEPS only takes a tiny volume of sample and a small quantity of solvent. Large numbers of several samples may be handled quickly by automated MEPS processes, which improves analysis efficiency and accuracy. In PSME research, the ability of gas-tight microliter syringes to self-modify is still crucial. MEPS sorbent is reusable, unlike SPE, which is only meant to be used once. Whereas the solution flows in two ways (up and down) in PSME, it flows in one direction (up to down) in SPE. Additionally, a novel type of PSME device is called SPDE. Both MEPS and SPDE employ the same methodology; the only differences are that in SPDE, a one-way microvalve was introduced, and sorbents with a lower size (5 mg) were utilized instead of 50 mg [95].

1.13 Micro solid phase extraction

In 2006, microSPE (μ-SPE), or membrane-protected μ-SPE, was presented as a substitute for multistep SPE. A scaled-down variant of conventional SPE, μ-SPE is a method of sample preparation. It uses column beds or coatings with sorbent amount ranging from micrograms to milligrams to extract target analytes from milliliter- or microliter-sized liquid samples. Because μ-SPE columns are meant to be used just once, they solve problems with traditional SPE that include cartridge reuse, sorbent drying, and carryover [96]. μ-SPE works on the same concept as SPE, using the analytes' affinity for a solid sorbent material in the liquid sample to extract and preconcentrate the analytes. The same physicochemical processes—such as reversed phase and ion exchange—are necessary for this. Because each extraction uses a separate disposable cartridge, the smaller format has the benefit of requiring less sample and solvent. Shorter flow pathways and diffusion lengths are caused by the tiny sorbent bed volumes in the microgram—milligram range, which improves kinetics for quick extraction. Conditioning, loading, washing, and elution are the same processes in the process, although on a smaller scale. The analytes that have been preconcentrated by μ-SPE can then be determined by analyzing the final eluent using the appropriate analytical equipment. High throughput analysis and small-volume bioanalytical applications are suitable uses for the downsized technique [97]. One of the many alluring advantages of μ-SPE is its significantly lower sample and solvent needs, which improve sustainability. The small quantities of the sorbent bed quicken the kinetics of extraction, enabling quicker sample preparation. Additionally, the smaller single-use cartridge format avoids the problems associated with traditional SPE in terms of drying, carryover, and reusing. But there are a few drawbacks as well. There are capacity issues with filthy or highly concentrated samples since there is a limited amount of sorbent available. Factors related to concentration may be limited by lower phase ratios between the sorbent and sample. Furthermore, because the performance of μ-SPE is sensitive to even small modifications in the sorbent, eluate, and physical parameters, validation and optimization processes are essential. To fully utilize the possibilities, intensive method development efforts are frequently required. However, in optimal circumstances, μ-SPE's high-throughput capacity, velocity, solvent reduction, and extraction efficiency makes it a compelling sustainable sample preparation method, particularly for small-scale bioanalytical applications where it addresses major problems with traditional SPE formats [98]. μ-SPE is anticipated to have significant expansion in the future by utilizing cutting-edge technology. Microchips and lab/organ-on-a-chips that integrate μ-SPE with analytical detectors for seamless analysis are predicted to continue miniaturizing. Unprecedented flexibility in μ-SPE device design and functioning might be achieved by 3D printing. Additionally, efforts are being made to optimize automation and improve selectivity using cutting-edge sorbent materials. Furthermore, it is anticipated that μ-SPE will continue to permeate high-value industries where sample sizes are decreasing, such as clinical diagnostics, pharmaceuticals, forensics/security, and environmental monitoring. There appears to be an increasing trend toward the use of portable instrumentation that depends on μ-SPE sample preparation, which might lead to greater options for on-site analysis.

1.14 Fabric phase sorptive extraction

Fabric phase sorptive extraction is a novel technique for sample preparation developed by Kabir and Furton in 2014. It cleverly combines SPE and SPME into a single technological platform. By simultaneously exerting an exhaustive extraction mechanism and equilibrium-driven extraction during the sample preparation process, FPSE achieves exhaustive or nearly exhaustive extraction even when the extraction is conducted under equilibrium extraction conditions (e.g., direct immersion extraction). Therefore, rather than being a new format for either SPME or SPE, FPSE is a true combination of both methodologies. Analytes may be extracted straight from sample matrices using the flexible FPSE, which doesn't need samples to be altered. Consequently, it does away with all pre- and postsample preparation procedures such as centrifugation, filtering, solvent evaporation, and sample reconstitution, thereby reducing the possibility of analyte loss, experimental mistakes, and sample preparation expenses. Not only FPSE has successfully combined the extraction techniques used in SPE and SPME, but it has also made available all of the sorbents that are exclusively used in SPE or SPME. For example, poly (dimethyl siloxane), or PDMS, is a popular sorbent coating used in SPME and FPSE. FPSE is the first sample preparation technique that utilizes the surface chemistry of the substrate. The organic polymer and one or more organically modified inorganic precursors are the origins of the FPSE membrane's extraction efficiency and selectivity. FPSE also benefits greatly from the sol—gel synthesis technique, which uses an inorganic or organically modified linker to chemically bond the organic polymer/ligand to the substrate. The FPSE membrane has an exceptionally good solvent, chemical, and thermal stability because of the chemical link between the polymer and substrate. The FPSE membrane may be utilized in an exhaustive extraction mode (as an SPE disk) or an equilibrium-based extraction mode (as in DI-SPME) because of its flat shape and open bed. Lakade et al. have shown that the FPSE membrane may be utilized as an SPE disk without sacrificing the quality of the analytical data, even if the application potential of the membrane has not yet been completely investigated [99].

1.14.1 Principles, advantages and disadvantages, and future trends of FPSE

The extraction mechanism of SPME and related microextraction methods, such as SBSE and TFME, is mimicked by FPSE in its traditional operating mode, direct immersion extraction. The FPSE process typically includes the following steps: the FPSE medium coated with sol–gel sorbent is first cleaned of any unwanted contaminants by immersing it in a suitable solvent. Subsequently, the material is washed with deionized water to eliminate any remaining organic solvent residue. The FPSE medium is then added to a glass vial containing a clean Teflon-coated magnetic stir bar, and a portion of the sample solution is transferred into the vial. For an appropriate extraction time, the sample is magnetically agitated so that the sorbent may sorb the analytes. The adsorbed analytes are then back-extracted into an eluting system when the FPSE device is taken out of the vial and placed in contact with the eluting solvent in a different vial, after that, an aliquot of the eluent is injected into the analytical apparatus (Fig. 4.9). Under equilibrium extraction conditions, the mass of the analyte(s) extracted by the FPSE membrane is proportional to the partition coefficient between the FPSE membrane and the sample matrix (K_{es}), the volume of the extracting phase (V_e), the volume of the sample (V_s), and the initial concentration of the

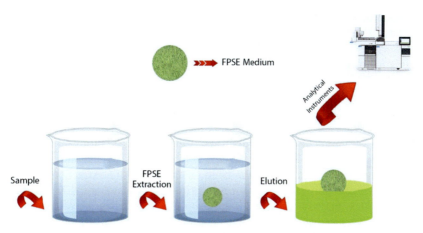

FIGURE 4.9 FPSE procedure steps.

analyte (C_o). This partition coefficient varies with different fabric substrates and the sorbent coatings on the substrate surface [100].

It is possible to express the mass of the analyte extracted by the FPSE membrane at equilibrium (n) as follows:

$$n = \frac{K_{es}V_eV_sC_o}{K_{es}V_e + V_s} \quad (4.9)$$

when the volume of the sample is greater than the volume of the sorbent used for extraction ($V_e \ll V_s$)

Eq. (4.3) is converted to:

$$n = K_{es}V_eC_o \quad (4.10)$$

Eq. (4.4) suggests that the mass of the analyte extracted by the FPSE membrane (n) is independent of sample volume and exactly proportional to the volume of the extracting sorbent. Therefore, if the starting concentration of the analyte(s) is maintained constant, the value of n can be raised by increasing the amount of the extracting sorbent. It must be noted that, the only microextraction method that provides the whole spectrum of sorbents, such as polar, medium polar, nonpolar, cation exchanger, anion exchanger, mixed mode, zwitterionic, and zwitterionic mixed-modal sorbents, is FPSE. It should be mentioned that any of these sorbents can be coated on the substrates that are polyester (hydrophobic) fiber glass (neutral), or 100% cotton cellulose (hydrophilic) [101]. FPSE has several key advantages that make it an attractive extraction technique. Firstly, FPSE provides very high extraction sensitivity and enrichment factors, allowing trace levels of analytes to be detected. Additionally, the fabric sorbent is stable and durable, allowing for multiple extractions and analyses without loss of performance. FPSE is also relatively fast, simple, and affordable compared to other extraction methods. Furthermore, the flexibility in the selection of sorbent kind allows FPSE to be optimized for different classes of analytes. However, FPSE also have some disadvantages. One of them is that the fabric sorbents can be limited in terms of selectivity and may coextract be interfering matrix components in some cases, requiring additional cleanup steps. Analyte carryover between extractions is a problem if the sorbent not properly conditioned. Finally, automation and high-throughput analysis capabilities are not as straightforward with FPSE compared to some other microextraction techniques [102].

A chemometric design of the experiment strategy or the traditional one-factor-at-a-time methodology might be used to carry out this technique development activity. The latter strategy is the more environmentally friendly and scientific one as it offers a thorough understanding the extraction process as a whole and clarifies whether or not various elements interact. To identify the variables that have the most impact on the total extraction efficiency, a screening design can be used. The optimal values of the most important components can then be determined by using a response surface model design.

1.15 Air-assisted solid phase extraction

In order to enhance the efficiency of SPE technique, which has drawbacks such as a lengthy extraction time and relatively low extraction recovery of, the notion of air-assisted SPE has arisen. During the extraction process, in air-assisted SPE turbulence is created and agitates the particles by introducing compressed air flow through the sorbent bed. Within the packed bed, this additional air stream creates regulated fluidization. In order to reduce the barrier to mass transfer, the dynamic motion breaks apart the diffusion layers that surround sorbent sites. Furthermore, the strong mixing increases the interaction between the solution and the sorbent, which improves the analytes and penetration into interior pores. This enables faster absorption throughout the whole sorbent binding regions. This approach allows for effective elution and analyte recovery by removing trapped compounds more quickly in addition to its capacity to quickly redirect airflow. In general, air-assisted SPE speeds up the rates of sorption and desorption to increase the productivity of extraction. This sets the method apart from conventional vacuum- or passive-based SPE procedures [103].

1.15.1 Principles of air-assisted SPE

Overall, during the extraction process, compressed air flow is passed over densely packed sorbent beds to provide controlled turbulence and particle fluidization. This is how air-assisted SPE works. In addition to promoting convection currents via pores rather than only surface adsorption, the improved hydrodynamics and mixing also increase the number of dynamic interfaces between the sorbent and sample, activate new binding sites, and lower stationary diffusion barriers. When combined, these effects improve the kinetics of inflow and sorption, make full use of the binding capacity of interior particles, and hasten the clearance of analytes that have been collected. When compared to conventional passive techniques that rely only on gravity or vacuum pressure differentials, this convection-powered agitation and circulation increases mass transfer rates to drive quick and effective extraction. The additional air stream overcomes transport resistances that usually impede extraction completeness and throughput, enabling high-performance SPE [103]. In air-assisted SPE wide variety of sorbent materials are used to extract and concentrate analytes from a gaseous sample and with no doubt it impacts the extraction efficacy.

Common absorbents used in air-assisted SPE include:

   I. PDMS: PDMS is a commonly used sorbent material for extracting VOCs from air samples.
   II. Tenax: Tenax is another popular sorbent material for trapping VOCs in air samples due to its high adsorption capacity and thermal stability.
   III. Carboxen: Carboxen is a highly porous carbon-based sorbent material that is effective for trapping a wide range of VOCs and SVOCs in air samples.

108 Green Analytical Chemistry

IV. Graphitized carbon black: Graphitized carbon black is a versatile sorbent material used for trapping a variety of organic compounds in air samples. These absorbents are packed into cartridges or tubes, and the air sample is passed through them to extract the compounds of interest. The adsorbed analytes can then be eluted from the sorbent material for further analysis [104].

1.15.2 Advantages and disadvantages air-assisted SPE

In comparison to traditional gravity-flow SPE, air-assisted SPE provides a number of noteworthy benefits, but there are also some particular drawbacks to take into account. The primary advantages of air-assisted SPE are its ability to automate processes via the use of vacuum manifolds or pressure processors, and its ability to provide quicker processing times through greater flow rates driven by air pressure or vacuum. By modifying airflow parameters, air-assisted SPE may also aid in the selective elution of analytes from the sorbent bed. Conversely, drawbacks include the need for expensive specialized vacuum/pressure equipment and more complicated technique development because of the additional parameters that need to be optimized. Due to the high air fluxes, there is also a chance that volatile analytes may evaporate. Furthermore, very strong or warm air flows may cause sorbent materials to overdry, which can reduce the ability to adsorb particular analytes. Although these possible drawbacks need to be considered, proper experimental design may often prevent them. Overall, when designed correctly, air-assisted SPE outperforms normal techniques in terms of throughput and lab productivity increases, although at the expense of more expensive equipment. As a result of the greater automation, consistency is also improved and less manual labor is needed [105].

1.15.3 Future trends of air-assisted SPE

The forthcoming trajectory of air-assisted SPE is anticipated to concentrate on progressions in technology and methodology in order to enhance the efficiency, sensitivity, and selectivity of the extraction process. There are several potential future trends in air-assisted SPE, including:

1. The advancement of innovative sorbent materials: Researchers are continuously engaged in the development of new sorbent materials that possess augmented adsorption capacity, selectivity, and stability to capture a broader range of analytes from air samples.
2. The process of reducing size and automating: There is an emerging inclination toward reducing the size of air-assisted SPE systems and integrating them with automated sample preparation platforms to enhance throughput, diminish the occurrence of sample handling errors, and improve reproducibility.

3. The online coupling with analytical techniques: The amalgamation of air-assisted SPE with analytical techniques such GC−MS or LC−MS has the potential to enable real-time analysis of the extracted analytes, leading to quicker and more precise outcomes.
4. The forthcoming advancements in air-assisted SPE may concentrate on eco-friendly methodologies, including the utilization of sustainable sorbent materials, diminished solvent consumption, and energy-efficient extraction techniques.
5. The adaptation of air-assisted SPE to target emerging contaminants such as pharmaceuticals, personal care products, and microplastics in the environment may be a potential future trend. This is due to the growing apprehension surrounding these new analytes.

In general, the prospective development of air-assisted SPE is expected to entail technological and methodological innovations. These innovations will aim to address current analytical challenges and meet the evolving requirements of environmental monitoring, food safety, and other domains necessitating air quality analysis.

1.16 Summary and outlook

SPE has become a well-established technique for preparing samples, and its applications are wide range. The methodology is constantly evolving, primarily in terms of its format, in order to simplify the sampling process or enable automation. There is anticipation for advancements in sorbent chemistry, specifically in the development of sorbents that are tailored to isolate and purify analytes in complex matrices. Additionally, there is an expectation for the use of computer-aided method development strategies to replace the currently employed trial-and-error procedures. Automated SPE systems are also anticipated to progress, in order to meet the demands of high sample throughput and continuous process monitoring in certain laboratories. Moreover, there are opportunities for the utilization of SPE in passive sampling, storage and preservation of extracts, field sampling, and as a biometric indicator, which present new avenues for research and development.

Abbreviations

CE Capillary electrophoresis
CEC Capillary electrochromatography
DI-SPME Direct immersion solid phase microextraction
DSPE Dispersive solid phase extraction
DVB Divinyl benzene
EF Enrichment factor
GAC Green analytical chemistry
GC−MS Gas chromatography−mass spectrometry

HLB Hydrophilic—lipophilic balanced particles
HPLC High-performance liquid chromatography
HS Headspace
LC—MS Liquid chromatography coupled to mass spectrometry
LLE Liquid—liquid extraction
LOD Limit of detection
MAE Microwave-assisted extraction
MIP Molecularly imprinted polymers
MNP Magnetic nanoparticle
MSPE Magnetic solid phase extraction
MDSPE Matrix dispersive solid phase extraction
ODS Octadecyl silane
PA Polyacrylate
PAH Polycyclic aromatic hydrocarbon
PAN Polyacrylonitrile
PCB Polychlorinated biphenyl
PDA Polydopamine
PDMDPS Poly (dimethyl diphenyl siloxane)
PDMS Poly dimethyl silane
PLE Pressurized liquid extraction
PS-DVB Poly (styrene-co-divinylbenzene)
QuEChERS Quick-Easy-Cheap-Effective-Rugged-Safe
RDSPE Rotating disk solid extraction
SBSE Stir-bar sorptive extraction
SFE Supercritical fluid extraction
SFODME Solidified floating organic drop microextraction
SPE Solid phase extraction
SPME Solid-phase microextraction
TFME Thin film microextraction
VOC Volatile organic compound

References

[1] J. Pezzatti, J. Boccard, S. Codesido, Y. Gagnebin, A. Joshi, D. Picard, V. González-Ruiz, S. Rudaz, Implementation of liquid chromatography—high-resolution mass spectrometry methods for untargeted metabolomic analyses of biological samples: a tutorial, Analytica Chimica Acta 1105 (2020) 28—44.

[2] M.N. Kalojeh, R.H. Sabet, M.A. Farajzadeh, M.R. Mogaddam, E.M. Khosrowshahi, M. Tuzen, Application of ZIF-67 coated by N-doped carbon dots in dispersive solid phase extraction of several pesticides from fruit juices and its combination with dispersive liquid—liquid microextraction followed by GC-FID, Journal of Food Composition and Analysis 121 (2023) 105372.

[3] E.V. Maciel, A.L. de Toffoli, E.S. Neto, C.E. Nazario, F.M. Lancas, New materials in sample preparation: recent advances and future trends, TrAC, Trends in Analytical Chemistry 119 (2019) 115633.

[4] B.H. Fumes, M.R. Silva, F.N. Andrade, C.E. Nazario, F.M. Lanças, Recent advances and future trends in new materials for sample preparation, TrAC, Trends in Analytical Chemistry 71 (2015) 9—25.

[5] Y. Wen, L. Chen, J. Li, D. Liu, L. Chen, Recent advances in solid-phase sorbents for sample preparation prior to chromatographic analysis, TrAC, Trends in Analytical Chemistry 59 (2014) 26–41.

[6] S. Dugheri, N. Mucci, A. Bonari, G. Marrubini, G. Cappelli, D. Ubiali, M. Campagna, M. Montalti, G. Arcangeli, Liquid phase microextraction techniques combined with chromatography analysis: a review, Acta Chromatographica 32 (2) (2020) 69–79.

[7] N. Patsos, K. Lewis, F. Picchioni, M.N. Kobrak, Extraction of acids and bases from aqueous phase to a pseudoprotic ionic liquid, Molecules 24 (5) (2019) 894.

[8] B. Yan, T.A. Abrajano, R.F. Bopp, D.A. Chaky, L.A. Benedict, S.N. Chillrud, Molecular tracers of saturated and polycyclic aromatic hydrocarbon inputs into Central Park Lake, New York City, Environmental Science & Technology 39 (18) (2005) 7012–7019.

[9] R. Mandrioli, M. Cirrincione, P. Mladěnka, M. Protti, L. Mercolini, Green analytical chemistry (GAC) applications in sample preparation for the analysis of anthocyanins in products and by-products from plant sources, Advances in Sample Preparation 3 (2022) 100037.

[10] J. Robles-Molina, B. Gilbert-López, J.F. García-Reyes, A. Molina-Díaz, Comparative evaluation of liquid–liquid extraction, solid-phase extraction and solid-phase microextraction for the gas chromatography–mass spectrometry determination of multiclass priority organic contaminants in wastewater, Talanta 117 (2013) 382–391.

[11] S. Pedersen-Bjergaard, A. Gjelstad, T.G. Halvorsen, Sample Preparation, Bioanalysis of Pharmaceuticals: Sample Preparation, Separation Techniques, and Mass Spectrometry, Wiley, 2015, pp. 73–122, https://doi.org/10.1002/9781118716830.ch6.

[12] M. Sargazi, S.H. Hashemi, M. Kaykhaii, Modern sample preparation techniques: a brief introduction, Sample Preparation Techniques for Chemical Analysis 9 (2021), https://doi.org/10.5772/intechopen.100715.

[13] E. Largy, B. Alies, G. Condesse, A. Gaubert, T. Livingston, K. Gaudin, Teaching with simulation tools to introduce the basics of analytical chemistry instrumentation, Analytical and Bioanalytical Chemistry 414 (23) (2022) 6709–6721.

[14] M. Faraji, Y. Yamini, M. Gholami, Recent advances and trends in applications of solid-phase extraction techniques in food and environmental analysis, Chromatographia 82 (8) (2019) 1207–1249.

[15] C.F. Poole, Core concepts and milestones in the development of solid-phase extraction, in: Solid-Phase Extraction, Elsevier, 2020, pp. 1–36, https://doi.org/10.1016/B978-0-12-816906-3.00001-7.

[16] M.A. Jouned, J. Kager, C. Herwig, T. Barz, Event driven modeling for the accurate identification of metabolic switches in fed-batch culture of *S. cerevisiae*, Biochemical Engineering Journal 180 (2022) 108345.

[17] W. Alahmad, S.I. Kaya, A. Cetinkaya, P. Varanusupakul, S.A. Ozkan, Green chemistry methods for food analysis: overview of sample preparation and determination, Advances in Sample Preparation 5 (2023) 100053.

[18] S. Ndwabu, M. Malungana, P. Mahlambi, Efficiency comparison of extraction methods for the determination of 11 of the 16 USEPA priority polycyclic aromatic hydrocarbons in water matrices: sources of origin and ecological risk assessment, Integrated Environmental Assessment and Management (2024), https://doi.org/10.1002/ieam.4904. Article (in press).

[19] O. Aydın Urucu, S. Dönmez, E. Kök Yetimoğlu, Solidified floating organic drop microextraction for the detection of trace amount of lead in various samples by electrothermal atomic absorption spectrometry, Journal of Analytical Methods in Chemistry 2017 (2017), https://doi.org/10.1155/2017/6268975.

[20] M.M. Daghi, M. Nemati, A. Abbasalizadeh, M.A. Farajzadeh, M.R. Afshar Mogaddam, A. Mohebbi, Combination of dispersive solid phase extraction using MIL−88A as a sorbent and deep eutectic solvent−based dispersive liquid−liquid microextraction for the extraction of some pesticides from fruit juices before their determination by GC−MS, Microchemical Journal 183 (2022) 107984.

[21] J. Płotka-Wasylka, N. Szczepańska, M. de La Guardia, J. Namieśnik, Modern trends in solid phase extraction: new sorbent media, TrAC, Trends in Analytical Chemistry 77 (2016) 23−43.

[22] A. Andrade-Eiroa, M. Canle, V. Leroy-Cancellieri, V. Cerdà, Solid-phase extraction of organic compounds: a critical review (part I), TrAC, Trends in Analytical Chemistry 80 (2016) 641−654.

[23] I. Liška, Fifty years of solid-phase extraction in water analysis−historical development and overview, Journal of Chromatography A 885 (1−2) (2000) 3−16.

[24] V. Camel, Solid phase extraction of trace elements, Spectrochimica Acta Part B: Atomic Spectroscopy 58 (7) (2003) 1177−1233.

[25] C.F. Poole, Principles and practice of solid-phase extraction, in: Comprehensive Analytical Chemistry, 37, Elsevier, 2002, pp. 341−387.

[26] M.E. Badawy, M.A. El-Nouby, P.K. Kimani, L.W. Lim, E.I. Rabea, A review of the modern principles and applications of solid-phase extraction techniques in chromatographic analysis, Analytical Sciences 38 (12) (2022) 1457−1487.

[27] N. Fontanals, F. Borrull, R.M. Marcé, New materials in sorptive extraction techniques for polar compounds, Journal of Chromatography A 1152 (1−2) (2007) 14−31.

[28] M.K. Murtada, F. de Andrés, Á. Ríos, M. Zougagh, A simple poly (styrene-co-divinylbenzene)-coated glass blood spot method for monitoring of seven antidepressants using capillary liquid chromatography-mass spectrometry, Talanta 188 (2018) 772−778.

[29] M.E. Leon-Gonzalez, L.V. Perez-Arribas, Chemically modified polymeric sorbents for sample preconcentration, Journal of Chromatography A 902 (1) (2000) 3−16.

[30] J.C. Nadal, K.L. Anderson, S. Dargo, I. Joas, D. Salas, F. Borrull, P.A. Cormack, R.M. Marcé, N. Fontanals, Microporous polymer microspheres with amphoteric character for the solid-phase extraction of acidic and basic analytes, Journal of Chromatography A 1626 (2020) 461348.

[31] E. Boyacı, Á. Rodríguez-Lafuente, K. Gorynski, F. Mirnaghi, É.A. Souza-Silva, D. Hein, J. Pawliszyn, Sample preparation with solid phase microextraction and exhaustive extraction approaches: comparison for challenging cases, Analytica Chimica Acta 873 (2015) 14−30.

[32] A. Andrade-Eiroa, M. Canle, V. Leroy-Cancellieri, V. Cerdà, Solid-phase extraction of organic compounds: a critical review. part ii, TrAC, Trends in Analytical Chemistry 80 (2016) 655−667.

[33] N. Reyes-Garces, E. Gionfriddo, G.A. Gómez-Ríos, M.N. Alam, E. Boyacı, B. Bojko, V. Singh, J. Grandy, J. Pawliszyn, Advances in solid phase microextraction and perspective on future directions, Analytical chemistry 90 (1) (2017) 302−360.

[34] A.R. Türker, New sorbents for solid-phase extraction for metal enrichment, Clean−Soil, Air, Water 35 (6) (2007) 548−557.

[35] A. Spietelun, A. Kloskowski, W. Chrzanowski, J. Namieśnik, Understanding solid-phase microextraction: key factors influencing the extraction process and trends in improving the technique, Chemical Reviews 113 (3) (2013) 1667−1685.

[36] C.F. Poole, New trends in solid-phase extraction, TrAC, Trends in Analytical Chemistry 22 (6) (2003) 362−373.

[37] F. Augusto, L.W. Hantao, N.G. Mogollón, S.C. Braga, New materials and trends in sorbents for solid-phase extraction, TrAC, Trends in Analytical Chemistry 43 (2013) 14−23.

[38] J. Płotka-Wasylka, N. Szczepańska, M. de La Guardia, J. Namieśnik, Miniaturized solid-phase extraction techniques, TrAC, Trends in Analytical Chemistry 73 (2015) 19−38.

[39] C.F. Poole, A.D. Gunatilleka, R. Sethuraman, Contributions of theory to method development in solid-phase extraction, Journal of Chromatography A 885 (1−2) (2000) 17−39.

[40] C.A. Impellitteri, Effects of pH and competing anions on the speciation of arsenic in fixed ionic strength solutions by solid phase extraction cartridges, Water Research 38 (5) (2004) 1207−1214.

[41] A.C. Gallo-Molina, H.I. Castro-Vargas, W.F. Garzón-Méndez, J.A. Ramírez, Z.J. Monroy, J.W. King, F. Parada-Alfonso, Extraction, isolation and purification of tetrahydrocannabinol from the *Cannabis sativa* L. plant using supercritical fluid extraction and solid phase extraction, The Journal of Supercritical Fluids 146 (2019) 208−216.

[42] M. Sajid, Porous membrane protected micro-solid-phase extraction: a review of features, advancements and applications, Analytica Chimica Acta 965 (2017) 36−53.

[43] A. Roszkowska, M. Tascon, B. Bojko, K. Goryński, P.R. Dos Santos, M. Cypel, J. Pawliszyn, Equilibrium ex vivo calibration of homogenized tissue for in vivo SPME quantitation of doxorubicin in lung tissue, Talanta 183 (2018) 304−310.

[44] S. Balasubramanian, S. Panigrahi, Solid-phase microextraction (SPME) techniques for quality characterization of food products: a review, Food and Bioprocess Technology 4 (2011) 1−26.

[45] E. Carasek, L. Morés, J. Merib, Basic principles, recent trends and future directions of microextraction techniques for the analysis of aqueous environmental samples, Trends in Environmental Analytical Chemistry 19 (2018) e00060.

[46] V. Jalili, A. Barkhordari, A. Ghiasvand, A comprehensive look at solid-phase microextraction technique: a review of reviews, Microchemical Journal 152 (2020) 104319.

[47] J. Zheng, Y. Kuang, S. Zhou, X. Gong, G. Ouyang, Latest improvements and expanding applications of solid-phase microextraction, Analytical Chemistry 95 (1) (2023) 218−237.

[48] M. Sajid, M.K. Nazal, M. Rutkowska, N. Szczepańska, J. Namieśnik, J. Płotka-Wasylka, Solid phase microextraction: apparatus, sorbent materials, and application, Critical Reviews in Analytical Chemistry 49 (3) (2019) 271−288.

[49] A. Spietelun, Ł. Marcinkowski, M. de la Guardia, J. Namieśnik, Recent developments and future trends in solid phase microextraction techniques towards green analytical chemistry, Journal of Chromatography A 1321 (2013) 1−3.

[50] G. Ouyang, D. Vuckovic, J. Pawliszyn, Nondestructive sampling of living systems using in vivo solid-phase microextraction, Chemical Reviews 111 (4) (2011) 2784−2814.

[51] A. Zhakupbekova, N. Baimatova, B. Kenessov, A critical review of vacuum-assisted headspace solid-phase microextraction for environmental analysis, Trends in Environmental Analytical Chemistry 22 (2019) e00065.

[52] J. Pawliszyn, Theory of solid-phase microextraction, in: Handbook of Solid Phase Microextraction, Elsevier, 2012, pp. 13−59.

[53] J. Ai, Headspace solid phase microextraction. Dynamics and quantitative analysis before reaching a partition equilibrium, Analytical Chemistry 69 (16) (1997) 3260−3266.

[54] C. Lancioni, C. Castells, R. Candal, M. Tascon, Headspace solid-phase microextraction: fundamentals and recent advances, Advances in Sample Preparation (2022) 100035.

[55] J. Feng, J. Feng, X. Ji, C. Li, S. Han, H. Sun, M. Sun, Recent advances of covalent organic frameworks for solid-phase microextraction, TrAC, Trends in Analytical Chemistry 137 (2021) 116208.

[56] I.D. Tegladza, T. Qi, T. Chen, K. Alorku, S. Tang, W. Shen, D. Kong, A. Yuan, J. Liu, H.K. Lee, Direct immersion single-drop microextraction of semi-volatile organic compounds in environmental samples: a review, Journal of Hazardous Materials 393 (2020) 122403.

[57] I. Pacheco-Fernández, M. Rentero, J.H. Ayala, J. Pasán, V. Pino, Green solid-phase microextraction fiber coating based on the metal−organic framework CIM-80 (Al): analytical performance evaluation in direct immersion and headspace using gas chromatography and mass spectrometry for the analysis of water, urine and brewed coffee, Analytica Chimica Acta 1133 (2020) 137−149.

[58] Q.Z. Su, P. Vera, C. Nerín, Direct immersion−solid-phase microextraction coupled to gas chromatography−mass spectrometry and response surface methodology for nontarget screening of (semi-) volatile migrants from food contact materials, Analytical Chemistry 92 (7) (2020) 5577−5584.

[59] Z. Zhang, M.J. Yang, J. Pawliszyn, Solid-phase microextraction. A solvent-free alternative for sample preparation, Analytical Chemistry 66 (17) (1994), 844A-53A.

[60] M. Mei, X. Huang, D. Yuan, Multiple monolithic fiber solid-phase microextraction: a new extraction approach for aqueous samples, Journal of Chromatography A 1345 (2014) 29−36.

[61] M. Háková, L.C. Havlíková, J. Chvojka, J. Erben, P. Solich, F. Švec, D. Šatínský, A comparison study of nanofiber, microfiber, and new composite nano/microfiber polymers used as sorbents for on-line solid phase extraction in chromatography system, Analytica Chimica Acta 1023 (2018) 44−52.

[62] H. Kataoka, Automated sample preparation using in-tube solid-phase microextraction and its application − a review, Analytical and Bioanalytical Chemistry 373 (2002) 31−45.

[63] Y. Moliner-Martinez, R. Herráez-Hernández, J. Verdú-Andrés, C. Molins-Legua, P. Campíns-Falcó, Recent advances of in-tube solid-phase microextraction, TrAC, Trends in Analytical Chemistry 71 (2015) 205−213.

[64] H. Saadaoui, A. Atayat, M. Necibi, F. Boujelbane, N. Mzoughi, Comparative evaluation of solid-phase extraction and in-tube liquid−liquid extraction for determination of triazole pesticides in water samples, International Journal of Environmental Analytical Chemistry 102 (19) (2022) 8385−8401.

[65] H. Kataoka, In-tube solid-phase microextraction: current trends and future perspectives, Journal of Chromatography A 1636 (2021) 461787.

[66] P. Rocío-Bautista, V. Termopoli, Metal−organic frameworks in solid-phase extraction procedures for environmental and food analyses, Chromatographia 82 (8) (2019) 1191−1205.

[67] R.A. Pérez, B. Albero, J.L. Tadeo, Matrix solid phase dispersion, in: Solid-Phase Extraction, Elsevier, 2020, pp. 531−549.

[68] A.K. El-Deen, An overview of recent advances and applications of matrix solid-phase dispersion, Separation and Purification Reviews 53 (1) (2024) 100−117.

[69] M.E. Queiroz, I.D. de Souza, C. Marchioni, Current advances and applications of in-tube solid-phase microextraction, TrAC, Trends in Analytical Chemistry 111 (2019) 261−278.

[70] D. Naviglio, P. Scarano, M. Ciaravolo, M. Gallo, Rapid solid−liquid dynamic extraction (RSLDE): a powerful and greener alternative to the latest solid-liquid extraction techniques, Foods 8 (7) (2019) 245.

[71] O. Zuloaga, N. Etxebarria, B. González-Gaya, M. Olivares, A. Prieto, A. Usobiaga, Stir-bar sorptive extraction, in: Solid-Phase Extraction, Elsevier, 2020, pp. 493−530, https://doi.org/10.1016/B978-0-12-816906-3.00018-2.

[72] S. Marín-San Román, J.M. Carot-Sierra, I.S. de Urturi, P. Rubio-Bretón, E.P. Pérez-Álvarez, T. Garde-Cerdán, Optimization of stir bar sorptive extraction (SBSE) and multi-stir bar sorptive extraction (mSBSE) to improve must volatile compounds extraction, LWT — Food Science and Technology 172 (2022) 114182.
[73] C.K. Hasan, A. Ghiasvand, T.W. Lewis, P.N. Nesterenko, B. Paull, Recent advances in stir-bar sorptive extraction: coatings, technical improvements, and applications, Analytica Chimica Acta 1139 (2020) 222—240.
[74] E. Velasco, J.J. Ríos-Acevedo, R. Sarria-Villa, M. Rosero-Moreano, Green method to determine triazine pesticides in water using Rotating Disk Sorptive Extraction (RDSE), Heliyon 7 (9) (2021), https://doi.org/10.1016/j.heliyon.2021.e07878.
[75] A. Fashi, M. Cheraghi, H. Ebadipur, H. Ebadipur, A. Zamani, H. Badiee, S. Pedersen-Bjergaard, Exploiting agarose gel modified with glucose-fructose syrup as a green sorbent in rotating-disk sorptive extraction technique for the determination of trace malondialdehyde in biological and food samples, Talanta 217 (2020) 121001.
[76] H. Tabani, M. Alexovič, J. Sabo, M.R. Payán, An overview on the recent applications of agarose as a green biopolymer in micro-extraction-based sample preparation techniques, Talanta 224 (2021) 121892.
[77] Y. Zhou, J.Y. Lin, Y. Bian, C.J. Ren, N. Xiao-Li, C.Y. Yang, X. Xiao-Xue, X.S. Feng, Non-steroidal anti-inflammatory drugs (NSAIDs) in the environment: updates on pretreatment and determination methods, Ecotoxicology and Environmental Safety 267 (2023) 115624.
[78] I. Hagarová, Magnetic solid phase extraction as a promising technique for fast separation of metallic nanoparticles and their ionic species: a review of recent advances, Journal of Analytical Methods in Chemistry 2020 (2020).
[79] C. Herrero-Latorre, J. Barciela-García, S. García-Martín, R.M. Peña-Crecente, J. Otárola-Jiménez, Magnetic solid-phase extraction using carbon nanotubes as sorbents: a review, Analytica Chimica Acta 892 (2015) 10—26.
[80] M. Wierucka, M. Biziuk, Application of magnetic nanoparticles for magnetic solid-phase extraction in preparing biological, environmental and food samples, TrAC, Trends in Analytical Chemistry 59 (2014) 50—58.
[81] E.M. KhosrowshahI, M.R. Afshar Mogaddam, M.A. Farajzadeh, M. Nemati, Magnetic silicon carbide nanocomposite as a sorbent in magnetic dispersive solid phase extraction followed by dispersive liquid—liquid microextraction in the gas chromatographic determination of pesticides, Microchemical Journal 181 (2022) 107786.
[82] A. Azzouz, S.K. Kailasa, S.S. Lee, A.J. Rascón, E. Ballesteros, M. Zhang, K.H. Kim, Review of nanomaterials as sorbents in solid-phase extraction for environmental samples, TrAC, Trends in Analytical Chemistry 108 (2018) 347—369.
[83] Y.A. Olcer, M. Tascon, A.E. Eroglu, E. Boyacı, Thin film microextraction: towards faster and more sensitive microextraction, TrAC, Trends in Analytical Chemistry 113 (2019) 93—101.
[84] L. Cai, J. Dong, Y. Wang, X. Chen, A review of developments and applications of thin-film microextraction coupled to surface-enhanced Raman scattering, Electrophoresis 40 (16—17) (2019) 2041—2049.
[85] V. Jalili, A. Barkhordari, A. Ghiasvand, Bioanalytical applications of microextraction techniques: a review of reviews, Chromatographia 83 (2020) 567—577.
[86] F.S. Mirnaghi, D. Hein, J. Pawliszyn, Thin-film microextraction coupled with mass spectrometry and liquid chromatography—mass spectrometry, Chromatographia 76 (2013) 1215—1223.

[87] B. Olayanju, A. Kabir, K.G. Furton, Development of a universal sol–gel sorbent for fabric phase sorptive extraction and its application in tandem with high performance liquid chromatography-ultraviolet detection for the analysis of phthalates in environmental and drinking water samples, Microchemical Journal 196 (2024) 109619.

[88] N. Yahaya, S.M. Ishak, A.H. Mohamed, S. Kamaruzaman, N.N. Zain, M.N. Waras, Y. Hassan, N. Abd Aziz, M. Miskam, W.N. Abdullah, Recent applications of electrospun nanofibres in microextraction based-sample preparation techniques for determination of environmental pollutants, Current Opinion in Environmental Science & Health 26 (2022) 100323.

[89] R.V. Emmons, R. Tajali, E. Gionfriddo, Development, optimization and applications of thin film solid phase microextraction (TF-SPME) devices for thermal desorption: a comprehensive review, Separations 6 (3) (2019) 39.

[90] M.R.A. Mogaddam, M.A. Farajzadeh, S.A. Damirchi, M. Nemati, Dispersive solid phase extraction combined with solidification of floating organic drop–liquid–liquid microextraction using in situ formation of deep eutectic solvent for extraction of phytosterols from edible oil samples, Journal of Chromatography A 1630 (2020) 461523.

[91] S. Büyüktiryaki, R. Keçili, C.M. Hussain, Functionalized nanomaterials in dispersive solid phase extraction: advances and prospects, TrAC, Trends in Analytical Chemistry 127 (2020) 115893.

[92] M. Ghorbani, M. Aghamohammadhassan, M. Chamsaz, H. Akhlaghi, T. Pedramrad, Dispersive solid phase microextraction, TrAC, Trends in Analytical Chemistry 118 (2019) 793–809.

[93] L. Yang, R. Said, M. Abdel-Rehim, Sorbent, device, matrix and application in microextraction by packed sorbent (MEPS): a review, Journal of Chromatography B 1043 (2017) 33–43.

[94] J.A. Pereira, J. Gonçalves, P. Porto-Figueira, J.A. Figueira, V. Alves, R. Perestrelo, S. Medina, J.S. Câmara, Current trends on microextraction by packed sorbent – fundamentals, application fields, innovative improvements and future applications, Analyst 144 (17) (2019) 5048–5074.

[95] M. Abdel-Rehim, Recent advances in microextraction by packed sorbent for bioanalysis, Journal of Chromatography A 1217 (16) (2010) 2569–2580.

[96] A. Salemi, N. Khaleghifar, N. Mirikaram, Optimization and comparison of membrane-protected micro-solid-phase extraction coupled with dispersive liquid–liquid microextraction for organochlorine pesticides using three different sorbents, Microchemical Journal 144 (2019) 215–220.

[97] M.N. Suseela, M.K. Viswanadh, A.K. Mehata, V. Priya, V.A. Setia, A.K. Malik, P. Gokul, J. Selvin, M.S. Muthu, Advances in solid-phase extraction techniques: role of nanosorbents for the enrichment of antibiotics for analytical quantification, Journal of Chromatography A (2023) 463937, https://doi.org/10.1016/j.chroma.2023.463937.

[98] N.N. Naing, S.C. Tan, H.K. Lee, Micro-solid-phase extraction, in: Solid-Phase Extraction, Elsevier, 2020, pp. 443–471, https://doi.org/10.1016/B978-0-12-816906-3.00016-9.

[99] R. Kumar, A.K. Malik, A. Kabir, K.G. Furton, Efficient analysis of selected estrogens using fabric phase sorptive extraction and high performance liquid chromatography-fluorescence detection, Journal of Chromatography A 1359 (2014) 16–25.

[100] V. Kazantzi, A. Anthemidis, Fabric sol–gel phase sorptive extraction technique: a review, Separations 4 (2) (2017) 20.

[101] A. Kabir, V. Samanidou, Fabric phase sorptive extraction: a paradigm shift approach in analytical and bioanalytical sample preparation, Molecules 26 (4) (2021) 865.

[102] N. Fontanals, F. Borrull, R.M. Marcé, Fabric phase sorptive extraction for environmental samples, Advances in Sample Preparation 5 (2023) 100050.
[103] A. Chisvert, S. Cárdenas, R. Lucena, Dispersive micro-solid phase extraction, TrAC, Trends in Analytical Chemistry 112 (2019) 226–233.
[104] J. Płotka-Wasylka, N. Jatkowska, M. Paszkiewicz, M. Caban, M.Y. Fares, A. Dogan, S. Garrigues, N. Manousi, N. Kalogiouri, P.M. Nowak, V.F. Samanidou, Miniaturized solid phase extraction techniques for different kind of pollutants analysis: state of the art and future perspectives—PART 1, TrAC, Trends in Analytical Chemistry (2023) 117034, https://doi.org/10.1016/j.trac.2023.117034.
[105] T. Liu, L. Ma, Y. Jiang, Y. Xiao, Y. Wu, K. Wang, Y. Yang, Simultaneous determination of four quinolones in honey by nickel/aluminum-layered double hydroxide-based air-assisted dispersive solid-phase extraction and high-performance liquid chromatography, Journal of Food Measurement and Characterization 16 (6) (2022) 5023–5031.

Chapter 5

Membrane-based extraction techniques

Abuzar Kabir and Basit Olayanju
Department of Chemistry and Biochemistry, Florida International University, Miami, FL, United States

1. Introduction

Green analytical chemistry (GAC) has emerged from the broad philosophy of Green Chemistry to emphasize the significant role of analytical chemists in making laboratory practices more environmentally friendly and sustainable. The primary goal of the so-called "Green Analytical Chemistry" movement is not only to improve the quality of chemical analysis but also to reduce the overall negative impact of chemical analysis on the environment as well as to implement sustainable development principles to analytical laboratories. Among the 12 guiding principles of GAC [1], a few principles are directly related to the sample preparation. For example, GAC demands that the sample size and the volume of samples should be kept at minimum, derivatization should be avoided, generation of large volume of analytical waste should be averted, the sample preparation workflow should be simplified, among others. To satisfy these requirements, the sample preparation technique should be field deployable, miniaturized, and simple in implementation. Membrane-based extraction techniques perfectly fit with the GAC principles and consequently have seen broader acceptance by practicing analytical chemists and numerous research and review articles have been published during the last couple of decades [2–15].

A major objective of analytical sample preparation is to eliminate or minimize the matrix interferents from the sample so that the upstream instrumental analysis (primarily chromatographic/electrophoretic separation) becomes less challenging, and the overall efficiency of the instrument is not adversely impacted by the presence of unwanted matrix interferents introduced into the system as an integral part of the sample to be analyzed. This problem becomes more critical when dealing with complex sample matrices such as blood, urine, saliva, and environmental water. To minimize the matrix interferents from the samples and to protect the extracting sorbents from contamination, membrane-based sample preparation techniques were introduced [10]. During the last 3 decades, many modifications of membrane-based extraction techniques have been introduced which can be broadly classified as (a) membrane-based solid phase microextraction (SPME) and membrane-based liquid-phase microextraction techniques based on the physical state of the extracting phases. Fig. 5.1 presents a classification scheme of membrane-based extraction techniques. As can be seen from the classification scheme, a few techniques such as fabric phase sorptive extraction (FPSE), biofluid sampler (BFS), carbon nanomaterial (CNM)-based solid phase extraction, and thin film microextraction (TFME), the sorbents are not protected by the membranes. These techniques have exploited the flat geometry of the membrane substrate to expand the primary contact surface area of the sample preparation device so that rapid mass transfer of the target analytes occurs between the extracting device and the bulk sample, resulting in fast extraction equilibrium.

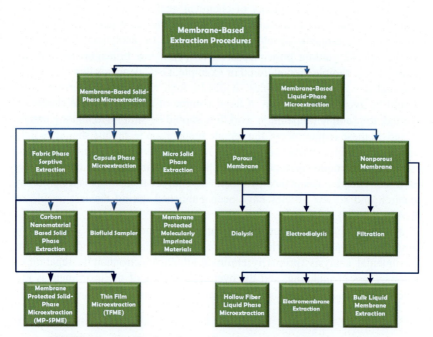

FIGURE 5.1 Classification scheme of membrane-based extraction techniques.

2. Membrane-based solid-phase microextraction and associated techniques

2.1 Fabric phase sorptive extraction

Fabric phase sorptive extraction, a new generation, porous fabric membrane-supported sample preparation technology, was invented by Kabir and Furton [16]. FPSE has addressed most of the shortcomings of SPE and SPME and combined these competing technologies into a single sample preparation platform. Consequently, FPSE has emerged as a universal sample preparation technique [17]. FPSE employs a fabric membrane (cellulose, polyester, and fiberglass) as the substrate and creates a highly porous, spongelike sol-gel sorbent coating on the substrate surface. The sol-gel sorbent coatings are chemically bonded during the sol-gel synthesis to the substrate, resulting in a bonded phase possessing extraordinarily high pH and sorbent stability. When porous fiberglass substrate is coated with sol-gel sorbent, the resulting FPSE membrane offers high thermal stability and allows thermal desorption of the extracted analytes via rapid heat shock (thermal desorption unit, TDU). For extracting the target analytes, sol-gel sorbent-coated FPSE membranes can be directly inserted into the pristine sample matrix, such as whole blood, urine, whole milk, and environmental water, without applying any sample pretreatment procedure. The extraction rate can be improved by diffusing the sample using a Teflon-coated magnetic stir bar. After the extraction, the analytes are eluted into a small volume of organic solvent. The spongelike porous morphology of the sol-gel sorbents allows speedy diffusion of the aqueous sample for rapid analyte extraction and the organic solvent for quick elution of the analytes. Selection of a suitable/compatible elution solvent allows injecting an aliquot of the sample into gas chromatography (GC), liquid chromatography (LC), capillary electrophoresis (CE), and inductively coupled plasma-mass spectrometry (ICP-MS) to maximize the analytical information.

Fabric phase sorptive extraction membranes are built using the following building blocks: (i) A porous fabric substrate; (ii) a sol-gel inorganic precursor/organically modified inorganic precursor; (iii) a sol-gel active inorganic/inorganic polymer. The selectivity and affinity of the sol-gel sorbent can be further improved by adding high surface area carbonaceous particles such as carboxen, graphene, and carbon nanotubes in the sol solution. Fig. 5.2 presents a schematic representation of a sol-gel sorbent-coated FPSE membrane.

Solid sorbent-based sample preparation techniques are primarily classified into two major classes based on their inherent extraction mechanism: (a) exhaustive extraction, as in solid-phase extraction (SPE) and (b) equilibrium extraction, as in SPME. Exhaustive extraction techniques, including SPE, microextraction by packed sorbent (MEPS), and disposable pipette tips extraction (DPTE), require liquid samples to percolate through a sorbent bed, and the analytes are exhaustively extracted during the percolation

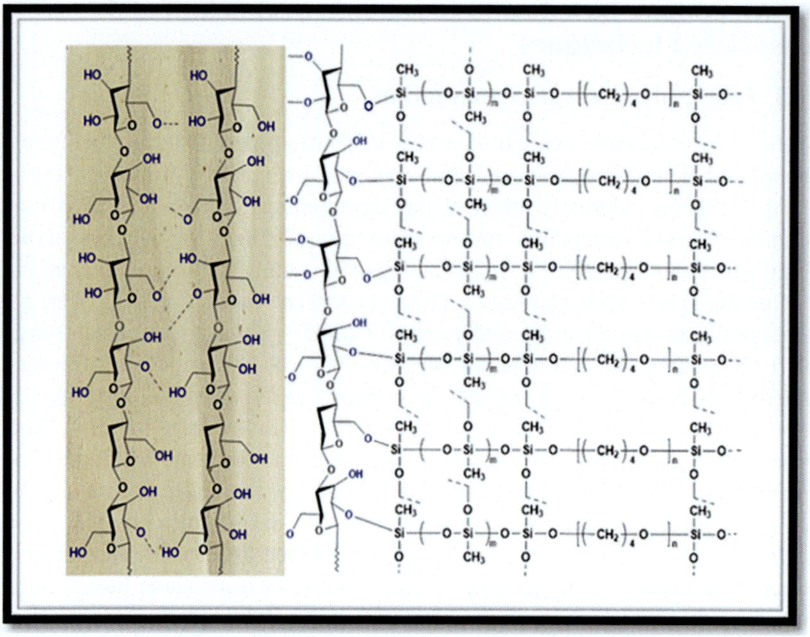

FIGURE 5.2 Schematic representation of an FPSE membrane.

through the sorbent bed. However, many unwanted matrix interferents/entities from the sample matrix during the percolation also become adsorbed. Removal of these unwanted materials is necessary before injecting the sample into the chromatographic instrument and is often accomplished by incorporating a washing step into the workflow before eluting the analytes. The washing step may result in substantial analyte loss and should be avoided.

Equilibrium extraction techniques such as SPME, in-tube solid-phase microextraction (IT-SPME), and SBSE require positioning the device inside the sample matrix (direct immersion extraction) or in the headspace of the sample (headspace extraction so that the analytes partition between the sorbent and the bulk of the sample and mass-transfer of analytes continues until the equilibrium is reached. Due to the high selectivity of the extracting sorbent, accumulation of unwanted matrix interferents is minimal, but the viscous nature of the extracting phases requires a substantially longer time to reach the extraction equilibrium. In addition, small sorbent masses (especially in SPME) often fail to accumulate a high mass of the target analyte during extraction, resulting in lower overall method sensitivity.

On the other hand, FPSE integrates both extraction techniques. When the FPSE membrane is immersed into the aqueous sample matrix during extraction, it mimics the direct immersion SPME. The fabric substrate is inherently permeable and remains intact even after the sol-gel sorbent coating. When the aqueous solution is diffused during the extraction with the help of an external Teflon-coated bar magnet, the aqueous sample matrix permeates through the porous bed of the FPSE membrane, mimicking an SPE disk, and the permeation continues thousands of times during the entire extraction period. Thanks to their unique material property, sol-gel sorbents are inherently superior in selectivity. As such, analytes are extracted by the FPSE membrane almost exhaustively without amassing any matrix interferents. Due to the spongelike porous architecture of the sol-gel sorbents, the presence of numerous functional groups to facilitate sorbent-analyte interactions via diverse intermolecular interactions, high sorbent loading, superior thermal, solvent, and pH stability, and combination of both the exhaustive and equilibrium extraction mechanism have positioned FPSE as a universal sample preparation technique indispensable for a modern analytical/bioanalytical/environmental/pharmaceutical laboratory.

2.1.1 Different implementations of fabric phase sorptive extraction

In the classical implementation of FPSE, a piece of sol-gel sorbent-coated FPSE membrane is immersed into the aqueous sample matrix, and a Teflon-coated magnetic stir bar is used to diffuse the sample matrix in order to expedite the analyte mass transfer from the solution to the FPSE membrane. During magnetic stirring, the FPSE membrane keeps randomly rotating inside the sample, occasionally becomes static on the sidewall of the extraction vial, and, in some cases, the FPSE membrane becomes coiled, resulting in reduced mass transfer and prolonged extraction kinetic. Due to the randomness of these processes, extraction reproducibility is also impacted. To improve analyte mass transfer from the solution to the FPSE membrane and to ensure that the entire primary contact surface area of the FPSE membrane remains accessible during extraction, several new implementations have been developed in recent years, including:

1. Stir fabric phase sorptive extraction (stir FPSE),
2. Stir-bar fabric phase sorptive extraction (Stir-bar FPSE),
3. Dynamic fabric phase sorptive extraction (DFPSE) and
4. Magnet-integrated fabric phase sorptive extraction (MI-FPSE).

2.1.1.1 Stir fabric phase sorptive extraction

Stir fabric phase sorptive extraction (stir FPSE) has an integrated FPSE membrane with a magnetic stirring mechanism. The integrated device spins on a magnetic stirrer during extraction, resulting in a faster extraction kinetic with improved reproducibility in replicate analysis. The rotation per min of stir FPSE on the magnetic stirrer can be precisely controlled, and the extraction equilibrium time can be substantially reduced [18].

The device was fabricated using a section of a polypropylene SPE cartridge, an FPSE membrane, a magnetic iron wire, and a section of a pipette tip. The new device has demonstrated improved extraction efficiency and reduced total extraction time for the determination of seven triazine herbicides in environmental water.

Huang et al. presented a modified form of stir FPSE known as magnetic stir FPSE (magnetic stir-FPSE) and demonstrated its application for extracting three brominated fire retardants in environmental water. Extraction was carried out for 15 min at 400 rpm. The device can be used many times and has reported excellent reproducibility and extraction sensitivity [19]. Fig. 5.3A represents the workflow of stir fabric phase sorptive extraction.

2.1.1.2 Stir-bar fabric phase sorptive extraction (stir bar-FPSE)

Roldan-Pijuan et al. developed a stir bar-FPSE that allows the FPSE membrane to be stirred during the entire extraction time. The device was created by cutting FPSE membranes into the shape of a house, clamping them, and fixing them with a stir bar. It requires only 10 min to reach extraction equilibrium. Stir bar-FPSE is recommended for large sample volumes [20]. A schematic presentation of stir-bar fabric phase sorptive extraction is presented in Fig. 5.3B.

2.1.1.3 Magnet integrated fabric phase sorptive extraction

Magnet-integrated fabric phase extraction (MI-FPSE) is the newest implementation of FPSE, which integrates a Teflon-coated bar magnet into the FPSE membrane. The MI-FPSE device is created by stitching two FPSE disks with a pocket at the center to insert the bar magnet into the device. The magnet can be introduced into the device (during extraction) and removed from the device (during elution). Due to the continuous circular motion during extraction, MI-FPSE has substantially reduced extraction equilibrium time with improved reproducibility [21].

2.1.1.4 Dynamic fabric phase sorptive extraction

Dynamic fabric phase sorptive extraction (DFPSE) mimics solid phase extraction in its disk format. The FPSE membranes can be cut into 47 cm and loaded on an SPE device or regular filtration assembly to carry out the extraction. Like SPE, DFPSE can substantially reduce the overall sample preparation time. Lakade et al. also compared the extraction efficiency of FPSE and DFPSE and observed almost similar extraction efficiency in both extraction modes for extracting residual pharmaceutical and personal care products in environmental water. This is indeed a manifestation of FPSE as a universal sample preparation technique that can be used in both SPME (equilibrium) and SPE (exhaustive) extraction modes [22].

Membrane-based extraction techniques **Chapter | 5** **125**

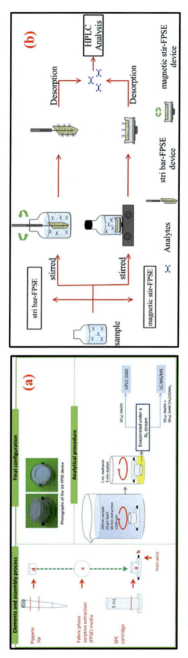

FIGURE 5.3 Schematic representation of (A) stir fabric phase sorptive extraction [20]; (B) stir bar-fabric phase sorptive extraction [19].

2.1.1.5 Fabric phase sorptive extraction interfacing with ion mobility spectrometry

Ion mobility spectrometry (IMS) has become very popular recently due to its suitability for field deployment, rapid on-site analysis, and portability. However, sampling and sample preparation remains a challenging task that limits the application of IMS. Recently, Sun et al. combined the power of IMS as the detection device and FPSE as the sampling and sample preparation device. Sol-gel poly dimethyl siloxane (sol-gel PDMS)-coated fiberglass FPSE membrane was used for sampling and sample preparation of polycyclic aromatic hydrocarbons (PAHs) in environmental water. Subsequently, the FPSE membrane was introduced into the IMS inlet for thermal desorption of the extracted analytes. The PAHs were analyzed in the positive ion mode with a corona discharge (CD) ionization source. The detection limits for phenanthrene, benzo[a] anthracene, and benzo[a] pyrene were found at 5, 8, and 10 ng/mL, respectively. FPSE-IMS opens new possibilities for rapid water quality monitoring in the field that can be easily used to establish an early warning system [23]. The concept can be easily expanded to other application areas.

2.1.2 Applications of fabric phase sorptive extraction

Fabric phase sorptive extraction has noticeably simplified the sample preparation process by eliminating several sample pretreatment and posttreatment steps from the sample preparation workflow. The unique combination of SPE and SPME mechanism into FPSE, an adaptation of a compelling sol-gel coating technology, and the flexible extraction membrane that supports the direct insertion of the FPSE membrane into the unmodified, original sample matrix in the presence of matrix interferents have encouraged many researchers to embrace FPSE as a green sample preparation technique. FPSE has been successfully applied to numerous samples, including environmental water [18,19,22,24–29], food [30–33], biological samples, including whole blood [34,35], and air [36,37]. FPSE has also been extensively used in heavy metal analysis from different sample matrices [38,39]. Several researchers have thoroughly reviewed the applications of FPSE in different sample matrices [17,40,41].

Among all these applications, FPSE has profoundly impacted the analysis of biological samples, especially whole blood. Traditionally, analysis of whole blood is carried out by converting it into plasma or serum, followed by protein precipitation and centrifugation to remove proteins and other cellular debris before applying any sample preparation technique. These processes incur a substantial analyte loss. After extracting the analytes, elution is carried out by employing a large volume of organic solvent. As a result, solvent evaporation and sample reconstitution are frequently applied to the eluant, resulting in another round of analyte loss. As such, the analytical data obtained from plasma/serum may not represent the actual concentration of the analytes present in the whole blood. FPSE allows the extraction of analytes directly from whole blood. The viscosity of the whole blood can be reduced by diluting it with deionized water. After the extraction, the FPSE membrane is exposed to a small volume of organic solvent. Elution of the analytes and protein precipitation cooccur at this step. The solvent quantitatively scavenges all the analytes from the FPSE membrane and cellular materials. Finally, the eluant is centrifuged to remove all the particulate materials before chromatographic analysis. Since a small volume of organic solvent quantitatively back-extracts all the analytes extracted by the FPSE membrane, no solvent evaporation and sample reconstitution are needed. A suitable solvent compatible with GC and LC can be used to elute the analytes from the FPSE membrane. Subsequently, it can be injected into GC and LC to obtain holistic analytical information from the whole blood sample.

2.2 Capsule phase microextraction

Conventional sample preparation techniques for the determination of pesticides in water samples include solid-phase extraction (SPE), and liquid-liquid extraction (LLE). However, since the introduction of GAC emerged from Green Chemistry, continuous efforts have been made to develop green, environmentally friendly analytical methods as sustainable alternatives to classical sample preparation protocols. As result, a wide variety of microextraction and miniaturized extraction techniques, such as solid-phase microextraction (SPME), dispersive liquid-liquid extraction (DLLME), magnetic solid-phase extraction (MSPE), stir bar sorptive extraction (SBSE), and stir fabric phase sorptive extraction (SFPSE) have arisen and used for the extraction of triazine herbicides from environmental and drinking water samples. However, in the sample above preparation techniques, sample filtration is typically required prior to the analysis in turbid samples containing solid particles. As a result, the potential application of these techniques for in situ sampling and sample preparation still needs to be improved. They often involve multiple steps, leading to irreversible analyte loss.

To overcome this limitation, capsule phase microextraction (CPME) was developed. CPME is a sample preparation technique introduced in 2017 by Kabir and Furton [16]. This sample preparation technique utilizes microextraction capsules with a built-in filtration step. The microextraction capsules consist of three main parts, that is, a magnetic rod, a high-performance sol-gel hybrid organic—inorganic sorbent coated onto a cellulose fiber substrate, and a porous tubular membrane. For the fabrication of the extraction device, two polypropylene tubes are used: one to accommodate the cylindrical magnetic rod and one to accommodate the sol-gel sorbent. Subsequently, the two tubes are welded to each other and combined to create a powerful extraction device known as microextraction capsules. As a result, the microextraction capsules (MECs) can spin themselves when placed on a magnetic stirrer, resulting in rapid extraction kinetics and high extraction efficiency.

Moreover, no sample filtration is required before the sample preparation since the filtration mechanism is incorporated in the extraction device, and the sol-gel sorbent is protected [43,44]. Therefore, CPME can be employed to directly analyze samples containing particulates, debris, and insoluble matrix interferants, which are commonly found in environmental samples. Until now, CPME has been successfully employed for the extraction of personal care products from environmental water samples [42], for the extraction of sulfonamides from milk [44], and for the extraction of acidic and basic compounds from environmental waters [43].

Sol-gel technology is a powerful and versatile material synthesis approach that not only allows the creation of a broad range of inorganic and hybrid inorganic—organic sorbents but also provides a facile pathway to in situ creation of surface coating on a suitable host substrate or a monolithic bed which cannot be created using other surface coating technologies traditionally used in fabricating different extraction and microextraction devices/sorbents. Microextraction capsules are typically fabricated by inserting precoated fibers using sol-gel coating technology [45], which involves several steps and may incur long assembly times and high production costs. To simplify the bulk production process, a superior approach might be to employ the in situ creation of a monolithic bed within the lumen of the microextraction capsules. As the sol-gel synthesis begins with a low-viscosity sol solution, the solution quickly permeates through the microextraction capsules' walls and fills the microextraction capsules' lumen. As the polycondensation progresses, the sol solution becomes a monolithic bed inside the microextraction capsules. In addition, a significant mass of the sol-gel sorbent gets trapped in the thick sponge-like walls of the microextraction capsules, leading to a very high sorbent loading on the CME device. The approach is simple and allows in situ creation of a broad range of high-performance sol-gel sorbents, including polar, nonpolar, ion exchanger, and mixed-mode sorbents. The schematic representation of CPME workflow is presented in Fig. 5.4.

2.3 Micro solid phase extraction

Micro solid-phase extraction was introduced by Basheer et al. [47] as a viable and greener alternative to classical solid phase extraction. μ-SPE utilizes a few milligrams of sorbent, packed inside a sheet of porous polymer membrane and the edges of the sheet are heat sealed to transform it into a small bag. The resulting bag is considered as a micro SPE device. After fabricating the device, it is cleaned and conditioned prior to deploying to analyte extraction. After

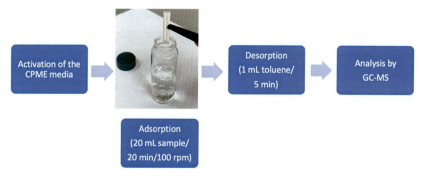

FIGURE 5.4 Workflow of capsule phase microextraction (CPME). *Adapted from N. Manousi, A. Kabir, K.G. Furton, E. Rosenberg, G.A. Zachariadis, Capsule phase microextraction of selected polycyclic aromatic hydrocarbons from water samples prior to their determination by gas chromatography-mass spectrometry, Microchemical Journal 166 (2021). https://doi.org/10.1016/j.microc.2021.106210 with permission from Elsevier.*

exposing to aqueous sample for extracting the target analytes, the μ-SPE device is dried and exposed to a small volume of organic solvent for eluting the extracted analytes. Subsequently, an aliquot of the eluent is injected into GC or LC. Fig. 5.5 represents different steps involved in fabricating μ-SPE device.

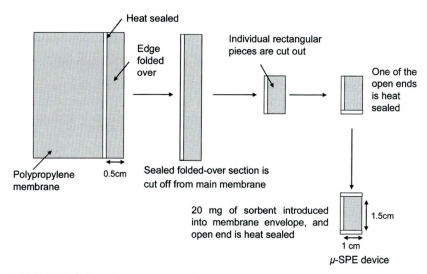

FIGURE 5.5 Schematic representation of μ-SPE device fabrication. *Adapted from C. Basheer, H.G. Chong, T.M. Hii, H.K. Lee, Application of porous membrane-protected micro-solid-phase extraction combined with HPLC for the analysis of acidic drugs in wastewater, Analytical Chemistry 79 (17) (2007) 6845–6850. https://doi.org/10.1021/ac070372r with permission from American Chemical Society.*

2.4 Carbon nanomaterial-based solid phase extraction

During the last 2 decades, CNMs have been enjoying increased popularity due to their low density, large specific surface area, high chemical activity, chemical stability, and easy surface functionalization. CNMs can be used individually or in combination with other materials depending on the analytical need. However, CNMs tend to aggregate into bundles and consequently limit their adsorption capacity significantly. To take full advantage of CNMs, membranes have been used to minimize agglomeration. Membranes offer several advantages such as porosity, high internal surface area, high loading capacity of caron nanomaterials, high transport rates, easy accessibility to the active sites, and of course operational flexibility. Membrane-protected CNMs can be used similar to the SPE in its disk format [4].

2.5 Biofluid sampler

Among all the biofluids (blood, urine, saliva, sweat), whole blood is the most information rich as well as the most complex sample matrix. It provides a temporal snapshot of the body and reveals the overall physiological condition. Blood contains ~55% aqueous fraction (plasma) and ~45% solid materials (formed elements). Although blood is the primary sample in most clinical chemistry-based investigations, due to the lack of available sample preparation technology that can handle whole blood directly, it is usually converted into plasma or serum prior to downstream processing. Plasma contains over 100 different dissolved solutes, including nutrients, gases, hormones, wastes, products of cell activity, ions, trace residues of exogenous substances, and proteins. The formed elements of blood consist of erythrocytes (red blood cells), leukocytes (white blood cells), and platelets.

Exogenous substances in the blood, such as drugs, poisons, and pollutants, enter the bloodstream via different routes (food, water, inhalation, ingestion, etc.), undergo metabolism, and the residues remain in equilibrium between the liquid part (plasma) and the suspended solid particles (formed elements). When blood is converted into plasma, a significant loss of the exogenous analytes could occur. The concentration of analytes in plasma also remains in equilibrium between the aqueous solution and the dissolved proteins. As such, when plasma undergoes protein precipitation prior to deploying any extraction technique, such as solid phase extraction, another round of analyte loss happens.

Often, after the solid phase extraction (the analytes are selectively extracted into a small volume of solid sorbent bed), the analytes are eluted in organic solvents. This point is also highlighted by one work published by Tartaglia and coworkers [48], where it was clearly reported that by the mean of the selective MIP stationary phase, the SPE procedure requires a large amount of solvents. Complete recovery of the retained analytes from the SPE sorbent generally requires a high organic solvent volume such as eluent. Consequently, an additional step of solvent evaporation and sample reconstitution often follows the SPE process, incurring another round of potential analyte loss.

In order to minimize the steps involved in blood analysis as well as to reduce the sample volume, dried blood spot cards (DBS cards) have been introduced [49]. DBS cards utilize 100% cellulose cards (Whatman 903 protein saver card) or glass microfiber filter paper (Agilent et al.). The cards are supplied with multiple 0.5-inch diameter printed circles for blood sample collection. Generally, blood samples are added to the marked circles, allowed to air dry and stored in plastic bags. For the analysis, a section of the blood spot is punched out and exposed to organic solvent for analyte extraction. Finally, the sample solution is centrifuged and subsequently injected into the analytical instrument for separation and detection.

Convenience in using, ability to handle whole blood, minimally invasive sample collection opportunity, and ability to ship the dried samples without expensive temperature control process and other advantages of DBS cards have catalyzed the exponential growth of this technique in numerous applications including newborn screening, toxicology, preclinical and clinical drug development, therapeutic drug monitoring (TDM), drug and sports doping screening, medical screening, and nutrition [49].

However, DBS cards suffer from several shortcomings that substantially limit the potential of this unique technology: (a) DBS cards are prepared using cellulose or glass microfiber, which offer only weak hydrophilic interaction toward the analytes. As such, the retention of the analytes is primarily governed by physical adsorption. Lack of strong affinity between the DBS card and the analyte may cause analyte loss, especially in highly volatile compounds. (b) Absence of viable intermolecular interactions between the DBS card and the analyte may lead to substantial loss during shipping at regular conditions; (c) due to the hematocrit issue, distribution of blood over the surface is not homogeneous, resulting in variability in the analyte concentration depending on the punching out location of the DBS card; (d) DBS cards cannot be used as an extraction device to reduce the matrix interferents; (f) cannot be applied to higher sample volume when sensitivity of the analysis needs to be scaled up; and (g) due to the reduced volume of blood used in the final analysis, only highly sensitive chromatographic instruments such as LC-MS/MS can be used in dried blood spots analyses.

Biofluid sampler and sample preparation device (BFS) is specifically designed to eliminate all major shortcomings of DBS cards and their different modifications in a rational scientific way. Advantages of BFSs include: (a) capable of retaining low to high sample volume (10—1000 µL); (b) each BFS is a standalone sampling and sample preparation device; therefore, hematocrit problem is not an issue in BFS (entire BFS is used for instrumental analysis, not a punched-out segment of the dried blood spot as in DBS cards); (c) spot homogeneity is not an issue in BFS as the entire device is exposed to back-extraction; (d) the BFSs can be created in small size (1/4″) to big size (2″) depending on the sample volume requirement; (e) unlike DBS card which primarily utilize physical adsorption as the retention mechanism, BFSs utilize a plethora of intermolecular interactions such as London dispersion, hydrogen bonding, dipole-dipole interaction, π-π stacking interaction; (f) sponge-like porous architecture of sol-gel sorbent allows rapid dissipation of the biofluid homogeneously throughout the device; (g) strong intermolecular interactions between the BFS and the biofluid minimizes the analyte loss during regular transportation and shipping; (h) when matrix interferents complicates the downstream separation, BFSs can be used as the extraction device. In this case, BFS is introduced into diluted biofluid, and analytes are extracted by magnetic stirring or orbital shaking. After the predetermined extraction period, the BFS is removed from the biofluid sample, wash with deionized water, dry in air, and subsequently subjected to back-extraction for chromatographic analysis; (i) BFSs can be stored after sampling for a prolonged period at laboratory room temperature without any discernible analyte loss; (j) sol-gel sorbent coating technology is a precisely controllable chemical coating process that ensures high batch-to-batch sorbent coating reproducibility; (k) as the U.S. Department of Transportation and the World Health Organization consider DBS specimen a nonregulated, nonbiohazard freight, BFSs can be shipped nationally and internationally after the sampling process.

Traditionally, blood samples are collected and analyzed in the same geographical area. As a result, the overall cost of analysis may be excessively high compared to other geographical areas, and the analyses may take longer due to overload or instrument breakdown. The preservation of sample integrity during this waiting period may be challenging. This unintended inconvenience may be easily addressed by using *Mail-in-Analysis*. Due to the availability of fast shipment throughout the world, blood samples can be collected on BFSs and shipped anywhere in the world in a very short period. Establishing a robust *mail-in-analysis* program may substantially reduce the cost and time of drug discovery and development, population-based research, monitoring epidemic diseases, forensic cases in big metropolitan cities with unsustainable caseloads, managing catastrophic events such as radiation exposure, and many others. Fig. 5.6 demonstrates the images of BFS.

2.6 Membrane-protected molecularly imprinted materials

Molecularly imprinted polymers (MIPs) are tailor made materials that are created using a specific template molecule or a group of template molecules that create nano cavities on the imprinted particles complementary to the shape, size, and functional composition of the template molecules. The resulting imprinted polymer can rebind the template molecules with improved selectivity. Although MIPs mimic antibodies used in clinical investigations, MIPs are stable, robust, and resistant to organic solvents. As such there has

FIGURE 5.6 Biofluid sampler (A) wet biofluid sampler (BFS) immediately after sampling blood, and (B) dry BFS after air drying for 1 h. *Adapted from M. Locatelli, A. Tartaglia, F. D'Ambrosio, P. Ramundo, H.I. Ulusoy, K.G. Furton, A. Kabir, Biofluid sampler: a new gateway for mail-in-analysis of whole blood samples, Journal of Chromatography B 1143 (2020) 122055. https://doi.org/10.1016/j.jchromb.2020.122055 with permission from Elsevier.*

been a large spike in the interest and applications of MIPs in recent years. Like other solid sorbents, MIPs may be susceptible to contamination and often requires rigorous sample cleanup when the MIPs are used in SPE cartridge format. To eliminate the sample cleanup exercises from the sample preparation workflow membrane protected molecular imprinted micro solid phase extraction is proposed. Fig. 5.7 demonstrates the application of membranes to protect MIPs.

2.7 Membrane-protected solid phase microextraction

Solid-phase microextraction, developed by Pawliszyn and coworkers, was a breakthrough invention in analytical/bioanalytical sample preparation [50]. SPME is a solvent-free sample preparation technique that substantially simplifies the overall sample preparation workflow. Due to its green attributes, ease in operation, and portability, SPME has gained enormous popularity in recent years. However, the SPME fiber is prone to be contaminated when the technique is used in direct immersion extraction mode (DI extraction mode) to extract the target analytes from biological or environmental samples containing high volume of matrix interferents. As such, many applications have been developed using headspace extraction mode (HS extraction mode). For low-volatile target analytes, this approach is not practical. To protect the SPME fiber coating from irreversible adsorption of matrix interferents,

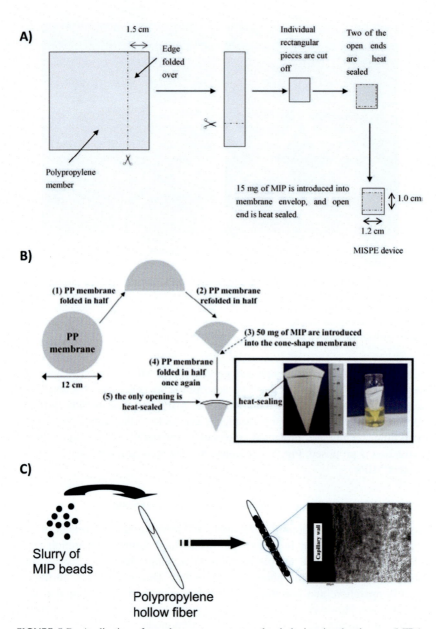

FIGURE 5.7 Application of membranes to protect molecularly imprinted polymers (MIPs). *Adapted from A. Martín-Esteban, Membrane-protected molecularly imprinted polymers: towards selectivity improvement of liquid-phase microextraction, TRAC Trends in Analytical Chemistry 138 (2021). https://doi.org/10.1016/j.trac.2021.116236 with permission from Elsevier.*

membrane-protected SPME was proposed [2]. There are two forms of membranes have been reported: membrane based on tubular hollow cellulose with molecular weight cut-off (MWCO) value of 18,000 Da [51] and membrane based on porous hollow propylene fiber [52]. However, membrane protected SPME has not gained much popularity as demonstrated by the limited number of published papers on this technique. Fig. 5.8 represents a membrane protected SPME and its operation.

2.8 Thin film microextraction

Thin film microextraction is another format of SPME that utilizes a thin sheet of a polymeric membrane holding the extracting sorbent. Due to the large surface area of TFME device, the sorbent coating assumes a large primary contact surface area that substantially boosts the mass transfer rate from the bulk sample to the extracting phase, resulting in very fast extraction equilibrium. However, there is no protection for the extracting sorbent in this system and the direct exposure of the TFME device to the complex sample may risk contamination of the extracting phase [2].

3. Membrane-based liquid phase microextraction

Membrane-based liquid phase microextraction techniques have eloquently addressed the shortcomings of classical liquid—liquid extraction (LLE) and its miniaturized version, liquid-liquid microextraction (LLME). These devices are simple, easy to design and use, and they allow fine tuning the selectivity by applying different modes of trapping the target analytes in the acceptor phase

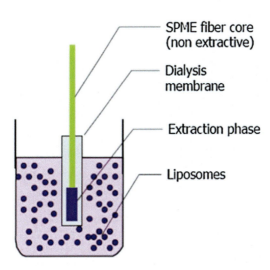

FIGURE 5.8 Schematic representation of membrane protected solid phase microextraction (SPME) and its operation. *Adapted from E. Carasek, J. Merib, Membrane-based microextraction techniques in analytical chemistry: a review, Analytica Chimica Acta 880 (2015) 8—25. https://doi.org/10.1016/j.aca.2015.02.049 with permission from Elsevier.*

(extracting liquid phase) [3]. Based on the presence or absence of porosity of the membrane, membrane-based liquid phase microextraction techniques can be classified as (i) porous membrane-based liquid phase microextraction techniques and (ii) nonporous membrane based liquid phase microextraction techniques.

3.1 Porous membrane–based liquid phase microextraction techniques

Porous membrane–based extraction techniques utilize a porous membrane through which the target analytes migrate from the donor phase to the acceptor phase. The pores of the membrane are bigger than the diameter of the molecules, and therefore, the molecules can easily permeate through the pores. Dialysis, electrodialysis (ED), and filtration are among the different implementations of porous membrane–based liquid phase microextraction. Physico-chemical properties existing between the porous membrane and the target analytes is believed to be the primary driving force toward the selective extraction via porous membrane [53]. Fig. 5.9 presents the schematics of different porous membrane-based microextraction techniques.

3.1.1 Dialysis

In dialysis, a semipermeable porous membrane possessing a well-defined molecular weight cut-off (MWCO) value is utilized. The membranes are made-up of cellulose or regenerated cellulose. Due to the well-defined molecular weight cut-off value of the membrane, the extraction is very selective in the dialysis process. The analytes are transported to the membrane via

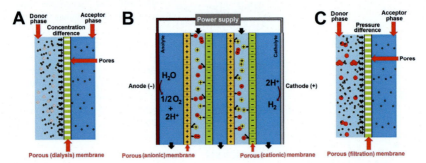

FIGURE 5.9 Schematic representation of (A) dialysis, (B) electrodialysis, and (C) filtration. *Adapted from H. Tabani, S. Nojavan, M. Alexovic, J. Sabo, Recent developments in green membrane-based extraction techniques for pharmaceutical and biomedical analysis, Journal of Pharmaceutical and Biomedical Analysis 160 (2018) 244–267. https://doi.org/10.1016/j.jpba.2018.08.002 with permission from Elsevier.*

spontaneous diffusion where the analytes migrate from the high concentration area (donor phase) through the porous membrane to the low concentration area (acceptor phase). The analyte mass transfer between the two phases continues until the concentration of the analytes is equal on both sides of the porous membrane. To improve the extraction recovery of the target analytes, the acceptor phase can be constantly replaced with fresh acceptor phase so that the concentration of the analytes in the acceptor phase can never reach the equilibrium. Subsequently, the concentration of the analyte in the acceptor phase can be further enhanced by evaporating the acceptor phase followed by reconstituting the sample in a small volume of solvent [15]. Dialysis is extensively used for the extraction of neural and charged species from different sample matrices including biological fluids. Fig. 5.9A represents the dialysis process.

3.1.2 Electrodialysis

Electrodialysis is an improved form of dialysis where ions are transported from the donor phase to the acceptor phase through the semipermeable membrane under the applied electric potential difference. The membrane, generally made up of regenerated cellulose, is highly hydrophilic and resistant to organic solvents. Compared to dialysis, mass transfer in ED is significantly faster due to the applied electric potential difference and can be controlled by adjusting the applied voltage. However, applying too high voltage may lead to temperature increase (joule heating) and bubble formation that slow down the mass transfer rate through the membrane. The electric charge of the membrane can be either neutral, positively charged or negatively charged. The schematic presented in Fig. 5.9B represents an ED cell consisting of a donor phase compartment, an acceptor phase compartment, a cation exchange membrane, an anion exchange membrane, a cathode, and an anode.

3.1.3 Filtration

Filtration is a one-phase extraction procedure that exploits the pressure gradient across the solid membrane placed between the donor phase and the acceptor phase. Pressure gradient across the membrane is created by applying positive or negative pressure or centrifugal force. Due to the applied pressure, the analytes are forced to move toward the membrane and permeate through it to the acceptor phase. Depending on the pore size of the membrane, membrane filters can be classified as microfilters (pore size rages from 0.1 to 1 μm) or nano filters (pore size ranges from 1 to 10 nm). Fig. 5.9C represents a filtration membrane. Filtration is extensively used in sample preparation as a sample pretreatment to eliminate unwanted matrix interferents and primarily used in treating environmental samples [54].

3.2 Nonporous membranes

In nonporous membrane-based liquid phase microextraction techniques (MLPME), a thin layer of liquid or a liquid mixture immobilized in the pores of a porous polymeric membrane is used. The primary building blocks of MLPME includes (i) an inert membrane that functions as the selective filter to inhibit particles, debris, macromolecules, and other matrix interferents to contaminate the acceptor phase (AP) (ii) the donor phase (DP) that contains the analytes, primarily the aqueous sample matrix and (iii) the acceptor phase (AP) that selectively accumulates the target analytes during the extraction process. The membrane, an inert permeable material, is located between DP and the AP and does not interact with these two phases. These nonporous membrane-based techniques employ two or three different phases with well-defined phase boundaries. Among the many formats of MLPME, hollow-fiber liquid phase microextraction (HF-LPME), electromembrane extraction (EME), and bulk liquid membrane extraction (BLME) are noteworthy and will be discussed in the following sections.

3.3 Hollow-fiber based liquid phase microextraction

Hollow-fiber liquid phase microextraction was invented by Pedersen-Bjergaard and Rasmussen in 1999 [55]. In HF-LPME, a small volume of organic solvent (~ 10 μL) is immobilized in the pores of a porous polymeric material that retains the organic solvent and serves as the support for the supported liquid membrane (SLM). Unlike porous membrane, the organic solvent serves as the supported liquid membrane. The membrane assembly is inexpensive and is recommended to be disposed to minimize the risk of carry-over and potential cross-contamination. During the extraction, analytes are transported from the DP to the AP based on the composition of the acceptor phase and the physico-chemical properties of the analytes such as partition coefficient and pKa values. The primary driving forces for the analytes from the DP to AP are the concentration and pH gradients.

HF-LPME can be classified as a 2-phase system and 3-phase system, based on the number of phases used in the process. Fig. 5.10 demonstrates the schematics of (A) a 2-phase HF-LPME system, and (B) a 3-phase LPME system.

3.4 Electromembrane extraction

Electron membrane extraction developed by Pedersen-Bjergaard and Rasmussen [56] is an improved implementation of hollow fiber L PME in HFL PME extraction time is relatively larger due to the reliance of the I spontaneous diffusion of the target analytes from the bulk the donor face to the acceptor phase this through mass transfer is substantially increased in Electro

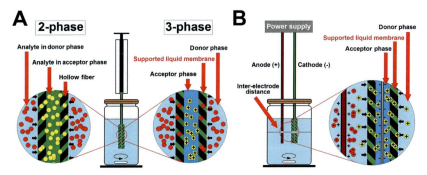

FIGURE 5.10 Schematic presentation of (A) hollow-fiber liquid phase microextraction (HF-LPME) and (B) electromembrane extraction (EME). *Adapted and modified from H. Tabani, S. Nojavan, M. Alexovic, J. Sabo, Recent developments in green membrane-based extraction techniques for pharmaceutical and biomedical analysis, Journal of Pharmaceutical and Biomedical Analysis 160 (2018) 244–267. https://doi.org/10.1016/j.jpba.2018.08.002 with permission from Elsevier.*

membrane extraction by applying electric field across the membrane the overall setup of EME is very similar to that of HFL PME except that there are electrodes immersed in the donor phase and the accepted phase and the power supply to create the voltage across the membrane in EME only charged analytes can be transferred from the donor phase to the acceptor phase under the influence of the applied voltage EMEA is considered as an ideal technique to clean up the analytes of interest from biological samples without requiring any sample pretreatment. Compared to HF-LPME, EME decreased the extraction time by 5–15 fold [57].

3.5 Bulk membrane extraction

Unlike supported liquid membrane extraction, bulk liquid membrane extraction technique does not use any porous polymeric support to retain the organic phase. Instead, in bulk liquid membrane extraction, the donor phase (aqueous sample matrix) is detached from the acceptor phase (organic solvent) by a solid flat or cylindrical barrier. The mass transfer of the target analytes from the donor phase to the acceptor phase is directed by the concentration gradient between the two phases. The rate of mass transfer or extraction kinetic can be expedited by stirring both the phases during the extraction process [54].

4. Conclusions

Membrane extraction techniques have been proven as green implementations of classical SPME and liquid–liquid microextraction techniques. The integration of membrane into the microextraction techniques protects the sorbent/acceptor phase from contamination. Membrane also allows sample preparation without any sample pretreatment exercises. As a result, sample preparation becomes lean and potential loss of the target analytes during sample pretreatment can be prevented or minimized. Due to the obvious advantages in membrane-based extraction, it is expected that these techniques will find many new applications in years to come.

References

[1] A. Galuszka, Z. Migaszewski, J. Namiesnik, The 12 principles of green analytical chemistry and the SIGNIFICANCE mnemonic of green analytical practices, TRAC Trends in Analytical Chemistry 50 (2013) 78−84, https://doi.org/10.1016/j.trac.2013.04.010.

[2] E. Carasek, J. Merib, Membrane-based microextraction techniques in analytical chemistry: a review, Analytica Chimica Acta 880 (2015) 8−25, https://doi.org/10.1016/j.aca.2015.02.049.

[3] L. Chimuka, M. Michel, E. Cukrowska, B. Buszewski, Advances in sample preparation using membrane-based liquid-phase microextraction techniques, TRAC Trends in Analytical Chemistry 30 (11) (2011) 1781−1792, https://doi.org/10.1016/j.trac.2011.05.008.

[4] C. Dal Bosco, M.G. De Cesaris, N. Felli, E. Lucci, S. Fanali, A. Gentili, Carbon nanomaterial-based membranes in solid-phase extraction, Microchimica Acta 190 (5) (2023), https://doi.org/10.1007/s00604-023-05741-y.

[5] A. Esrafili, M. Baharfar, M. Tajik, Y. Yamini, M. Ghambarian, Two-phase hollow fiber liquid-phase microextraction, TRAC Trends in Analytical Chemistry 108 (2018) 314−322, https://doi.org/10.1016/j.trac.2018.09.015.

[6] M. Ghambarian, Y. Yamini, A. Esrafili, Developments in hollow fiber based liquid-phase microextraction: principles and applications, Microchimica Acta 177 (3−4) (2012) 271−294, https://doi.org/10.1007/s00604-012-0773-x.

[7] A. Gjelstad, S. Pedersen-Bjergaard, Electromembrane extraction-Three-phase electrophoresis for future preparative applications, Electrophoresis 35 (17) (2014) 2421−2428, https://doi.org/10.1002/elps.201400127.

[8] C.X. Huang, Z.L. Chen, A. Gjelstad, S. Pedersen-Bjergaard, X.T. Shen, Electromembrane extraction, TRAC Trends in Analytical Chemistry 95 (2017) 47−56, https://doi.org/10.1016/j.trac.2017.07.027.

[9] C.X. Huang, H. Jensen, K.F. Seip, A. Gjelstad, S. Pedersen-Bjergaard, Mass transfer in electromembrane extraction-The link between theory and experiments, Journal of Separation Science 39 (1) (2016) 188−197, https://doi.org/10.1002/jssc.201500905.

[10] J. Jönsson, L. Mathiasson, Membrane-based techniques for sample enrichment, Journal of Chromatography A 902 (1) (2000) 205−225, https://doi.org/10.1016/s0021-9673(00)00922-5.

[11] A. Martín-Esteban, Membrane-protected molecularly imprinted polymers: towards selectivity improvement of liquid-phase microextraction, TRAC Trends in Analytical Chemistry 138 (2021), https://doi.org/10.1016/j.trac.2021.116236.

[12] L.N. Moskvin, T.G. Nikitina, Membrane methods of substance separation in analytical chemistry, Journal of Analytical Chemistry 59 (1) (2004) 2−16, https://doi.org/10.1023/B:JANC.0000011661.47796.b2.

[13] S. Pedersen-Bjergaard, K.E. Rasmussen, Liquid-phase microextraction with porous hollow fibers, a miniaturized and highly flexible format for liquid-liquid extraction, Journal of Chromatography A 1184 (1−2) (2008) 132−142, https://doi.org/10.1016/j.chroma.2007.08.088.

[14] M. Sajid, Porous membrane protected micro-solid-phase extraction: a review of features, advancements and applications, Analytica Chimica Acta 965 (2017) 36−53, https://doi.org/10.1016/j.aca.2017.02.023.

[15] N.C. van de Merbel, Membrane-based sample preparation coupled on-line to chromatography or electrophoresis, Journal of Chromatography A 856 (1−2) (1999) 55−82, https://doi.org/10.1016/s0021-9673(99)00581-6.

[16] A. Kabir, K.G. Furton, Fabric Phase Sorptive Extractors, United States Patents and Trademark Office, 2016.

[17] A. Kabir, R. Mesa, J. Jurmain, K. Furton, Fabric phase sorptive extraction explained, Separations 4 (2) (2017) 21.

[18] R.L.M. Roldan-Pijuan, S. Cardenas, M. Valcarcel, A. Kabir, G. Kenneth, Furton Stir fabric phase sorptive extraction for the determination of triazine herbicides in environmental water by using ultra-high performance liquid chromatography-UV detection, Joural of Chromatography A (2014) under review.

[19] G. Huang, S. Dong, M. Zhang, H. Zhang, T. Huang, Fabric phase sorptive extraction: two practical sample pretreatment techniques for brominated flame retardants in water, Water Research 101 (2016) 547−554, https://doi.org/10.1016/j.watres.2016.06.007.

[20] M. Roldan-Pijuan, R. Lucena, S. Cardenas, M. Valcarcel, A. Kabir, K.G. Furton, Stir fabric phase sorptive extraction for the determination of triazine herbicides in environmental waters by liquid chromatography, Journal of Chromatography A 1376 (2015) 35−45, https://doi.org/10.1016/j.chroma.2014.12.027.

[21] G. Antoniou, V. Alampanos, A. Kabir, T. Zughaibi, K.G. Furton, V. Samanidou, Magnet integrated fabric phase sorptive extraction for the extraction of resin monomers from human urine prior to HPLC analysis, Separations 10 (4) (2023), https://doi.org/10.3390/separations10040235.

[22] S.S. Lakade, F. Borrull, K.G. Furton, A. Kabir, R.M. Marce, N. Fontanals, Dynamic fabric phase sorptive extraction for a group of pharmaceuticals and personal care products from environmental waters, Journal of Chromatography A 1456 (2016) 19−26, https://doi.org/10.1016/j.chroma.2016.05.097.

[23] T. Sun, D. Wang, Y. Tang, X. Xing, J. Zhuang, J. Cheng, Z. Du, Fabric-phase sorptive extraction coupled with ion mobility spectrometry for on-site rapid detection of PAHs in aquatic environment, Talanta 195 (2019) 109−116, https://doi.org/10.1016/j.talanta.2018.11.018.

[24] R. Kumar, H. Gaurav, A.K. Malik, A. Kabir, K.G. Furton, Efficient analysis of selected estrogens using fabric phase sorptive extraction and high performance liquid chromatography-fluorescence detection, Journal of Chromatography A 1359 (0) (2014) 16−25, https://doi.org/10.1016/j.chroma.2014.07.013.

[25] R. Kumar, Gaurav, A. Kabir, K.G. Furton, A.K. Malik, Development of a fabric phase sorptive extraction with high-performance liquid chromatography and ultraviolet detection method for the analysis of alkyl phenols in environmental samples, Journal of Separation Science 38 (18) (2015) 3228−3238, https://doi.org/10.1002/jssc.201500464.

[26] S.S. Lakade, F. Borrull, K.G. Furton, A. Kabir, N. Fontanals, R. Maria Marcé, Comparative study of different fabric phase sorptive extraction sorbents to determine emerging contaminants from environmental water using liquid chromatography-tandem mass spectrometry, Talanta (2015), https://doi.org/10.1016/j.talanta.2015.08.009. TALD1501278.

[27] S. Montesdeoca-Esponda, Z. Sosa-Ferrera, A. Kabir, K.G. Furton, J.J. Santana-Rodriguez, Fabric phase sorptive extraction followed by UHPLC-MS/MS for the analysis of benzotriazole UV stabilizers in sewage samples, Analytical and Bioanalytical Chemistry 407 (26) (2015) 8137−8150, https://doi.org/10.1007/s00216-015-8990-x.

[28] I. Racamonde, R. Rodil, J.B. Quintana, B.J. Sieira, A. Kabir, K.G. Furton, R. Cela, Fabric phase sorptive extraction: a new sorptive microextraction technique for the determination of non-steroidal anti-inflammatory drugs from environmental water samples, Analytica Chimica Acta 865 (0) (2015) 22−30, https://doi.org/10.1016/j.aca.2015.01.036.

[29] A. Castinñeira-Landeira, L. Vazquez, A.M. Carro, M. Celeiro, A. Kabir, K.G. Furton, T. Dagnac, M. Llompart, Fabric phase sorptive extraction as a sustainable sample preparation procedure to determine synthetic musks in water, Microchemical Journal 196 (2024), https://doi.org/10.1016/j.microc.2023.109542.

[30] V. Samanidou, L.-D. Galanopoulos, A. Kabir, K.G. Furton, Fast extraction of amphenicols residues from raw milk using novel fabric phase sorptive extraction followed by high-performance liquid chromatography-diode array detection, Analytica Chimica Acta 855 (0) (2015) 41−50, https://doi.org/10.1016/j.aca.2014.11.036.

[31] E. Karageorgou, N. Manousi, V. Samanidou, A. Kabir, K.G. Furton, Fabric phase sorptive extraction for the fast isolation of sulfonamides residues from raw milk followed by high performance liquid chromatography with ultraviolet detection, Food Chemistry 196 (2016) 428−436, https://doi.org/10.1016/j.foodchem.2015.09.060.

[32] M. Aznar, S. Úbeda, C. Nerin, A. Kabir, K.G. Furton, Fabric phase sorptive extraction as a reliable tool for rapid screening and detection of freshness markers in oranges, Journal of Chromatography A 1500 (2017) 32−42, https://doi.org/10.1016/j.chroma.2017.04.006.

[33] E. Agadellis, A. Tartaglia, M. Locatelli, A. Kabir, K.G. Furton, V. Samanidou, Mixed-mode fabric phase sorptive extraction of multiple tetracycline residues from milk samples prior to high performance liquid chromatography-ultraviolet analysis, Microchemical Journal 159 (2020) 10, https://doi.org/10.1016/j.microc.2020.105437.

[34] M. Locatelli, A. Kabir, D. Innosa, T. Lopatriello, K.G. Furton, A fabric phase sorptive extraction-High performance liquid chromatography-Photo diode array detection method for the determination of twelve azole antimicrobial drug residues in human plasma and urine, Journal of Chromatography B 1040 (2017) 192−198, https://doi.org/10.1016/j.jchromb.2016.10.045.

[35] A. Kabir, K.G. Furton, N. Tinari, L. Grossi, D. Innosa, D. Macerola, A. Tartaglia, V. Di Donato, C. D'Ovidio, M. Locatelli, Fabric phase sorptive extraction-high performance liquid chromatography-photo diode array detection method for simultaneous monitoring of three inflammatory bowel disease treatment drugs in whole blood, plasma and urine, Journal of Chromatography B 1084 (2018) 53−63, https://doi.org/10.1016/j.jchromb.2018.03.028.

[36] V. Kazantzi, A. Anthemidis, Fabric sol−gel phase sorptive extraction technique: a review, Separations 4 (2) (2017) 20.

[37] M.C. Alcudia-León, R. Lucena, S. Cárdenas, M. Valcárcel, A. Kabir, K.G. Furton, Integrated sampling and analysis unit for the determination of sexual pheromones in environmental air using fabric phase sorptive extraction and headspace-gas chromatography−mass spectrometry, Journal of Chromatography A 1488 (2017) 17−25, https://doi.org/10.1016/j.chroma.2017.01.077.

[38] V. Kazantzi, A. Kabir, K.G. Furton, A. Anthemidis, Fabric fiber sorbent extraction for on-line toxic metal determination by atomic absorption spectrometry: determination of lead and cadmium in energy and soft drinks, Microchemical Journal (2018), https://doi.org/10.1016/j.microc.2017.11.006.

[39] A. Kabir, K.G. Furton, Fabric Phase Sorptive Extraction: A New Genration, Green Sample Preparation Approach, Elsevier Science Bv, Amsterdam, 2020, https://doi.org/10.1016/b978-0-12-816906-3.00013-3.
[40] S. Montesdeoca-Esponda, R. Guedes-Alonso, S. Santana-Viera, Z. Sosa-Ferrera, J. Santana-Rodríguez, Applications of fabric phase sorptive extraction to the determination of micropollutants in liquid samples, Separations 5 (3) (2018) 35.
[41] V.F.S.A. Kabir, Fabric phase sorptive extraction in pharmaceutical analysis, Pharmaceutica Analytica Acta 6 (7) (2015) 1−3, https://doi.org/10.4172/21532435.1000e177.
[42] S.S. Lakade, F. Borrull, K.G. Furton, A. Kabir, R.M. Marce, N. Fontanals, Novel capsule phase microextraction in combination with liquid chromatography-tandem mass spectrometry for determining personal care products in environmental water, Analytical and Bioanalytical Chemistry 410 (12) (2018) 2991−3001, https://doi.org/10.1007/s00216-018-0984-z.
[43] J.C. Nadal, F. Borrull, K.G. Furton, A. Kabir, N. Fontanals, R.M. Marce, Selective monitoring of acidic and basic compounds in environmental water by capsule phase microextraction using sol-gel mixed-mode sorbents followed by liquid chromatography-mass spectrometry in tandem, Journal of Chromatography A 1625 (2020) 9, https://doi.org/10.1016/j.chroma.2020.461295.
[44] D.-E. Georgiadis, A. Tsalbouris, A. Kabir, K.G. Furton, V. Samanidou, Novel capsule phase microextraction in combination with high performance liquid chromatography with diode array detection for rapid monitoring of sulfonamide drugs in milk, 2019, https://doi.org/10.1002/jssc.201801283.
[45] A. Kabir, K.G. Furton, A. Malik, Innovations in sol-gel microextraction phases for solvent-free sample preparation in analytical chemistry, TRAC Trends in Analytical Chemistry 45 (2013) 197−218, https://doi.org/10.1016/j.trac.2012.11.014.
[46] N. Manousi, A. Kabir, K.G. Furton, E. Rosenberg, G.A. Zachariadis, Capsule phase microextraction of selected polycyclic aromatic hydrocarbons from water samples prior to their determination by gas chromatography-mass spectrometry, Microchemical Journal 166 (2021), https://doi.org/10.1016/j.microc.2021.106210.
[47] C. Basheer, H.G. Chong, T.M. Hii, H.K. Lee, Application of porous membrane-protected micro-solid-phase extraction combined with HPLC for the analysis of acidic drugs in wastewater, Analytical Chemistry 79 (17) (2007) 6845−6850, https://doi.org/10.1021/ac070372r.
[48] A. Tartaglia, A. Kabir, S. Ulusoy, H.I. Ulusoy, G.M. Merone, F. Savini, C. D'Ovidio, U. de Grazia, S. Gabrielli, F. Maroni, P. Bruni, F. Croce, D. Melucci, K.G. Furton, M. Locatelli, Novel MIPs-parabens based SPE stationary phases characterization and application, Molecules 24 (18) (2019) 15, https://doi.org/10.3390/molecules24183334.
[49] M. Locatelli, A. Tartaglia, F. D'Ambrosio, P. Ramundo, H.I. Ulusoy, K.G. Furton, A. Kabir, Biofluid sampler: a new gateway for mail-in-analysis of whole blood samples, Journal of Chromatography B 1143 (2020) 122055, https://doi.org/10.1016/j.jchromb.2020.122055.
[50] C.L. Arthur, J. Pawliszyn, Solid phase microextraction with thermal desorption using fused silica optical fibers, Analytical Chemistry 62 (19) (1990) 2145−2148, https://doi.org/10.1021/ac00218a019.
[51] Z.Y. Zhang, J. Poerschmann, J. Pawliszyn, Direct solid phase microextraction of complex aqueous samples with hollow fibre membrane protection, Analytical Communications 33 (7) (1996) 219−221, https://doi.org/10.1039/ac9963300219.

[52] C. Basheer, H.K. Lee, Hollow fiber membrane-protected solid-phase microextraction of triazine herbicides in bovine milk and sewage sludge samples, Journal of Chromatography A 1047 (2) (2004) 189−194, https://doi.org/10.1016/j.chroma.2004.06.130.

[53] N. Jakubowska, Z. Polkowska, J. Namiesnik, A. Przyjazny, Analytical applications of membrane extraction for biomedical and environmental liquid sample preparation, Critical Reviews in Analytical Chemistry 35 (3) (2005) 217−235, https://doi.org/10.1080/10408340500304032.

[54] H. Tabani, S. Nojavan, M. Alexovic, J. Sabo, Recent developments in green membrane-based extraction techniques for pharmaceutical and biomedical analysis, Journal of Pharmaceutical and Biomedical Analysis 160 (2018) 244−267, https://doi.org/10.1016/j.jpba.2018.08.002.

[55] S. Pedersen-Bjergaard, K.E. Rasmussen, Liquid-liquid-liquid microextraction for sample preparation of biological fluids prior to capillary electrophoresis, Analytical Chemistry 71 (14) (1999) 2650−2656, https://doi.org/10.1021/ac990055n.

[56] S. Pedersen-Bjergaard, K.E. Rasmussen, Electrokinetic migration across artificial liquid membranes: new concept for rapid sample preparation of biological fluids, Journal of Chromatography A 1109 (2) (2006) 183−190, https://doi.org/10.1016/j.chroma.2006.01.025.

[57] A. Gjelstad, K.E. Rasmussen, S. Pedersen-Bjergaard, Electrokinetic migration across artificial liquid membranes: tuning the membrane chemistry to different types of drug substances, Journal of Chromatography A 1124 (1) (2006) 29−34, https://doi.org/10.1016/j.chroma.2006.04.039.

Chapter 6

Introduction and overview of applications related to green solvents used in sample preparation

Seçkin Fesliyan[1], Hameed Ul Haq[2], Nail Altunay[1], Mustafa Tuzen[3], Ebaa Adnan Azooz[4] and Zainab Hassan Muhamad[5]

[1]*Sivas Cumhuriyet University, Faculty of Science, Department of Chemistry, Sivas, Turkey;* [2]*Gdansk University of Technology, Faculty of Civil and Environmental Engineering, Department of Sanitary Engineering, Gdansk, Poland;* [3]*Tokat Gaziosmanpaşa University, Faculty of Science and Arts, Chemistry Department Tokat, Turkey;* [4]*Medical Laboratory Technology Department, College of Medical Technology, The Islamic University, Najaf, Iraq;* [5]*Department of Computer Science, The Gifted Students School, Gifted Guardianship Committee, Ministry of Education, Najaf, Iraq*

1. Introduction

Working areas for scientists are increasing day by day. There is no doubt that chemists also have their share of this increase. Increasing work areas have brought about intense labor, chemicals, information resources and energy requirements for chemists, as well as exposure to various chemicals. To solve all these problems and meet the requirements, the green chemistry approach was put forward by researchers in the 1990s. According to this approach, Green Chemistry refers to all efforts to dispose of or minimize the use and production of various reagents, solvents and products that are risky to the sphere and human health [1]. Green chemistry was initially used for industrial-scale products, pharmaceuticals, etc. Although it is used for medical purposes and the production of synthetic substances, it has begun to be discussed more broadly in the future [2]. The fact that analytical chemists are exposed to chemicals in all analytical processes such as taking samples, preparing them for analysis, and being at risk for health has laid the foundations of Green analytical chemistry. In parallel with this situation, 12 principles of green analytical chemistry were presented to researchers by Anastas and Warner in 1998 [3]. Items such as minimizing the amount of samples, the quantity of chemicals used and the amount of energy required, using more minimal methods and avoiding waste generation are the most striking of the 12 principles and are the ones that guide today's studies the most.

Samples to be chemically analyzed generally have a complex internal structure. However, the analyte concentration to be determined chemically may be at trace levels in the specimen, and species. Such reasons make sample preparation techniques necessary before chemical analysis. Separation and enrichment methods are also necessary for effective measurement in analytical devices. A well-dissolved and enriched sample to a lower concentration forms the basis for accurate, reliable and sensitive chemical analysis.

Traditional extraction techniques don't provide the principles of green analytical chemistry because they are time-consuming, needed high energy consumption [4]. Due to the various disadvantages they present, the interest of researchers has shifted in favor of more minimal, environmentally friendly and simpler microextraction process, in which the levels of organic solvents used are reduced to microliters, as an alternating to customary extraction techniques. These modern analytical techniques can generally be categorized into two groups: solid phase micro-extraction (SPME) [5–7] and liquid phase micro-extraction (LPME) [8–10].

Sort of solvent is one of the significant variables in microextraction processes. To perform extraction with high yield, it is needful to select the correct solvent. By choosing the right solvent, analyte-solvent interaction can be maximized and effective mass transfer can be achieved. In recent years, dichloromethane, toluene, chloroform, etc., which are highly volatile and dangerous to the environment and health, have been used. There has been a leaning toward the utilize of new generation green extraction solvents that are easier to prepare, more green friendly and less costly as an alternating to organic solvents such as dichloromethane, toluene, chloroform etc. [11,12]. In this context, the developed analytical procedures focus on these solvents. Green solvents utilized in various sample preparation process can be classified under three main headings: 1-amphiphilic solvents, 2-ionic liquids and 3-deep eutectic solvents.

2. Amphiphilic solvents

Surfactants are among the solvents frequently used by analytical chemists in sample preparation, attracting attention with their low toxicity, and are a member of the amphiphilic solvent type. Surfactant molecules have the ability to show activity between two phases. Additionally, these molecules have the feature of having two different groups, which makes them quite interesting. One of these groups is the hydrophilic head part of the molecule and the other is the hydrophobic tail part [13]. Toward both hydrophilic and hydrophobic properties in the same molecule offers a wide range of analytes to sample preparation techniques using amphiphilic solvents. Surfactants form different clustered structures in the environment hinging on the structure of the molecules, surfactant and electrolyte concentration, and temperature. If there is surfactant in the liquid phase above the critical concentration, the space required for all molecules to take place becomes limited. In this case, the surfactants begin to self-assemble into clusters called micelles (Fig. 6.1A) and reverse micelles (in organic solvents) (Fig. 6.1B). The increase in surfactant concentration causes the molecules to shift from a micelle structure to a different long and cylindrical cluster structure (Fig. 6.1C). If the concentration increases further, these cylindrical clusters come together and form a liquid crystalline structure (Fig. 6.1D). In some cases, increasing the concentration causes the molecules to come together to form a different crystalline liquid or layered structure (Fig. 6.1E). Liposome structures are concentric, multilayered hollow spherical special structures (Fig. 6.1F).

The most extensively utilized surfactants in sample preparation are non-ionic surfactants and ionic surfactants. Triton X-100, Triton X-114 and PONPE 7.5 (polyethyleneglycol mono-p-nonylphenyl ether) can be given as examples of non-ionic surfactants. The most commonly used ionic surfactants are sodium dodecyl sulfate (SDS) and cetyltrimethylammonium bromide (CTAB). The most commonly used ionic and non-ionic surfactants are depicted in Fig. 6.2.

These solvents, which are low-cost, low-toxicity and allow interaction with many analytes, have found frequent application in the literature [14]. Another advantage of this solvent group is that it can be combined with paramagnetic materials. Due to this feature, the phase containing the analyte can be easily separated by the effect of a magnetic field, thus saving energy and time [15]. Different micelle shapes (A) and micelle-assisted extraction steps (B) are presented in Fig. 6.3.

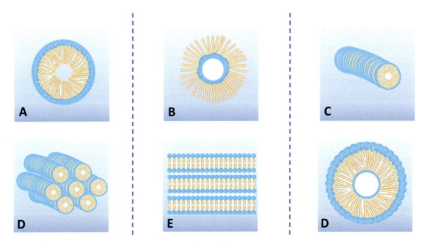

FIGURE 6.1 Different structures formed by surfactants in phase.

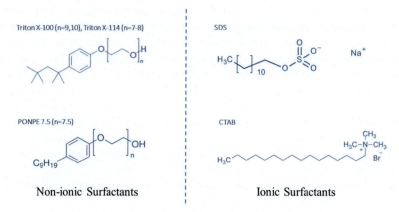

FIGURE 6.2 The most commonly used non-ionic and ionic surfactants.

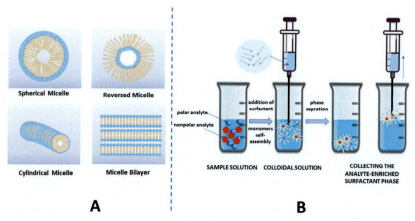

FIGURE 6.3 Illustration of different micelle shapes (A) and steps (B) of micelle-assisted extraction.

The authors were [16,17] used dispersive solid phase microextraction in the technique they improved to extract 4-nitrophenol and 4-chlorophenol, which are carcinogenic pollutants. In this study, the formation of mixed hemimicelles was benefited and magnetic particle supported CTAB was used. Experimental conditions were evaluated with central compound design, and the optimum pH was recorded as 10 and CTAB concentration as 0.86 mmol L^{-1}. As a result of the study, two analytes were successfully extracted simultaneously. Su et al. In a study by Ref. [18] on the extraction of pectin from orange peels, a process based on microwave system assisted extraction was developed and Tween-80 was used as a surfactant in this method. Variables such as pH and irradiation duration were optimized using the Box-Behnken design. It was emphasized that in this developed method, pectin extraction was achieved with 17% more efficiency compared to other extraction methods. Before the identification of five different triazole fungicides in water specimens by HPLC, a method was enhanced for the separation of these analytes in which magnetic nanocomposite material supported with surfactant was used in dispersive solid phase extraction. Effective hydrophobic and electrostatic interactions occurred between the sorbent material against fungicides and a good recovery of 90% −104% was achieved [19]. Surfactants are also used effectively in sample preparation before the identification of inorganic substances. In a study in which a procedure based on surfactant-assisted dispersive LLM was improved before the spectrophotometric identification of trace levels of vanadium in fresh juice specimens, CTAB was used as a surfactant and tannic acid as a sequestering agent. The tannic acid-vanadium complex, which was made more hydrophobic under current conditions, was efficiently extracted with 1-undecanol under optimum conditions. Finally, the developed method has been proven to be green, friendly to earth and safe using the Analytical Greenness Calculator (AGREE) [20]. In a study conducted by Bişgin [21], a method involving surfactant-assisted DLLME was investigated for the extraction of Cu (II) from food and water specimens before its determination by FAAS. In the study, Triton X-15 was used for extraction, Triton X-114 as the dispersing solvent, and dithizone as the complexing agent. As a result, a low-cost and environmentally friendly method was obtained, providing recovery of between 96% and 101% for Cu (II).

3. Supramolecular solvents

The term supramolecular solvent (SUPRAS) describes nano-structured liquids formed by self-assembly of amphiphiles. Self-assembly of amphiphiles occurs in two stages. One of these processes is on the molecular scale, while the second is on the nanoscale. First of all, above the critical micelle concentration, amphiphiles molecules self-assemble depending on the solvent and amphiphiles structure and form three-dimensional aggregates with different structures. These aggregates are aqueous micelles (3−6 nm), reverse micelles (5−8 nm) and vesicles (30−500 nm). These nano-structured aggregates self-assemble again to form aggregates with different size scales in nano and micro structures by changing the parameters of the system (temperature, pH, electrolyte, etc.) and are then collected from the solution in a further separation process [22,23]. The formation of SUPRAS via self-assembly (A) and the micrograph of a specific SUPRAS with hexagonal arrangement of amphiphile molecules (B) are shown in Fig. 6.4.

SUPRAS have many important physicochemical properties that make them alternatives to classical extraction solvents. Access to amphiphiles, which are the components of supramolecular solvents, is quite easy and low-cost. The fact that the amphiphile has both hydrophilic and hydrophobic properties

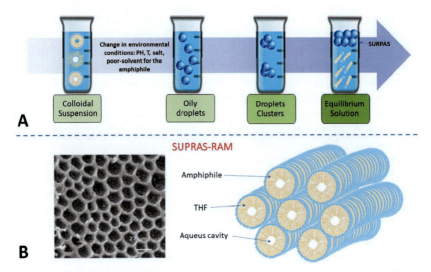

FIGURE 6.4 The self-assembly process of amphiphiles that takes place in SUPRAS formation (A), the hexagonal arrangement of amphiphilic molecules in a SUPRAS, and a micrograph of alkanol-based SUPRAS obtained by cryo-scanning electron microscopy (B).

provides easy solubility to various organic and inorganic compounds. Additionally, supramolecular solvents can be adjusted to suit different properties by changing the polarities of the hydrophobic and hydrophilic functional groups of the amphiphile. Non-volatility and non-flammability are other advantages of supramolecular solvents [24].

Nonpolar compounds can be efficiently extracted by SUPRAS depending on their octanol-water constants. The reason for the extraction yield is that the hydrocarbon chain of the amphiphiles formed in SUPRAS is non-polar. The forces involved in the separation of nonpolar analytes are dispersive forces, dipole-dipole forces and dipole-induced dipole forces. As can be easily understood, the extraction efficiency of SUPRAS when extracting polar compounds is determined via the polar section in the SUPRAS conformation. The bonding in extraction results from ionic bond, hydrogen bond, π −cation, and π - π interactive relations. If SUPRAS has a benzene ring in its structure, the bonding occurs with delocalized electrons in the π orbitals. The most important factor from the point of extraction efficiency is the amount of amphiphiles. In SUPRAS containing alkylcarboxylic acid-tetrabutylammonium alkylcarboxylate vesicles, the amphiphile concentration increases according to; decanoic > dodecanoic > tetradecanoic. SUPRAS are also very effective in extracting compounds such as surfactants, various drugs and pesticides by forming mixed amphiphiles. Recently, the tendency to use alkanol-based SUPRAS in microextraction techniques has increased. Alkanol-based SUPRAS are solutions formed by tetrahydrofuran (THF)/water mixture by

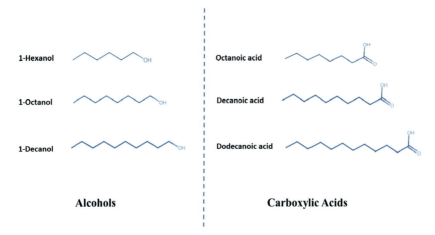

FIGURE 6.5 Some alcohols and carboxylic acids used to obtain SUPRAS.

forming reverse micelles of alkanols. In these solvents, the hydrocarbons of alkanols dissolve in THF and polar groups self-assemble. The dimension of the aggregates can be affected by adjusting the ratio dispersive solvent/water at a certain rate [24,25]. The different alcohols and carboxylic acids used to obtain SUPRAS are shown in Fig. 6.5.

Researchers frequently benefit from the superior solubility properties that SUPRAS provide to various analyte groups. In the ultrasound-assisted technique developed for the assignation of phthalates in water and cosmetic products, SUPRAS, which forms nanostructured reverse micellar aggregates with the combination of decanoic acid and THF, was used. After optimizing the parameters affecting the extraction, an enrichment ranging from 176 to 412 times was achieved under optimum conditions. With the sonication effect, a better mass transfer was achieved and the extraction time was shortened [26]. A study conducted in shops, restaurants, etc. It focuses on the change of bisphenol A (BPA) in the dust in public places such as. In this study, alkanol (1-hexanol, 1-octanol, 1-decanol) based SUPRAS were created with a mixture of amphiphile (alkanol), THF and water. After extraction, BPAs were analyzed by LC-MS/MS. While the study was quite efficient for the detection of these analytes at very low concentrations, it also revealed that these pollutants have the potential to pass into the environment [27]. SUPRAS, containing vesicular aggregates, was used in a method developed to enrich tetracycline antibiotics (TC) from milk, egg and honey matrix using liquid phase microextraction before analysis by HPLC. SUPRAS was obtained by adding NaCl to the mixture of didecyldimethylammonium bromide (DDAB) and dodecyltrimethylammonium bromide (DTAB), which belong to the class of cationic surfactants. This process ensured high enrichment factors (48−198) and good sensitivity [28]. In a study on the determination of organophosphorus pesticides (OPPs) in tea beverages, microextraction was applied to enrich OPPs prior to their assignment through high-performance liquid chromatography (HPLC-UV) with UV detector. SUPRAS, which consists of a mixture of Di-(2-ethylhexyl)phosphoric acid (DEHPA), THF and HCl, was used in the enrichment stage. Under optimum conditions, the technique was effectually performed for the mentioned pesticides and recoveries between 82% and 106% were achieved (Table 6.1) [29].

4. Hydrophilicity switchable solvent

Hydrophilicity switchable solvent (SHS) refers to solvents that are hydrophobic in aqueous solutions in the absence of CO_2, but switch to hydrophilic form in the presence of CO_2 [60]. In 2005, Jessop et al. [61] by adding CO_2 to the 1:1 mixture of non-ionic DBU (1,8-diazabicyclo-[5.4.0]-undec-7-ene) and 1-hexanol at 1 atm pressure and room temperature, the liquid admixture was converted to ionic (They discovered that the system gains polar) properties and that the system returns to its previous properties when CO_2 is removed. CO_2 is here called the trigger for the transition between two different (hydrophilic and hydrophobic) properties. The reason why CO_2 is used as a trigger is that it can be easily removed and is benign. Two different transition processes via CO_2 can be seen in Fig. 6.6.

The condition for a solvent to act as SHS is as follows; If the solvent mixes with water and forms a single phase without the addition of CO_2, the system is single phase, this solvent cannot be SHS. If the solvent forms a two-phase system both before and after the spiking of CO_2, it is called a two-phase system. If the mixture of solvent and water forms a two-phase system before the addition of CO_2, and if the solvent mixes with water and forms a single liquid phase with the addition of CO_2, the resulting mixture is SHS [62].

Amines, which are initially immiscible with water, that is, form a binary phase system, become miscible with water with the adding of CO_2, according to the chemical Eq. (6.1) below. Here, adding CO_2 to the amine -water mixture is

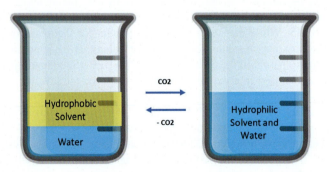

FIGURE 6.6 Transition process to water and SHS mixture triggered by CO_2.

TABLE 6.1 Some applications of amphiphilic solvents in sample preparation process.

Green solvent	Solvent composition	Sample preparation technique	Analytes	Matrix	Analytical method	Enrichment factor	Linearity	LOD	Refs.
SUPRAS	Decanoic acid-THF	Microextraction	Chlorophenols	Environmental waters	HPLC	94–102	5.0–400.0 ng/mL	1.5–2.0 ng/mL	[30]
SUPRAS	Ethanediyl-1,2-bis (dimethyl tetradecyl ammonium bromide) - 1 propanol	LPME	Parabens	Cosmetic, beverage and water samples	HPLC-UV	98–156	5–200 µg/L	0.5 µg/L	[31]
SUPRAS	Decyl hydrogen phosphonate–tetrahexylammonium	LLME	Polycyclic aromatic hydrocarbons	Tap water, natural waters	LC-fluorimetry	—	0.1–1000 µg/L	0.6–7.1 ng/L	[32]
SUPRAS	1-Tetradecanol-THF	Microextraction	Cadmium	Food and environmental samples	FAAS	165, 133	50–3500 µg kg^{-1}, 1–400 ng mL^{-1}	15 µg kg^{-1}, 0.3 ng mL^{-1}	[33]
Surfactant	Triton X-405, 1-Pentanol	CPE	Curcumin	Spices herbal teas	UV-Vis	—	—	0.012–0.027 µg mL^{-1}	[34]
Surfactant	Cationic methyltrioctyl ammonium chloride (aliquat 336), Anionic sodium dodecyl sulfate (SDS), Nonionic trioctyl phosphine oxide (TOPO).	Microextraction	Fluoroquinolone	Chicken livers	HPLC-fluorescence	32–189	—	9–21 µg/kg 5–23 µg/kg 9–20 µg/kg	[35]
Surfactant	Triton X-100, Triton X-114	DLLME	Nickel	Water samples	UV-Vis	32	≥100 µg/L	0.24 µg/L	[36]

Continued

TABLE 6.1 Some applications of amphiphilic solvents in sample preparation process.—cont'd

Green solvent	Solvent composition	Sample preparation technique	Analytes	Matrix	Analytical method	Enrichment factor	Linearity	LOD	Refs.
SUPRAS	Heptanol/THF	Microextraction	Carbaryl	Water, fruits and vegetables	LC-MS/MS	15	—	30 μg/L	[37]
20 μL SUPRA	SDS/tetrabutylammonium bromide (1:4)/AlCl3	Microextraction	Phenoxy acid herbicides	Water and rice	HPLC	37–149	10–100 μg/L	1–2 μg/L	[38]
SUPRA	200 μL undecanol/1.0 mL THF/10% NaCl	Microextraction	Fluorine-containing pesticides	Water	UPLC-Q-Orbitrap HRMS	23.5–26.7	1–500 μg/L	0.125–0.25 μg/L	[39]
SUPRA	50 μL 1-decanol/500 μL THF/8.7% NaCl	Microextraction	Carbendazim, fipronil and picoxystrobin	Water	HPLC-DAD	21.43–71.36	0.87–1000 μg/L	0.23–0.45 μg/L	[40]
SUPRA	400 mg decanoic acid/15% THF	Microextraction	Phenylurea herbicides	Water and rice	HPLC	—	0.3–20 μg/L	10–30 μg/L	[41]
SUPRA	8 μL SUPRA (1-dodecanol/toluene (1:3)) and 10 mg of CLDH@Fe3O4@TA.	AA-DMSPE	Organophosphates, diazinon, metalaxyl	Orange, peach, grape and apple juices	GC-FID	500	0.6–2000 μg/L	0.2–0.8 μg/L	[42]
SUPRA	200 μL decanoic acid/1.0 mL THF/10% NaCl	Microextraction	Pesticide residues	Rice, cucumber, tomatoes	HPLC-UV	102–176	0.1–1500 μg/kg	0.05–0.20 μg/L	[43]
SUPRA	50 μL 1-decanol/100 μL THF/NaCl	Microextraction	Sugarcane herbicides	Water	HPLC-DAD	15–48.5	0.43–400 μg/L	0.13–1.45 μg/L	[44]
SUPRA	100 mg alkanol, 5% THF	Microextraction	Organophosphorus pesticides	Juices and water	HPLC-UV	—	1.65–500 μg/L	0.50–1.30 μg/L	[45]
SUPRA	100 μL Heptanol/400 μL THF	Microextraction	Malathion	Environmental samples	UHPLC-UV	40	—	1.40 μg/L	[46]

SUPRA	Decanol/THF/water	Microextraction	Chiral triadimefon and triadimenol	Beer	LC-MS/MS	–	0.5–100 µg/L	0.24–0.98 µg/L	[47]
SUPRA	Decanoic acid/magnetic nanoparticles	Microextraction	Triazine herbicides	Water	HPLC-UV	183–256	300–250000 µg/L	300–500 µg/L	[48]
SUPRA	Decanoic acid/THF	SM-DLLME	Malachite green	Food	UV-Vis	52	–	4.0 µg/L	[49]
SUPRA	1-Decanol/THF	VA-SMS-LLME	Mercury	Environmental and biological samples	UV-Vis	100	1.0–100 µg/L	0.3 µg/L	[50]
SUPRA	100 µL 1-dodecanol/450 µL THF	VA-SSMSFD	Arsenic	Water	UV-Vis	200	5.0–140 µg/L	0.4 µg/L	[51]
SUPRA	125 µL 1-octanol/1400 µL THF	Microextraction	Selenium	Food and environmental samples	GFAAS	58	0.4–100 µg/L	0.1 µg/L	[52]
SUPRA	Hexanol/water/THF	DSPE	Ionophores	Food	LC-MS/MS	–	1–50 µg/L	0.004–0.07 µg kg−1	[53]
Surfactant	NP4EO/THF/water	VA–SS–DLLME	Orthophosphate	Water	UV-Vis	50	0.5–28.0 µg/L	0.1 µg/L	[54]
SUPRA	450 µL 1-decanol/300 µL THF	UA–SS—LLME	Manganese/zinc	Vegetables	FAAS	107	0.1–200 µg/L	0.035 µg/L	[55]
SUPRA	Decanoic acid/THF	VA-LPME	Nitrite	Meat and chicken	UV-Vis	200	0.1–300 µg/L	0.035 µg/L	[56]
SUPRA	C$_{12}$minBr ionic liquid/SDS	Microextraction	Three synthetic food dyes	Foodstuff	HPLC-UV	68–86	0.2–500 µg/L	0.05–0.1 µg/L	[57]
SUPRA	200 µL 1-decanol/400 µL THF	LLME	Thorium	Water and soil	UV-Vis	40	2.0–100 µg/L	0.4 µg/L	[58]
SUPRA	1 mL THF/70 µL 1-octanol	D-µSPE	Copper	Water and food		280	0.5–15 µg/L	0.2 µg/L	[59]

AA-DMSPE, air-assisted dispersive micro solid phase extraction; *D-µSPE*, dispersive micro-solid phase extraction; *DSPE*, Dispersive solid phase extraction; *NP4EO*, non-ionic nonylphenol tetraethoxylate; *SM-DLLME*, supramolecular-based dispersive liquid-liquid microextraction; *UA–SS–LLME*, Ultrasonic-assisted supramolecular solvent liquid-liquid microextraction; *UPLC-Q-Orbitrap HRMS*, ultrahigh performance liquid chromatography quadrupole Orbitrap high resolution mass spectrometry; *VA-SMS-LLME*, vortex-assisted supramolecular solvent liquid–liquid microextraction; *VA-SSMSFD*, vortex-assisted supramolecular solvent microextraction based on solidification of floating drop.

based on the protonation of the relevant amine and formation of a water-soluble bicarbonate salt. If Ar, N_2 or air is passed through the system and CO_2 is removed, Eq. (6.1) is reversed and a two-phase system occurs. Removal of CO_2 from the environment can also be achieved by using other physical and chemical methods; Examples include the utilized sonication, heating, use of ion exchange resins, addition of acid or base, and use of triethanolamine (TEA). Here, the time needful for the amine to mix with water depends on the size of the sample, the sample container and shape, as well as the method of adding CO_2 to the medium. The possible reason for this situation is that the reaction rate is controlled via the mass transfer of CO_2 to the aqueous solution [60,62,63].

$$NR_3 + H_2O + CO_2 \overset{*}{\leftrightarrow} NR_3H^+ + HCO_3^- \tag{6.1}$$

In 2010 Jessop et al. The first SHS introduced by Ref. [64] was N,N,N'-tributylpentanamide, a liquid amidine. However, N,N,N'-tributylpentanamide was not commercially suitable and was very troublesome to synthesize. There were also reasons such as being hydrolytically unstable and causing bioaccumulation. For these reasons, the aforementioned researchers tested a number of amines to find suitable SHS candidates that are water immiscible in the absence of CO_2 and miscible with water regards 1 bar CO_2. While some of the amines (triethanolamine, N-ethylmorpholine, N-methylpiperidine) were soluble in water without the addition of CO_2, some of them (tripropylamine, trioctylamine, N,N-dimethyldodecylamine) were not proper due to the fact that they were immiscible with water with the addition of CO_2. Another point to consider here is that some amines are too volatile to be used as SHS, and some have toxic effects. In the light of the research conducted, some notable findings have been revealed regarding the selection of the amine to be used as SHS. In order for the amine to be used as SHS, its $logK_{ow}$ value must be between 1.2 and 2.5. Amines at values lower than the lower limit will combine with water in neutral form and thus form a single-phase system. Otherwise, at values higher than the upper limit, the amine will be too hydrophobic to mix with water even after the addition of CO_2 and will form a two-phase system. Another parameter regarding the amines to be selected for SHS is; The pKa value should be above 9.5. Otherwise, the basicity of the amine will be insufficient and the switching from a two-phase mixture to a single-phase mixture with carbonated water will not occur. Although these criteria are very important and necessary, some amines could not be used as SHS despite meeting these conditions. This means that the mentioned conditions are not sufficient to obtain SHS [60,62]. Unlike amine-containing solvents, saturated oleic acids such as hexanoic, octanoic, nonanoic, and decanoic acids can also be used to form SHS. In this case, sodium carbonate and sulfuric acid are required in stages such as single phase formation and separation of phases [63]. The molecular structures of some amine-based SHSs are presented in Fig. 6.7.

Incorporation with water in SHS procures a very large surface area that allows interaction with different species. Thus, the speed of separation process increases. Additionally, there is no need for distillation to separate the product and solvent. In parallel, the utilize of detrimental organic solvents and the formation of toxic vapors are prevented and energy is saved [63,65].

Many researchers have taken advantage of these wonderful advantages of SHSs in their sample preparation techniques. In an experiment on the identification of methadone and tramadol from urine samples, the separation of analytes was carried out by homogeneous liquid-liquid microextraction (HLLME). Dipropylamine, an SHS, was chosen as the extraction solvent. CO_2 was used as phase trigger. In this study, in which GC-FID was used as the analysis technique, analysis was successfully carried out and the preconcentration factors were recorded as 118 and 128 [66]. Another study using SHS aims to determine benz[a]anthracene from water specimens. Before fluorometric determination, the departure of benz[a]anthracene was carried out by liquid-liquid microextraction and (N,N-Dimethylcyclohexylamine) was utilized for extraction. For SHS formation, CO_2 was used to form the single phase and finally NaOH was used to form the binary phase. As a result of the study, good sensitivity was achieved and recoveries were recorded between 72%−100% [67]. SHSs can also be used in the microextraction of inorganic substances. For this purpose, effervescence-assisted SHS-based liquid-liquid micro-extraction was used to enrich Cu (II) prior to its identification by FAAS with micro injection system. Here, decanoic acid was used as SHS, CO_2 as phase trigger and H_2SO_4 as proton donor. Since decanoic acid is solid at room temperature, it facilitates phase distinction, and the proposed method provides sensitivity, repeatability and convenience under optimum conditions. The enhancement factor is stated as 52 [68]. In another study, a method based on vortex-assisted switchable hydrophilic solvent liquid phase microextraction (VA−SHS−LPME) was enhanced to enrich trace amounts of vanillin in food specimens prior to its determination with UV-Vis. While four different SHS (N,N-Dimethylbenzylamine, N,N-Dimethylcyclohexylamine, 1-ethylpiperidine and TEA) were used as solvents in the microextraction technique, NaOH was preferred as the switching-off trigger. The extraction recovery (97% ± 4%) and improvement factor (220) were recorded, and the developed procedure provided advantages such as good extraction yield, reduced chemical consumption and low matrix effect (Table 6.2) [69].

5. Ionic liquids

One of the most important green alternatives to classical extraction solvents is undoubtedly ionic liquids (ILs). The following explanation can be made to concretize the term ionic liquid; ionic liquids are liquids comprise entirely ions, not an ionic solution, which is a salt solution dissolved in a molecular solvent (Fig. 6.8.) [98]. Ethyl ammonium nitrate was the first IL introduced to

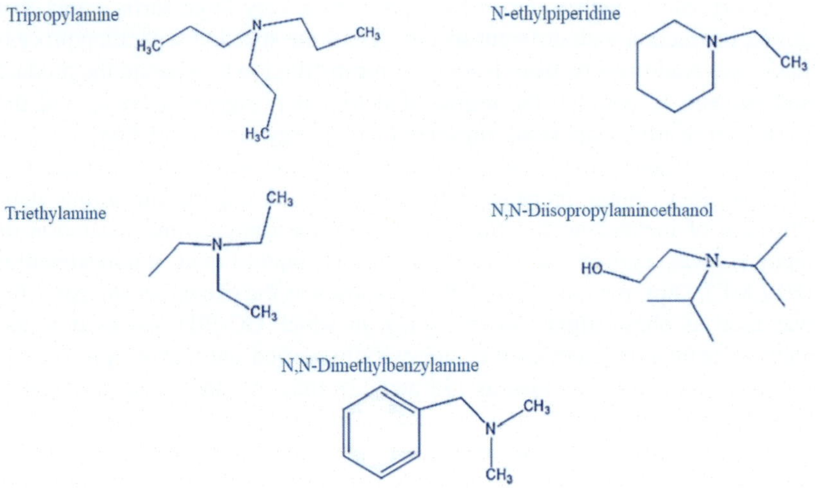

FIGURE 6.7 Some amine-based SHSs used.

the literature. It was Paul Walden who pioneered this in 1914 [99]. Although Paul Walden probably did not foresee that his invention would make a big splash almost a century later, ILs are among the green solvents that have been frequently taken advantage of by researchers in recent years with their extraordinary properties. The superior advantages provided by ILs arise from their physicochemical qualities. ILs have omissible vapor pressure, high thermal pertinacity, low melting point, superior solubilization ability for many analytes, low flammability, and good electrical conductivity, while their viscosity and hydrophilicity can be adjusted. Due to these properties, ILs are used in many fields such as electrochemistry, materials chemistry, energy technology, organic synthesis, as well as sample preparation techniques [100–102]. The difference between ionic liquid and ionic solution is concretized in Fig. 6.8.

ILs are green solvents because [103];

- They have low vapor pressure under ambient conditions and are not volatile. This situation is important for the environment and researcher health.
- It is liquid over a wide temperature range.
- Acidity, alkalinity, hydrophilic properties and viscosity can be adjusted.
- It is mostly colorless and polar.

The low melting point of ionic liquids is because of they consist of ions with a high diameter and salts with a low degree of symmetry of the cation. In this case, the crystal lattice energy and melting point of the salt decrease accordingly [104]. The density of ILs is inversely proportional to the magnitude of the ion size. The density of commonly utilized ILs is higher than that of water. Thanks to this feature, the IL phase precipitates quickly in phase separation appliances utilized in dispersive liquid phase micro-extraction. Adjustable viscosity can be used to suspend large drops at the needle tip when performing single drop microextraction, and this is an important advantage [100]. Some of the preferred cations and anions for ionic liquids are given in Fig. 6.9.

ILs can be classified structurally (Fig. 6.10). Task-specific ionic liquids are very important and attract great attention because they are among the ILs in which the researcher can adjust the anion and cation combination as desired in line with the conditions required by the study. The poisonousness and low biodegradability of alkyl imidazolium and alkyl benzimidazolium consisted ILs have accelerated the synthesis of low-toxicity, environmentally friendly, biodegradable and recyclable bio-ionic liquids. For this purpose, compounds containing various choline cations and counter anions with different amino acid contents have been synthesized. Recently, another group of ionic liquids called energetic ionic liquids has engaged the attention of researchers with their advantages such as high density, thermal pertinacity, easy synthesis, neglectable vapor pressure and low vapor toxicity [103].

A subclass of ionic liquids is magnetic ionic liquids (MIL). Magnetic ionic liquids have all the physicochemical characteristic of ionic liquids and also respond rapidly to an externally applied magnetic field. Members of this subclass include magnetic cations (such as ferrocenium) or magnetic anions ($FeCl_4^-$, $CoCl_4^{2-}$, $GdCl_6^{3-}$). Magnetic susceptibility in these solvents arises from the addition of high-spin transition metals or the addition of radical contents to parts of the anion or cation of the IL [105−107]. MILs can be configured to be task specific. MILs, which form the analyte-enriched extraction phase, eliminate back-extraction steps that cause analyte loss and can be injected directly from the injection chamber of the relevant device. In this case, time is also saved. MILs with low UV-Vis absorbance are frequently preferred in analytical determination systems where photometric detectors are used. In this way, it is possible to synthesize task-specific MIL that takes analytical devices into account. In general, as the increment in the lengthiness of the alkyl chain of the IL cation increases the viscosity, suitable MILs can be obtained by making some modifications. While viscous species may be preferred for the stationary phase of gas chromatography (GC), lower viscosity MILs can be synthesized to effectively pipet the extraction phase in microextraction. The magnetic susceptibility of MILs is inversely proportional to temperature. The feature of having strong magnetic susceptibility gives this solvent group the advantage of being effective in the microextraction [107].

Ionic liquids have found a wide interval of implementing in the literature thanks to their superior properties. In a study aiming to test the extractability of 45 pollutants, comprising benzene, toluene, ethylbenzene and xylene, by liquid phase microextraction before their determination by HPLC, two ILs with the structure of 1-alkyl-3 methylimidazolium hexafluorophosphate

TABLE 6.2 Some studies in the literature on sample preparation techniques using SHSs.

SHS	Switching-on trigger	Switching-off trigger	Sample preparation technique	Analyte	Matrix	Detection technique	Enrich-ment factor	Linearity	LOD	Refs.
DMCA	NaOH	–	SHS-LLME	Quercetin	Food samples	UV-Vis				[70]
DMCA, TEA, DMBA	CO_2	NaOH	SHS-UA-LLME/DSPME	Nitrite	Food samples	UV-Vis	250	0.3–600 µg/L	0.1 µg/L	[71]
DPA[3]	–	–	SHS-HLLME	Antidepressants	Biological and wastewater samples	GC-FID	178.7–194.8	5.0–1000 µg/L	≤1.0 µg/L	[72]
Hexanoic acid	Na_2CO_3	H_2SO_4	SHS-HLLME	Sulfonamides	Chicken meat	HPLC-UV				[73]
DMCA, DPA	HCl	–	SHS-HLLME	Synthetic drugs	Blood samples	LC-MS/MS				[74]
DPA	HCl	NaOH	SHS-UA-LPME	Ni	Medicinal plant	UV-Vis				[75]
DEBT	HNO_3	–	RP–SHS–LLME	Cu	Edible oil samples	Digital image colorimetry				[76]
DMBA	Dry ice	–	SHS-LPME	Mn	Soil samples	SQT-FAAS				[77]
DMBA	Dry ice	NaOH	SHS-LPE	Iron	Human hair samples	SQT-FAAS				[78]
DMCA	HAC	–	SHSE	Lead	Environmental samples	ICP-OES				[79]
DMCA	Dry ice	NaOH	SPE-HLLME	Chloramphenicol	Water	HPLC-UV	340	0.5–50 µg/L	0.1 µg/L	[80]

DMBA	CO$_2$	NaOH	SS-LPME	Fluoxetine, estrone, Pesticides, and endocrine disruptors	Wastewater	GC-MS	22–307	0.5–1000 µg/L	0.16–8.6 µg/L	[81]
DMCA	HCl	NaOH	LLME	11 drugs	Urine	GC-MS	–	5–2000 µg/L	0.36–12.5 µg/L	[82]
TEABC	CO$_2$	NaOH	LPME	Uranium	Water and soil	UV-Vis	40	–	0.3 µg/L	[83]
Triethylamine	CO$_2$	NaOH	AA-PLME	Palladium	Real samples	ETAAS	64	0.16–2.5 µg/L	0.07 µg/L	[84]
DMCA	CO$_2$	NaOH	LLME	Palladium	Water and soil	FAAS	37.5	15–1600 µg/L	4.28 µg/L	[85]
DMOABC	Dry ice	NaOH	LPME	Cobalt	Tobacco and food	FAAS	40	–	3.2 µg/L	[86]
1-Ethylpiperidine	Dry ice	NaOH	LPME	Nickel	Food and cigarette	FAAS	10	17–500 µg/L	5.2 µg/L	[87]
Triethylamine, TEABC	–	NaOH	UA-LPME	Cadmium, nickel, lead, cobalt	Water, urine, tea	FAAS	100	0.7–200 µg/L	0.24–0.76 µg/L	[88]
Triethylamine	Dry ice	NaOH	LPME	Nickel	Water	UV-Vis	–	0.05–0.60 µg/mL	0.02 µg/mL	[89]
DETA/MWCNT	–	CO$_2$	HSL-SDM	Arsenic	Water	HG-AAS	83	0.01–0.80	0.003 µg/L	[90]
TEABC	Dry ice	NaOH	LPME	Cadmium	Baby food	FAAS	54.2	0.1–60 µg/L	0.02 µg/L	[91]
Decanol/DBU	HNO$_3$	CO$_2$	Extraction	Aluminum	Blood	FAAS	25	10–50 µg/L	0.47 µg/L	[92]
Triethylamine	CO$_2$	NaOH	LPME	Lead and cadmium	Biological	GFAAS	49/52	0.01–2.0 µg/L	0.016 µg/L	[93]
DMBA	Dry ice	NaOH	VA-LPME	Vitamin B12 and cobalt	Egg yolk	SQT-FAAS	–	8–200 µg/L	2.3 µg/L	[94]

Continued

TABLE 6.2 Some studies in the literature on sample preparation techniques using SHSs.—cont'd

SHS	Switching-on trigger	Switching-off trigger	Sample preparation technique	Analyte	Matrix	Detection technique	Enrich-ment factor	Linearity	LOD	Refs.
Schiff base ligand	CO_2	NaOH	LLME	Palladium	Water	SQT-FAAS	34	50–750 µg/L	15 µg/L	[95]
Hexanoic acid	Na_2CO_3	H_2SO_4	Microextraction	Dyes	Foodstuffs	HPLC	62–66	4.0–3000 µg/L	0.001–0.005 µg/L	[96]
DMBA	Dry ice	NaOH	VA-LLME	Cadmium	Wastewater	SQT-FAAS	–	2.0–30 µg/L	0.7 µg/L	[97]

DBU, 1,8-diazabicyclo [5.4.0] undec-7-ene; *DEBT*, N,Ndiethyl-N'-benzoylthiourea; *DETA*, diethylenetriamine; *DMCA*, N, N-dimethylbenzylamine; *DMOABC*, N, N dimethyl-n-octylamine bicarbonate; *DPA*, Dipropylamine; *HAC*, Glacial Acetic Acid; *HSL-SDM*, hydrophilic switchable liquid-solid dispersive microextraction; *MWCNT*, Multiwall carbon nanotube; *SQT*, Slotted quartz tube; *TEABC*, triethylamine bicarbonate.

Introduction and overview of applications **Chapter | 6** **163**

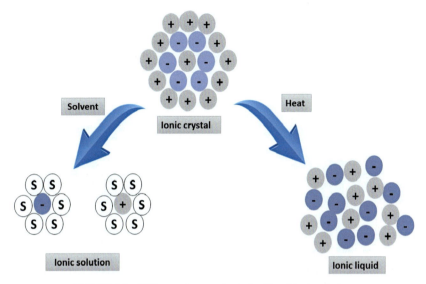

FIGURE 6.8 Difference between ionic liquid and ionic solution.

(n = 4.8) were used. Researchers stated that the enhanced technique is a very suitable and rapid method for the enrichment of the mentioned pollutants by using environmentally friendly and recyclable IL [108]. In another study, a new method called cold vapor ionic liquid assisted headspace single drop microextraction (CV-ILAHS-SDME) was enhanced for the identification of trace levels of inorganic and organomercury species. After the method in

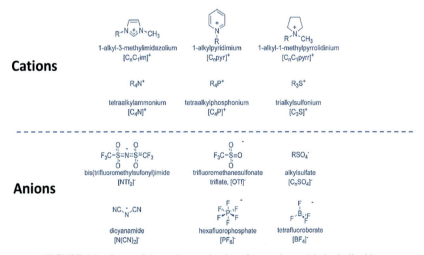

FIGURE 6.9 Some of the cations and anions frequently used in ionic liquids.

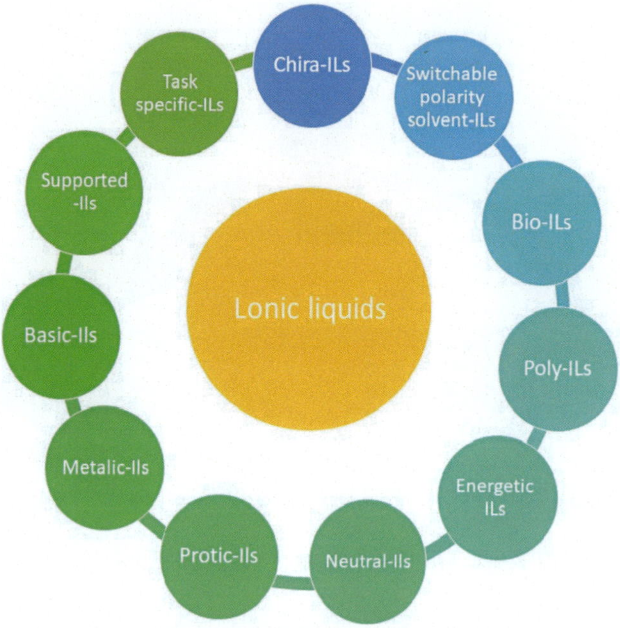

FIGURE 6.10 Structural classification of ionic liquids.

which tetradecyl(trihexyl)phosphonium chloride was used as IL, electrothermal atomic absorption spectrometry (ETAAS) was utilized for the determination of analytes. Experimental parameters affecting the extraction (temperature, sample and solvent volume, extraction time) were examined and the method showed a fast, low-cost and environmentally friendly activity for the identification of mercury species in various specimens such as sea water, fish tissues, hair and wine [109]. In a study in which the single drop microextraction method using magnetic liquid was combined with voltammetry for the first time, the determination of ascorbic acid from effervescent tablets and orange juice was aimed. In this study, magnetic ionic liquid containing aliquat tetrachloromanganate (II) was used, and the extraction phase containing ascorbic acid was separated by a magnet. This method, used for the first time in the literature, has been effectually performed for the identification of ascorbic acid from the mentioned matrices [110]. In a study aiming to investigate neurotransmitters in urine samples of patients with mild cognitive impairment, mild dementia and moderate dementia, ionic liquid-based liquid-liquid micro-extraction was used for separation and enrichment. Experimental parameters affecting microextraction were examined and optimized using Plackett-Burman scanning and Central Composite Design. After the enrichment step, 15 neurotransmitters were determined using tandem mass spectrometry. The improved method provided high throughput usability to examine neurotransmitters in various clinical biological fluids (Table 6.3) [111].

6. Deep eutectic solvents

In 2001, Abbott and colleagues reported that a mixture of choline chloride and zinc chloride could form a liquid at temperatures below 100°C [130]. In 2003, the same people introduced the concept of deep eutectic solvent (DES) for the first time. With this concept, the authors have shown that when salt or solid ionic liquid is utilized in a mixture, a large decrease in melting point is achieved and the resulting solvent can be called deep eutectic solvents [131,132]. Although DESs are seen as alternative solvents to ILs, as they eliminate the disadvantages such as high cost, toxicity, poor biodegradability exhibited by some types of ILs, and maintain the advantages such as low vapor pressure, tunability, easy preparation and superior solubilization ability, ILs are It can also be said that they are a subspecies [132,133]. DESs are formed by way of the combination of components called hydrogen bond donor (HBD) and hydrogen bond acceptor (HBA) by means of hydrogen bonding at a certain temperature and mole ratio (Fig. 6.11). This mixture is called eutectic mixture. The melting temperatures of DESs are lower than the melting temperatures of each of the constituents. The main reason for this situation is the charge delocalization that occurs due to hydrogen bond formation [134–137]. In addition, the process of formation of DES's components does not create any waste products [25,132,138]. In the components that make up DESs, quaternary ammonium salts are generally utilized as HBA, and amines, carboxylic acids, alcohols and carbohydrates are used as HBD [139].

By changing the components of DESs, new eutectic admixtures with various chemical and physical characteristics can be created. This is very important in terms of DES's ability to affect large-scale analyte groups. For example, when it is desired to extract the nonpolar analyte from aqueous solution, DES with a hydrophobic structure will make it easier to achieve this goal. Viscosity and density of DESs are very important parameters for mixing and phase separation. Since these parameters are directly related to the DES components, the effect of the replaceable properties of the components on the extraction yield is also encountered here [138]. The polarity of DESs also depends on the presence of water. In the absence of water, DES polarity follows the following order: organic acid-based DESs > amino acid-based DESs > sugar-based DESs > polyol-based DESs [140]. Some HBAs and HBDs used as DES components are given in Figs. 6.12 and 6.13, respectively.

Choi et al. [141] started by questioning the role of sugars, some amino acids, choline and some organic acids such as malic acid, citric acid, lactic acid and succinic acid in living cells and organisms, and then tested the usability of these compounds for the separation of natural products. For this

TABLE 6.3 Some studies using ILs in the literature.

Province	Matrix	Analyte	Microextraction technique	Detection technique	%Recovery	Enrichment factor	Linearity	LOD	Refs.
[C$_{16}$MIM][NTf$_2$]	Tap, river, sewage waters	DDT and its metabolites	HF-SPME	UHPLC-UV	64–112	–	0.5–50 µg/L	0.33–0.38 µg/L	[112]
[P$_{6,6,6,14}$][FeCl$_4$] (MIL)	Honey samples	Arsenic	DLLME	ETAAS	95.2–102	110	0.02–5.0 µg/L	0.012 µg/L	[113]
[P$^+_{6,6,6,14}$I$_2$][MnCl$_4^{2-}$] [Aliquat-I$_2$][MnCl$_4^{2-}$]	Human urine	Estrogens	DLLME	HPLC	67.5–115.6	28.9–33.4	5–500 µg/L	2.0 µg/L	[114]
Decylguanidinium chloride	Urine samples	PAHs	IL-ME	HPLC-FD	95.1–110	–	–	0.001 µg/L	[115]
[C$_6$mim][FAP]	Chocolate-based samples	Nickle, cobalt	VA-IL-DLLME	FAAS	94–97.2, 93.5–97.5	62.5	0.7–400 µg/L	0.3, 0.2 µg/L	[116]
[C$_4$mim][Tf$_2$N]	Vegetables	Aluminum, chromium	VA-IL-DLLME	FAAS	96.2–98.8, 96.7–98.1	95, 135	0.07–100 µg/L	0.02, 0.05 µg/L	[117]
[C$_4$MIM][PF$_6$]	Human saliva	Cortisol, cortison	IL-DLLME	LC-UV	83.3–115.8				[118]
[Emim]$_2$[Co(NCS)$_4$]	Water	Chlorobenzenes	HS-SDME	TD-GC-MS	82–114				[119]
[CnMIM][BF$_4$] (n = 4,6,8,10)	Water	Microcystins	Capillary microextraction	LC-MS	70.2–107	–	0.001–10 µg/L	0.0004–0.0014 µg/L	[120]
[C4MIm]Cl	Water	Organophosphorus pesticides	IL-DLLME	GC-MS	85–118	–	0.1–2.0 µg/L	0.005–0.016 µg/L	[121]
[N1,8,8,8][NTF2]	Rice	Endrin, chlordane, and dieldrin	HF-LPME	GC-MS	86.0–92.9	–	0.064–16.7 µg/L	0.019–0.029 µg/L	[122]

[VBC₁₆IM⁺][NTf₂⁻]	Water	Ultraviolet filters	HS-SPME	GC-MS	63.5–110	—	—	0.0028–0.026 µg/L	[123]
BMIMPF₆/acetonitrile	Rat brain, plasma, cell	15 neurotransmitters	IL-USADLLME	LC-MS/MS	81–128	—	0.04–200 µg/L	0.02–0.987 µg/L	[124]
[VC₆IM][Cl], [VC₁₆IM][Br], and [(VIM)₂C₁₀]₂[Br]	Plasma	Endocannabinoids	On-line in-tube SPME	UHPLC-MS/MS	—	—	0.1–100 µg/L	0.015 µg/L	[125]
10 HDIL	Water	Cadmium and lead	Extraction	UV-Vis	95.48–98.46	20	—	—	[126]
[OMIM][PF₆]	Water	Ag NPs	IL-DLLME	ICP-MS	71.0–90.9	—	0–1000 µg/L	0.01 µg/L	[127]
[C₈C₁Im][NTf₂]	Arnica infusions	Cadmium, lead and mercury	IL-DLLME	ICP-MS	94–107	89–127	1.5–100 µg/L	0.078–0.27 µg/L	[128]
[BMIM]PF₆/Fe₃O₄@SiO₂@ILs	Water, fruits and vegetables	Linuron	MSPE	UV-Vis	95.0–101.0	—	0.04–20.00 µg/L	0.005 µg/L	[129]

[Aliquat+I₂][MnCl₄²⁻], aliquat tetrachloromanganate (II); [C₁₆MIM][NTf₂], 1-Hexadecyl-3-methylimidazolium bis(trifluoromethylsulfonyl)imide; [C₄MIM][NTf₂], 1-butyl-3-methylimidazolium bis(trifluorosulfonyl)imide; [C4MIm]Cl, 1-butyl-3-methylimidazolium chloride; [C₄MIM][PF₆], 1-butyl-3-methylimidazolium hexafluorophosphate; [C₆mim][FAP], 1-hexyl-3-methylimidazolium tris(pentafluoroethyl)trifluorophosphate; [C₈C₁Im][NTf2], 1-octyl-3-methylimidazolium bis(trifluoromethanesulfonyl)imide; [CnMIM][BF4], 1-Butyl-3-methylimidazolium tetra fluoroborate; [Emim]₂[Co(NCS)₄], 1-ethyl-3-methylimidazolium tetraisothiocyanatocobaltate(II); [N1,8,8,8][NTF₂], 1-Ethyl-3-methylimidazolium bis(trifluoromethylsulfonyl)imide; [OMIM][PF6], 1-octyl-3-methylimidazolium hexafluorophosphate; [P₆,₆,₆,₁₄]FeCl₄(MIL), trihexyl(tetradecyl)phosphonium tetrachloroferrate (III); [P₆,₆,₆,₁₄]₂[MnCl₄²⁻], trihexyltetradecylphosphonium tetrachloromanganate (II); [VBC₁₆IM⁺][NTf₂⁻], 1-Vinylbenzyl-3-hexadecylimidazolium bis(trifluoromethyl)sulfonyl] imide; BMIMPF₆, 1-butyl-3-methylimidazolium hexafluorophosphate; HDIL, hydrophobic dicationic ionic liquids; HF-LPME, hollow fiber-protected liquid-phase microextraction.

purpose, they introduced a new sort of DES by mixing the mentioned compounds in different combinations and called it natural deep eutectic solvent (NADES). In the light of their research, they reported that the dissolubility of rutin, a flavonoid, in different NADES is 50−100 times higher than in water. Additionally, non-water-soluble paclitaxel and ginkgolide B showed high dissolubility in a NADES comprise glucose-choline chloride components. Researchers also noted that DNA (from male salmon), albumin, and amylase have good solubility in some NADES. NADES can be generated by means of mixing their components in a certain molar ratio and at a certain temperature. It has been reported that the number of HBA and HBD possessed by the components, the spatial structure of these groups and the position of the bonds are effective on the creation and pertinacity of NADES. For example, the excess number of carboxyl and hydroxyl groups participating in hydrogen bonding in a NADES, one or more of whose components are organic acids, is effective in increasing the stability of the NADES [142]. NADES, which have a natural composition, are frequently used safely in food, pharmaceutical, cosmetics and various other industries as an alternating to classical extraction solvents owing to their environmental friendliness and biodegradability and high dissolving abilities [143,144].

DESs can be combined with structures with paramagnetic properties and converted into a form that can react to the external magnetic field. In this way, magnetic DES allow quick and easy gathering of the analyte-containing extraction phase, while also saving processing time [15]. Typically, the preparation of magnetic DES involves two sequential stages. Initially, the appropriate molar ratio of the HBA and HBD is combined, and the resultant mixture is heated from 40 to 80°C until a clear liquid is obtained. Subsequently, as the temperature returns to room temperature, the second stage commences. In this phase, a specific metal chloride (such as $FeCl_3$, $CoCl_2$, $MnCl_2$) is introduced to the obtained DES and mixed at a defined mole ratio. A small amount of a volatile solvent (like dichloromethane) is then incorporated into the blend. Following approximately 24 h of stirring and subsequent evaporation under pressure, a magnetic DES is produced [145].

In microextraction techniques using DESs, the energy of ultrasonic sound waves and vortex formation can be used to increase the speed of movement of molecules. Ultrasonic effect and vortex effect increase the contact of the extraction solution with the analyte. In this case, effective mass transfer occurs. In techniques based on ultrasonic effect, extraction efficiency depends on the type and volume of extraction solvent, pH of the environment, ultrasonic conditions and properties of the sample. Since the ultrasonic effect accelerates and facilitates contact between phases, it can replace the dispersing liquid used in DLLME. Such physical techniques increase extraction efficiency and provide convenience while also saving time [138,146].

DESs are the pioneers of green solvents with the various advantages they offer to researchers. It is frequently used in sample preparation from various sample matrices. In a study aiming at the extraction of chlorogenic acids and flavonoids from coffee grounds, DES consisting of 1,6- hexanediol:choline chloride was used. The experimental parameters affecting this ultrasound-assisted extraction were optimized using a central composite design. As a result of the study, good recovery values were obtained [143]. In a study conducted to enrich curcumin from food samples before spectrophotometric determination, the DES based microextraction technique was developed. An alcohol-based DES consisting of betaine hydrochloride and glycerol was used as the extraction solvent. The method provided good efficiency in the detection and quantification of curcumin from food samples under optimum conditions [147]. A new extraction method via DES was tested to determine trace amounts of zinc in fish samples. Choline chloride and phenol were used as extraction solvents, and 8-hydroxyquinoline was used to chelate zinc ions. After optimizing the experimental conditions, the developed method was successfully used in the determination of zinc from fish samples under optimum conditions [148]. In one study, a method based on magnetic solid phase extraction was developed for the simultaneous separation and enrichment of Cu (II) and Pb (II) from real samples. DES-modified nano-Fe_3O_4 particles were used as adsorbent in the study. After optimizing the parameters affecting the experimental procedure, the method was effectively used for the simultaneous enrichment of copper and lead under optimum conditions. Determination of Cu (II) and Pb (II) was carried out by ICP-OES (Table 6.4) [149].

7. Conclusions and prospects for the future

GAC has motivated analytical researchers to create new sustainable environmental solvents, enhance methods, and create sophisticated equipment to raise the overall sustainability and greenness of the method of analysis while also improving analysis quality. Several environmentally friendly, highly promising replacements for traditional organic solvents are sustainable green solvents,

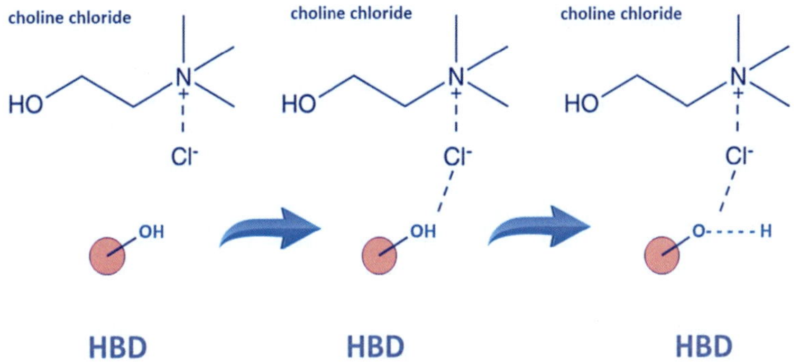

FIGURE 6.11 Interaction between choline chloride (HBA) and an HBD.

170 Green Analytical Chemistry

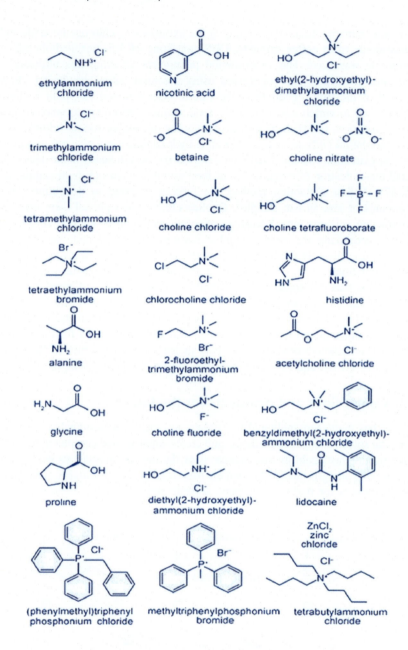

Hydrogen Bond Acceptor (HBA)

FIGURE 6.12 Some HBAs as DES component.

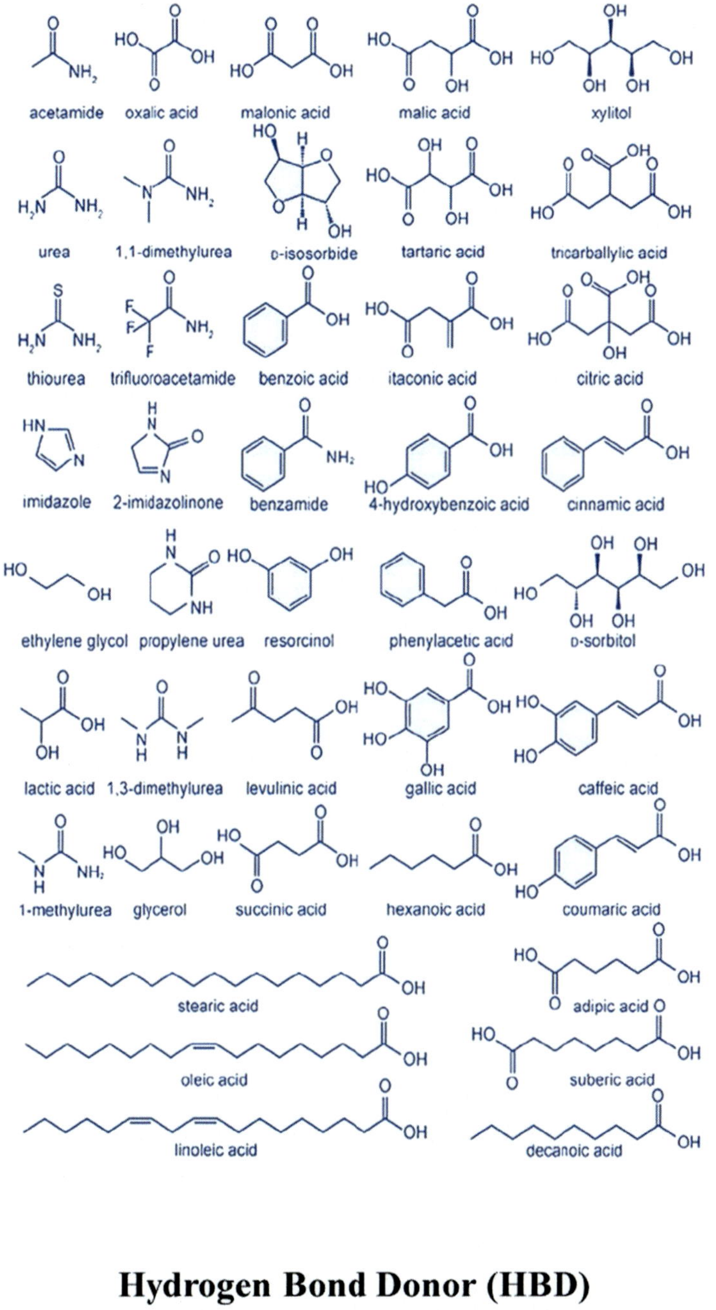

Hydrogen Bond Donor (HBD)

FIGURE 6.13 Some HBDs as DES component.

TABLE 6.4 Some of the studies using DESs in the literature.

DES/NADES composition	Molar ratio	Matrix	Analyte	Microextraction technique	Detection technique	Recovery %	Enrichment factor	Linearity	LOD	Refs.
Lactic acid-glucose-water	5:1:3	Flower petals of catharanthus roseus	Anthocyanins	LPME	HPLC-DAD					[150]
Choline chloride-ethylene glycol	1:4	Orange peel	Polyphenolic compounds	SLE	HPLC					[151]
l-menthol-acetic acid	1:1	Water samples	Phthalic acid esters	DLLME	HPLC-UV	71–120	—	4–425 µg/L	1.08–6.9 µg/L	[152]
Citric acid-sucrose	1:3	Vegetable samples	Lead, cadmium	HI-DES-ME	FAAS					[153]
Nonanoic acid- n-decanoic acid-undecanoic acid	2:1:1	Fruit juices	Lycopene, carotene	DES-ABLLME	HPLC	—	—	0.1–100 µg mL^{-1}	0.025–5.00, 0.05–0.002 µg mL^{-1}	[154]
Fructose-citric acid	1:1	Food samples	Gluten	UAME	Immunoassay				—	[155]
Choline chloride-chlorophenol	1:2	Water and urine samples	Volatile aromatic hydrocarbons	HSDME	GC					[156]
Choline chloride-lactic acid- water	1:1:5.5	Oil samples	Metals	M.E.	ICP-OES	35–100	—	0.1–1500 mg kg^{-1}	0.02–17 mg kg^{-1}	[157]
Sucrose-choline chloride-water	1:4:4	Safflower (Carthamus tinctorius)	Colorants	LPME	HPLC-DAD					[158]

Octanol/menthol	1:1	Beverages	Bisphenols	VA-DES-LLME	HPLC	91.33–103-95	23.60–27.23	5.0–1000 µg/L	0.44–0.96 µg/L	[159]
Choline chloride/urea	1:1	Water, vegetable, and commercial food	Chlorophyll (E140)	UA-DES-DLLME	UV-Vis	93–104	16.7	—	0.7 µg/L	[160]
Menthol: Octanoic acid	1:1	Chicken meat	Chloramphenicol and florfenicol	SH-DES–MAE-in–syringe HLLME	HPLC–DAD	77, 81	92, 97	$0.37–2 \times 10^3$, $0.43–2 \times 10^3$ ng/g	0.11, 0.13 ng/g	[161]
Phenyl salicylate/DL-menthol	1:1	Water and soil	Palladium	TC-DLLME	ETAAS	95.1–100.4	205	0.1–3.0 µg/L	0.03 µg/L	[162]
Thymol/decanoic acid	1:2	Water	Arsenic	LLE	ICP-MS	94–95.7	—	7.2–100 µg/L	1.0 µg/L	[163]
Choline chloride/ethylene glycol	1:1	Vine pruning residues	Polyphenols	SPE	HPLC	31–100	—	0.31–20.5 µg/L	0.03–1.0 µg/L	[164]
$N_{4444}Br$/dodecanal	1:2	Real samples	Cinnamon, cumin, fennel, clove, thyme, and nutmeg	HS-SDME	GC-MS	—	—	1–500 µg/g	0.47 µg/g	[165]
$N_{8888}Br$/1-dodecanol/decanoic	1:1:1	Water and tea beverage	Pyrethroids	LPME	HPLC	89.3–97.7	92–105	1.9–500 µg/L	0.56–1.24 µg/L	[166]
l-Menthol/butylated hydroxytoluene	3:1	Wine	Pesticides	DLLME	LC-MS	66–100	51–86	20–4000 µg/L	0.00070–1.6 µg/L	[167]
PChCl/fatty acid		Deionized water and exhaled breath condensate	Lung surfactants	AA-LLME	LC-MS/MS	86–97	71–86	0.39–1000 µg/L	0.12–0.23 µg/L	[168]

Continued

TABLE 6.4 Some of the studies using DESs in the literature.—cont'd

DES/NADES composition	Molar ratio	Matrix	Analyte	Microextraction technique	Detection technique	Recovery %	Enrichment factor	Linearity	LOD	Refs.
Monoterpene/fatty acid	1:1	Microbial fermented functional beverage and bottled water	Alkyl phenols, bisphenols, and alkyl phenol ethoxylates	VA-DLLME	UHPLC-MS	70.0–124.3	—	0.2–50 µg/L	0.1–2.99 µg/L	[169]

$N_{4444}Br$, tetrabutylammonium bromide; *SH-DES–based MAE and in–syringe HLLME*, switchable hydrophilicity deep eutectic solvent–based microwave–assisted extraction combined with in-syringe homogenous liquid–liquid microextraction; *TC-LLME*, temperature-controlled liquid–liquid microextraction.

such as amphiphilic solvents, DESs, ILs, surfactants, SUPRASs, and SHS. These solvents have useful chemical-based and thermal stability, higher extraction efficiency, and adjustable physical characteristics. This study offers an initial overview of how green solvents have been used in sample preparation methods throughout the last 20 years, with a focus on applications in various analysis classes. The field of study on green solvents will certainly continue to expand quickly because it is still in its early stages.

The future of green analytical procedures looks to be focused on some aspects, such as: Further investigation is required to explain the structure-property correlations and intermolecular interactions of green solvents. since these insights will be beneficial for particular applications and will help increase their use. It is also necessary to do a thorough investigation into factors including compatibility with various analytes and matrices, effectiveness of extraction, and solvent selection. This may encourage the use of green solvents that are better suited for particular sample preparation tasks. Further study and development of cheaper, polar and non-polar bio-based solvents is necessary. Also, the integration of microextraction and miniature detection equipment with sustainable green solvents would be essential for on-site and point-of-care analysis to advance the use of green analytical procedures. Furthermore, microextraction enters the nano-extraction range due to the confinement of environmentally friendly solvents within nanomaterials. This, combined with further reductions in sample volumes, raises the total relative greenness of microextraction techniques. This development makes it possible to easily modify the extraction of medium-polar or non-polar compounds with high effectiveness, excellent stability, and accuracy by customizing these nanoparticle fluids.

References

[1] P.T. Anastas, Green chemistry and the role of analytical methodology development, Critical Reviews in Analytical Chemistry 29 (3) (1999) 167–175.

[2] A. Gałuszka, Z. Migaszewski, J. Namieśnik, The 12 principles of green analytical chemistry and the SIGNIFICANCE mnemonic of green analytical practices, TrAC, Trends in Analytical Chemistry 50 (2013) 78–84.

[3] P.T. Anastas, J.C. Warner, Green chemistry, Frontiers 640 (1998) (1998) 850.

[4] S. Armenta, S. Garrigues, M. de la Guardia, The role of green extraction techniques in Green Analytical Chemistry, TrAC, Trends in Analytical Chemistry 71 (2015) 2–8.

[5] V.K. Balakrishnan, K.A. Terry, J. Toito, Determination of sulfonamide antibiotics in wastewater: a comparison of solid phase microextraction and solid phase extraction methods, Journal of Chromatography A 1131 (1–2) (2006) 1–10.

[6] A. Arcoleo, F. Bianchi, M. Careri, Helical multi-walled carbon nanotube-coated fibers for solid-phase microextraction determination of polycyclic aromatic hydrocarbons at ultratrace levels in ice and snow samples, Journal of Chromatography A 1631 (2020) 461589.

[7] M. Tuzen, B. Hazer, A. Elik, N. Altunay, Synthesized of poly (vinyl benzyl dithiocarbonate-dimethyl amino ethyl methacrylate) block copolymer as adsorbent for the

vortex-assisted dispersive solid phase microextraction of patulin from apple products and dried fruits, Food Chemistry 395 (2022) 133607.

[8] X. Li, H. Li, W. Ma, Z. Guo, X. Li, X. Li, Q. Zhang, Determination of patulin in apple juice by single-drop liquid-liquid-liquid microextraction coupled with liquid chromatography-mass spectrometry, Food Chemistry 257 (2018) 1−6.

[9] F. Vincenti, C. Montesano, L. Cellucci, A. Gregori, F. Fanti, D. Compagnone, M. Sergi, Combination of pressurized liquid extraction with dispersive liquid liquid micro extraction for the determination of sixty drugs of abuse in hair, Journal of Chromatography A 1605 (2019) 360348.

[10] J. Xiong, B. Hu, Comparison of hollow fiber liquid phase microextraction and dispersive liquid−liquid microextraction for the determination of organosulfur pesticides in environmental and beverage samples by gas chromatography with flame photometric detection, Journal of Chromatography A 1193 (1−2) (2008) 7−18.

[11] A. del Pilar Sánchez-Camargo, M. Bueno, F. Parada-Alfonso, A. Cifuentes, E. Ibáñez, Hansen solubility parameters for selection of green extraction solvents, TrAC, Trends in Analytical Chemistry 118 (2019) 227−237.

[12] B.K. Tiwari, Ultrasound: a clean, green extraction technology, TrAC, Trends in Analytical Chemistry 71 (2015) 100−109.

[13] F.A. Wannas, E.A. Azooz, R.K. Ridha, S.K. Jawad, Separation and micro determination of zinc(II) and cadmium(II) in food samples using cloud point extraction method, Iraqi Journal of Science 64 (3) (2023) 1046−1061, https://doi.org/10.24996/ijs.2023.64.3.2.

[14] A. Melnyk, J. Namieśnik, L. Wolska, Theory and recent applications of coacervate-based extraction techniques, TrAC, Trends in Analytical Chemistry 71 (2015) 282−292.

[15] M.Q. Farooq, I. Ocaña-Rios, J.L. Anderson, Analysis of persistent contaminants and personal care products by dispersive liquid-liquid microextraction using hydrophobic magnetic deep eutectic solvents, Journal of Chromatography A 1681 (2022) 463429.

[16] J. Płotka-Wasylka, M. Rutkowska, K. Owczarek, M. Tobiszewski, J. Namieśnik, Extraction with environmentally friendly solvents, TrAC, Trends in Analytical Chemistry 91 (2017) 12−25.

[17] A.A. Asgharinezhad, H. Ebrahimzadeh, A simple and fast method based on mixed hemimicelles coated magnetite nanoparticles for simultaneous extraction of acidic and basic pollutants, Analytical and Bioanalytical Chemistry 408 (2016) 473−486.

[18] D.L. Su, P.J. Li, S.Y. Quek, Z.Q. Huang, Y.J. Yuan, G.Y. Li, Y. Shan, Efficient extraction and characterization of pectin from orange peel by a combined surfactant and microwave assisted process, Food Chemistry 286 (2019) 1−7.

[19] K. Seebunrueng, S. Tamuang, S. Ruangchai, S. Sansuk, S. Srijaranai, In situ self-assembled coating of surfactant-mixed metal hydroxide on Fe3O4@ SiO2 magnetic composite for dispersive solid phase microextraction prior to HPLC analysis of triazole fungicides, Microchemical Journal 168 (2021) 106396.

[20] W.I. Mortada, H.E. Zedan, M.E. Khalifa, Spectrophotometric determination of trace vanadium in fresh fruit juice samples by ion pair-based surfactant-assisted microextraction procedure with solidification of floating organic drop, Spectrochimica Acta Part A: Molecular and Biomolecular Spectroscopy 302 (2023) 123107.

[21] A.T. Bişgin, Surfactant-assisted emulsification and surfactant-based dispersive liquid−liquid microextraction method for determination of Cu (II) in food and water samples by flame atomic absorption spectrometry, Journal of AOAC International 102 (5) (2019) 1516−1522.

[22] G.J. Shabaa, F.A. Semysim, R.K. Ridha, E.A. Azooz, E.A.J. Al-Mulla, Air-assisted dual-cloud point extraction coupled with flame atomic absorption spectroscopy for the separation and quantification of zinc in pregnant women's serum, Journal of the Iranian Chemical Society 20 (2023) 2277–2284, https://doi.org/10.1007/s13738-023-02834-6.

[23] M. Martinefski, N. Feizi, M.L. Lunar, S. Rubio, Supramolecular solvent-based high-throughput sample treatment platform for the biomonitoring of PAH metabolites in urine by liquid chromatography-tandem mass spectrometry, Chemosphere 237 (2019) 124525.

[24] A. Ballesteros-Gómez, M.D. Sicilia, S. Rubio, Supramolecular solvents in the extraction of organic compounds. A review, Analytica Chimica Acta 677 (2) (2010) 108–130.

[25] E. Carasek, G. Bernardi, D. Morelli, J. Merib, Sustainable green solvents for microextraction techniques: recent developments and applications, Journal of Chromatography A 1640 (2021) 461944.

[26] M. Moradi, Y. Yamini, M. Tayyebi, H. Asiabi, Ultrasound-assisted liquid-phase microextraction based on a nanostructured supramolecular solvent, Analytical and Bioanalytical Chemistry 405 (2013) 4235–4243.

[27] M.J. Dueñas-Mas, A. Ballesteros-Gómez, S. Rubio, Supramolecular solvent-based microextraction of emerging bisphenol A replacements (colour developers) in indoor dust from public environments, Chemosphere 222 (2019) 22–28.

[28] N. Gissawong, S. Boonchiangma, S. Mukdasai, S. Srijaranai, Vesicular supramolecular solvent-based microextraction followed by high performance liquid chromatographic analysis of tetracyclines, Talanta 200 (2019) 203–211.

[29] C. Vakh, S. Koronkiewicz, Surfactants application in sample preparation techniques: insights, trends, and perspectives, TrAC, Trends in Analytical Chemistry (2023) 117143.

[30] H. Karimiyan, M. Hadjmohammadi, Ultrasound-assisted supramolecular-solvent-based microextraction combined with high-performance liquid chromatography for the analysis of chlorophenols in environmental water samples, Journal of Separation Science 39 (24) (2016) 4740–4747.

[31] N. Feizi, Y. Yamini, M. Moradi, M. Karimi, Q. Salamat, H. Amanzadeh, A new generation of nano-structured supramolecular solvents based on propanol/gemini surfactant for liquid phase microextraction, Analytica Chimica Acta 953 (2017) 1–9.

[32] L. Algar, M.D. Sicilia, S. Rubio, Ribbon-shaped supramolecular solvents: synthesis, characterization and potential for making greener the microextraction of water organic pollutants, Talanta 255 (2023) 124227.

[33] N. Altunay, A. Elik, Ultrasound-assisted alkanol-based nanostructured supramolecular solvent for extraction and determination of cadmium in food and environmental samples: experimental design methodology, Microchemical Journal 164 (2021) 105958.

[34] M. Al-Nidawi, O. Ozalp, U. Alshana, M. Soylak, Synergistic cloud point microextraction prior to spectrophotometric determination of curcumin in food samples, Analytical Letters 56 (12) (2023) 1977–1988.

[35] D. Moema, S. Dube, M.M. Nindi, Greener surfactant-assisted microextraction method for the determination of fluoroquinolones in chicken livers by high-performance liquid chromatography, Analytical Chemistry Letters 12 (2) (2022) 185–197.

[36] Q. Deng, M. Chen, L. Kong, X. Zhao, J. Guo, X. Wen, Novel coupling of surfactant assisted emulsification dispersive liquid–liquid microextraction with spectrophotometric determination for ultra trace nickel, Spectrochimica Acta Part A: Molecular and Biomolecular Spectroscopy 104 (2013) 64–69.

[37] Z.A. Alothman, E. Yilmaz, M.A. Habila, B. Alhenaki, M. Soylak, A.Y.B.H. Ahmed, E.A. Alabdullkarem, Development of combined-supramolecular micro-extraction with

ultra-performance liquid chromatography-tandem mass spectrometry procedures for ultra-trace analysis of carbaryl in water, fruits and vegetables, International Journal of Environmental Analytical Chemistry (2020) 1−11, https://doi.org/10.1080/03067319.2020.1738419.

[38] K. Seebunrueng, P. Phosiri, R. Apitanagotinon, S. Srijaranai, A new environment-friendly supramolecular solvent-based liquid phase micro-extraction coupled to high performance liquid chromatography for simultaneous determination of six phenoxy acid herbicides in water and rice samples, Microchemical Journal 152 (2020) 104418, https://doi.org/10.1016/j.microc.2019.104418.

[39] H. Deng, H. Wang, M. Liang, X. Su, A novel approach based on supramolecular solvent micro-extraction and UPLC-Q-Orbitrap HRMS for simultaneous analysis of perfluorinated compounds and fluorine-containing pesticides in drinking and environmental water, Microchemical Journal 151 (2019) 104250, https://doi.org/10.1016/j.microc.2019.104250.

[40] G.L. Scheel, C.R. Teixeira Tarley, Simultaneous micro-extraction of carbendazim, fipronil and picoxystrobin in naturally and artificial occurring water bodies by water-induced supramolecular solvent and determination by HPLC-DAD, Journal of Molecular Liquids 297 (2020) 111897, https://doi.org/10.1016/j.molliq.2019.111897.

[41] S. Amir, J. Shah, M.R. Jan, Supramolecular solvent micro-extraction of phenylurea herbicides from environmental samples, Desalination and Water Treatment 148 (2019) 202−212, https://doi.org/10.5004/dwt.2019.23789.

[42] L. Adlnasab, M. Ezoddin, M. Shabanian, B. Mahjoob, Development of ferrofluid mediated CLDH@Fe3O4@Tanic acid- based supramolecular solvent: application in air-assisted dispersive micro solid phase extraction for pre-concentration of diazinon and metalaxyl from various fruit juice samples, Microchemical Journal 146 (2019) 1−11, https://doi.org/10.1016/j.microc.2018.12.020.

[43] S. Gorji, P. Biparva, M. Bahram, G. Nematzadeh, Rapid and direct micro-extraction of pesticide residues from rice and vegetable samples by supramolecular solvent in combination with chemometrical data processing, Food Analytical Methods 12 (2019) 394−408, https://doi.org/10.1007/s12161-018-1371-2.

[44] G.L. Scheel, C.R.T. Tarley, Feasibility of supramolecular solvent-based microextraction for simultaneous pre-concentration of herbicides from natural waters with posterior determination by HPLC-DAD, Michrochem, D-J Series 133 (2017) 650−657, https://doi.org/10.1016/j.microc.2017.03.007.

[45] M. Peyrovi, M. Hadjmohammadi, Alkanol-based supramolecular solvent microextraction of organophosphorus pesticides and their determination using high performance liquid chromatography, Journal of the Iranian Chemical Society 14 (2017) 995−1004, https://doi.org/10.1007/s13738-017-1049-5.

[46] Z.A. Alothman, E. Yilmaz, M.A. Habila, A.A. Ghfar, B. Alhenaki, M. Soylak, A.Y.B.H. Ahmed, Supramolecular solvent micro-extraction and ultra-performance liquid chromatography-tandem mass spectrometry combination for the pre-concentration and determination of malathion in environmental samples, Desalination and Water Treatment 144 (2019) 166−171, https://doi.org/10.5004/dwt.2019.23574.

[47] W. Zhao, J. Zhao, H. Zhao, Y. Cao, W. Liu, Supramolecular solvent-based vortexmixed micro-extraction: determination of chiral triazole fungicide in beer samples, Chirality 30 (2018) 302−309, https://doi.org/10.1002/chir.22798.

[48] M. Safari, Y. Yamini, E. Tahmasebi, B. Ebrahimpour, Magnetic nanoparticle assisted supramolecular solvent extraction of triazine herbicides prior to their determination by HPLC

with UV detection, Microchimica Acta 183 (2016) 203–210, https://doi.org/10.1007/s00604-015-1607-4.

[49] S. Jafarvand, F. Shemirani, Supramolecular-based dispersive liquid-liquid microextraction: a novel sample preparation technique utilizes coacervates and reverse micelles, Journal of Separation Science 34 (2011) 455–461.

[50] A.A. Gouda, A.M. Alshehri, R. El Sheikh, W.S. Hassan, S.H. Ibrahim, Development of green vortex-assisted supramolecular solvent-based liquid-liquid microextraction for preconcentration of mercury in environmental and biological samples prior to spectrophotometric determination, Microchemical Journal 157 (2020) 105108.

[51] A. Najafi, M. Hashemi, Vortex-assisted supramolecular solvent microextraction based on solidification of floating drop for preconcentration and speciation of inorganic arsenic species in water samples by molybdenum blue method, Microchemical Journal 150 (2019) 104102.

[52] M. Moradi, R. Kashanaki, S. Borhani, H. Bigdeli, N. Abbasi, A. Kazemzadeh, Optimization of supramolecular solvent microextraction prior to graphite furnace atomic absorption spectrometry for total selenium determination in food and environmental samples, Journal of Molecular Liquids 232 (2017) 243–250.

[53] S. Gonzalez-Rubio, D. García-Gomez, A. Ballestros-Gomez, S. Rubio, A new sample treatment strategy based on simultaneous supramolecular solvent and dispersive solid-phase extraction for the determination of ionophore coccidiostats in all legislated foodstuffs, Food Chemistry 326 (2020) 126987.

[54] A. Najafi, M. Hashemi, Feasibility of liquid phase microextraction based on a new supramolecular solvent for spectrophotometric determination of orthophosphate using response surface methodology optimization, Journal of Molecular Liquids 297 (2020) 111768.

[55] N. Altunay, K.P. Katin, Ultrasonic-assisted supramolecular solvent liquid-liquid microextraction for determination of manganese and zinc at trace levels in vegetables: experimental and theoretical studies, Journal of Molecular Liquids 310 (2020) 113192.

[56] N. Altunay, A. Elik, A green and efficient vortex-assisted liquid-phase microextraction based on supramolecular solvent for UV-VIS determination of nitrite in processed meat and chicken products, Food Chemistry 332 (2020) 127395.

[57] Q. Salamat, Y. Yamini, M. Moradi, M. Karimi, M. Nazraz, Novel generation of nanostructured supramolecular solvents based on an ionic liquid as a green solvent for microextraction of some synthetic food dyes, New Journal of Chemistry 42 (2018) 19252–19259.

[58] A.A. Gouda, M.S. Elmasry, H. Hashem, H.M. El-Sayed, Eco-friendly environmental trace analysis of thorium using a new supramolecular solvent-based liquid-liquid microextraction combined with spectrophotometry, Microchemical Journal 142 (2018) 102–107.

[59] R. Kashanaki, H. Ebrahimzadeh, M. Moradi, Metaleorganic framework based micro solid phase extraction coupled with supramolecular solvent microextraction to determine copper in water and food samples, New Journal of Chemistry 42 (2018) 5806–5813.

[60] P.G. Jessop, L. Kozycz, Z.G. Rahami, D. Schoenmakers, A.R. Boyd, D. Wechsler, A.M. Holland, Tertiary amine solvents having switchable hydrophilicity, Green Chemistry 13 (3) (2011) 619–623.

[61] P.G. Jessop, D.J. Heldebrant, X. Li, C.A. Eckert, C.L. Liotta, Reversible nonpolar-to-polar solvent, Nature 436 (7054) (2005) 1102.

[62] J.R. Vanderveen, J. Durelle, P.G. Jessop, Design and evaluation of switchable-hydrophilicity solvents, Green Chemistry 16 (3) (2014) 1187–1197.

[63] U. Alshana, M. Hassan, M. Al-Nidawi, E. Yilmaz, M. Soylak, Switchable-hydrophilicity solvent liquid-liquid microextraction, TrAC, Trends in Analytical Chemistry 131 (2020) 116025.

[64] P.G. Jessop, L. Phan, A. Carrier, S. Robinson, C.J. Dürr, J.R. Harjani, A solvent having switchable hydrophilicity, Green Chemistry 12 (5) (2010) 809−814.

[65] Y. Bazel, M. Rečlo, Y. Chubirka, Switchable hydrophilicity solvents in analytical chemistry. Five years of achievements, Microchemical Journal 157 (2020) 105115.

[66] H. Ahmar, M. Nejati-Yazdinejad, M. Najafi, K.S. Hasheminasab, Switchable hydrophilicity solvent-based homogeneous liquid−liquid microextraction (SHS-HLLME) combined with GC-FID for the quantification of methadone and tramadol, Chromatographia 81 (2018) 1063−1070.

[67] G. Lasarte-Aragonés, R. Lucena, S. Cárdenas, M. Valcárcel, Use of switchable solvents in the microextraction context, Talanta 131 (2015) 645−649.

[68] H.E. Pelvan, Ç. Arpa, An effervescence-assisted switchable hydrophobicity solvent microextraction before microsampling flame atomic absorption spectrometry for copper ions in vegetables, International Journal of Environmental Analytical Chemistry 103 (9) (2023) 1955−1970.

[69] A. Elik, N. Altunay, Optimization of vortex-assisted switchable hydrophilicity solvent liquid phase microextraction for the selective extraction of vanillin in different matrices prior to spectrophotometric analysis, Food Chemistry 399 (2023) 133929.

[70] M. Hassan, F. Uzcan, S.N. Shah, U. Alshana, M. Soylak, Switchable-hydrophilicity solvent liquid-liquid microextraction for sample cleanup prior to dispersive magnetic solid-phase microextraction for spectrophotometric determination of quercetin in food samples, Sustainable Chemistry and Pharmacy 22 (2021) 100480.

[71] A. Elik, A.Ö. Altunay, M.F. Lanjwani, M. Tuzen, A new ultrasound-assisted liquid-liquid microextraction method utilizing a switchable hydrophilicity solvent for spectrophotometric determination of nitrite in food samples, Journal of Food Composition and Analysis 119 (2023) 105267.

[72] M. Behpour, S. Nojavan, S. Asadi, A. Shokri, Combination of gel-electromembrane extraction with switchable hydrophilicity solvent-based homogeneous liquid-liquid microextraction followed by gas chromatography for the extraction and determination of antidepressants in human serum, breast milk and wastewater, Journal of Chromatography A 1621 (2020) 461041.

[73] X. Di, X. Wang, Y. Liu, X. Guo, Microwave assisted extraction in combination with solid phase purification and switchable hydrophilicity solvent-based homogeneous liquid-liquid microextraction for the determination of sulfonamides in chicken meat, Journal of Chromatography B 1118 (2019) 109−115.

[74] C. Scheid, S. Eller, A.L. Oenning, E. Carasek, J. Merib, T.F. de Oliveira, Application of homogeneous liquid−liquid microextraction with switchable hydrophilicity solvents to the determination of MDMA, MDA and NBOMes in postmortem blood samples, Journal of Analytical Toxicology 46 (7) (2022) 776−782.

[75] X. Yang, C. Yan, Y. Sun, Y. Liu, S. Yang, Q. Deng, X. Wen, Micro-spectrophotometric determination of nickel in Gentiana rigescens after switchable hydrophilicity solvent-based ultrasound-assisted liquid phase microextraction, Microchemical Journal 168 (2021) 106402.

[76] M. Al-Nidawi, U. Alshana, Reversed-phase switchable-hydrophilicity solvent liquid-liquid microextraction of copper prior to its determination by smartphone digital image colorimetry, Journal of Food Composition and Analysis 104 (2021) 104140.

[77] B.T. Zaman, N.B. Turan, E.G. Bakirdere, S. Topal, O. Sağsöz, S. Bakirdere, Determination of trace manganese in soil samples by using eco-friendly switchable solvent based liquid phase microextraction-3 holes cut slotted quartz tube-flame atomic absorption spectrometry, Microchemical Journal 157 (2020) 104981.

[78] N. Atsever, T. Borahan, E.G. Bakırdere, S. Bakırdere, Determination of iron in hair samples by slotted quartz tube-flame atomic absorption spectrometry after switchable solvent liquid phase extraction, Journal of Pharmaceutical and Biomedical Analysis 186 (2020) 113274.

[79] C. Yan, X. Yang, Z. Li, Y. Liu, S. Yang, Q. Deng, X. Wen, Switchable hydrophilicity solvent-based preconcentration for ICP-OES determination of trace lead in environmental samples, Microchemical Journal 168 (2021) 106529.

[80] X. Di, X. Wang, Y.P. Liu, X.J. Guo, Solid-phase extraction coupled with switchable hydrophilicity solvent based homogeneous liquid-liquid microextraction for chloramphenicol enrichment in environmental water samples: a novel alternative to classical extraction techniques, Analytical and Bioanalytical Chemistry 411 (4) (2019) 803−812, https://doi.org/10.1007/s00216-018-1486-8.

[81] S. Erarpat, A. Caglak, S. Bodur, S.D. Chormey, G.O. Engin, S. Bakirdere, Simultaneous determination of fluoxetine, estrone, pesticides, and endocrine disruptors in wastewater by gas chromatography-mass spectrometry (GC-MS) following switchable solvent- liquid phase microextraction (SS-LPME), Analytical Letters 52 (5) (2019) 869−878, https://doi.org/10.1080/00032719.2018.1505897.

[82] F.M. Xu, Q. Li, W.L. Wei, L.Y. Liu, H.B. Li, Development of a liquid-liquid microextraction method based on a switchable hydrophilicity solvent for the simultaneous determination of 11 drugs in urine by GC-MS, Chromatographia 81 (12) (2018) 1695−1703, https://doi.org/10.1007/s10337-018-3643-9.

[83] M. Soylak, M. Khan, E. Yilmaz, Switchable solvent based liquid phase microextraction of uranium in environmental samples: a green approach, Analytical Methods 8 (5) (2016) 979−986, https://doi.org/10.1039/C5AY02631H.

[84] M. Ezoddin, K. Abdi, N. Lamei, Development of air assisted liquid phase microextraction based on switchable-hydrophilicity solvent for the determination of palladium in environmental samples, Talanta 153 (2016) 247−252, https://doi.org/10.1016/j.talanta.2016.03.018.

[85] M. Reclo, E. Yilmaz, Y. Bazel, M. Soylak, Switchable solvent based liquid phase microextraction of palladium coupled with determination by flame atomic absorption spectrometry, International Journal of Environmental Analytical Chemistry 97 (14−15) (2017) 1315−1327, https://doi.org/10.1080/03067319.2017.1413185.

[86] Z.M. Memon, E. Yilmaz, M. Soylak, Switchable solvent based green liquid phase microextraction method for cobalt in tobacco and food samples prior to flame atomic absorption spectrometric determination, Journal of Molecular Liquids 229 (2017) 459464, https://doi.org/10.1016/j.molliq.2016.12.098.

[87] M. Reclo, E. Yilmaz, M. Soylak, V. Andruch, Y. Bazel, Ligandless switchable solvent based liquid phase microextraction of nickel from food and cigarette samples prior to its micro-sampling flame atomic absorption spectrometric determination, Journal of Molecular Liquids 237 (2017) 236241, https://doi.org/10.1016/j.molliq.2017.04.066.

[88] A. Habibiyan, M. Ezoddin, N. Lamei, K. Abdi, M. Amini, M. Ghazi-khansari, Ultrasonic assisted switchable solvent based on liquid phase microextraction combined with micro sample injection flame atomic absorption spectrometry for determination of some heavy

metals in water, urine and tea infusion samples, Journal of Molecular Liquids 242 (2017) 492496, https://doi.org/10.1016/j.molliq.2017.07.043.

[89] Y. Bazel, M. Reclo, J. Sandrejova, Using a switchable hydrophilicity solvent for the extraction spectrophotometric determination of nickel, Journal of Analytical Chemistry 72 (10) (2017) 10181023, https://doi.org/10.1134/S1061934817080032.

[90] J. Ali, M. Tuzen, T.G. Kazi, Determination of arsenic in water samples by using a green hydrophobic hydrophilic switchable liquid-solid dispersive microextraction method, Water, Air, & Soil Pollution 228 (1) (2017), https://doi.org/10.1007/s11270-016-3211-6.

[91] E. Vessally, E. Ghorbani-Kalhor, R. Hosseinzadeh- Khanmiri, M. Babazadeh, A. Hosseinian, F. Omidi, et al., Application of switchable solvent-based liquid phase microextraction for preconcentration and trace detection of cadmium ions in baby food samples, Journal of the Iranian Chemical Society 15 (2) (2018) 491498, https://doi.org/10.1007/s13738-017-1249-z.

[92] M.S. Arain, T.G. Kazi, H.I. Afridi, N. Khan, J. Ali, A innovative switchable polarity solvent, based on 1,8-diazabicyclo- 5.4.0 undec-7-ene and decanol was prepared for enrichment of aluminum in biological sample prior to analysis by flame atomic absorption spectrometry, Applied Organometallic Chemistry 32 (3) (2018), https://doi.org/10.1002/aoc.4157.

[93] S. Zhang, B.B. Chen, M. He, B. Hu, Switchable solvent based liquid phase microextraction of trace lead and cadmium from environmental and biological samples prior to graphite furnace atomic absorption spectrometry detection, Microchemical Journal 139 (2018) 380385, https://doi.org/10.1016/j.microc.2018.03.017.

[94] Z. Tekin, S. Erarpat, A. Sahin, D.S. Chormey, S. Bakirdere, Determination of Vitamin B12 and cobalt in egg yolk using vortex assisted switchable solvent based liquid phase microextraction prior to slotted quartz tube flame atomic absorption spectrometry, Food Chemistry 286 (2019) 500505, https://doi.org/10.1016/j.foodchem.2019.02.036.

[95] M. Firat, E.G. Bakirdere, An accurate and sensitive analytical strategy for the determination of palladium in aqueous samples: slotted quartz tube flame atomic absorption spectrometry with switchable liquid-liquid microextraction after preconcentration using a Schiff base ligand, Environmental Monitoring and Assessment 191 (3) (2019) 9, https://doi.org/10.1007/s10661-019-7252-3.

[96] M. Hemmati, M. Rajabi, Switchable fatty acid based CO2-effervescence ameliorated emulsification microextraction prior to high performance liquid chromatography for efficient analyses of toxic azo dyes in foodstuffs, Food Chemistry 286 (2019) 185190, https://doi.org/10.1016/j.foodchem.2019.01.197.

[97] M. Firat, S. Bodur, B. Tisli, C. Ozlu, D.S. Chormey, F. Turak, et al., Vortex-assisted switchable liquid-liquid microextraction for the preconcentration of cadmium in environmental samples prior to its determination with flame atomic absorption spectrometry, Environmental Monitoring and Assessment 190 (7) (2018) 8, https://doi.org/10.1007/s10661-018-6786-0.

[98] A. Stark, K.R. Seddon, Ionic liquids, Kirk-Othmer Encyclopedia of Chemical Technology 26 (2000) 836.

[99] Z. Lei, B. Chen, Y.M. Koo, D.R. MacFarlane, Introduction: ionic liquids, Chemical Reviews 117 (10) (2017) 6633−6635.

[100] V. Vičkačkaitė, A. Padarauskas, Ionic liquids in microextraction techniques, Central European Journal of Chemistry 10 (2012) 652−674.

[101] Q. Zhang, J.N.M. Shreeve, Energetic ionic liquids as explosives and propellant fuels: a new journey of ionic liquid chemistry, Chemical Reviews 114 (20) (2014) 10527−10574.

[102] A.R. Hajipour, F. Rafiee, Basic ionic liquids. A short review, Journal of the Iranian Chemical Society 6 (2009) 647–678.
[103] S.K. Singh, A.W. Savoy, Ionic liquids synthesis and applications: an overview, Journal of Molecular Liquids 297 (2020) 112038.
[104] M.J. Earle, K.R. Seddon, Ionic liquids. Green solvents for the future, Pure and Applied Chemistry 72 (7) (2000) 1391–1398.
[105] H. Yu, J. Merib, J.L. Anderson, Faster dispersive liquid-liquid microextraction methods using magnetic ionic liquids as solvents, Journal of Chromatography A 1463 (2016) 11–19.
[106] M.J. Trujillo-Rodriguez, V. Pino, J.L. Anderson, Magnetic ionic liquids as extraction solvents in vacuum headspace single-drop microextraction, Talanta 172 (2017) 86–94.
[107] M.S. Alves, L.C.F. Neto, C. Scheid, J. Merib, An overview of magnetic ionic liquids: from synthetic strategies to applications in microextraction techniques, Journal of Separation Science 45 (1) (2022) 258–281.
[108] J.F. Liu, Y.G. Chi, G.B. Jiang, Screening the extractability of some typical environmental pollutants by ionic liquids in liquid-phase microextraction, Journal of Separation Science 28 (1) (2005) 87–91.
[109] E.M. Martinis, R.G. Wuilloud, Cold vapor ionic liquid-assisted headspace single-drop microextraction: a novel preconcentration technique for mercury species determination in complex matrix samples, Journal of Analytical Atomic Spectrometry 25 (9) (2010) 1432–1439.
[110] Z. Jahromi, A. Mostafavi, T. Shamspur, M. Mohamadim, Magnetic ionic liquid assisted single-drop microextraction of ascorbic acid before its voltammetric determination, Journal of Separation Science 40 (20) (2017) 4041–4049.
[111] G.S. Zhou, Y.C. Yuan, Y. Yin, Y.P. Tang, R.J. Xu, Y. Liu, J.A. Duan, Hydrophilic interaction chromatography combined with ultrasound-assisted ionic liquid dispersive liquid–liquid microextraction for determination of underivatized neurotransmitters in dementia patients' urine samples, Analytica Chimica Acta 1107 (2020) 74–84.
[112] L. Pang, P. Yang, R. Pang, S. Li, Bis (trifluoromethylsulfonyl) imide-based frozen ionic liquid for the hollow-fiber solid-phase microextraction of dichlorodiphenyltrichloroethane and its main metabolites, Journal of Separation Science 40 (16) (2017) 3311–3317.
[113] E.F. Fiorentini, B.V. Canizo, R.G. Wuilloud, Determination of as in honey samples by magnetic ionic liquid-based dispersive liquid-liquid microextraction and electrothermal atomic absorption spectrometry, Talanta 198 (2019) 146–153.
[114] J. Merib, D.A. Spudeit, G. Corazza, E. Carasek, J.L. Anderson, Magnetic ionic liquids as versatile extraction phases for the rapid determination of estrogens in human urine by dispersive liquid-liquid microextraction coupled with high-performance liquid chromatography-diode array detection, Analytical and Bioanalytical Chemistry 410 (2018) 4689–4699.
[115] I. Pacheco-Fernández, V. Pino, J. Lorenzo-Morales, J.H. Ayala, A.M. Afonso, Salt-induced ionic liquid-based microextraction using a low cytotoxic guanidinium ionic liquid and liquid chromatography with fluorescence detection to determine monohydroxylated polycyclic aromatic hydrocarbons in urine, Analytical and Bioanalytical Chemistry 410 (2018) 4701–4713.
[116] N. Altunay, A. Elik, R. Gürkan, Vortex assisted-ionic liquid based dispersive liquid-liquid microextraction of low levels of nickel and cobalt in chocolate-based samples and their determination by FAAS, Microchemical Journal 147 (2019) 277–285.

[117] N. Altunay, E. Yıldırım, R. Gürkan, Extraction and preconcentration of trace Al and Cr from vegetable samples by vortex-assisted ionic liquid-based dispersive liquid–liquid microextraction prior to atomic absorption spectrometric determination, Food Chemistry 245 (2018) 586–594.

[118] F. Abujaber, A.I.C. Ricardo, Á. Ríos, F.J.G. Bernardo, R.C.R. Martín-Doimeadios, Ionic liquid dispersive liquid-liquid microextraction combined with LC-UV-Vis for the fast and simultaneous determination of cortisone and cortisol in human saliva samples, Journal of Pharmaceutical and Biomedical Analysis 165 (2019) 141–146.

[119] E. Fernández, L. Vidal, A. Canals, Hydrophilic magnetic ionic liquid for magnetic headspace single-drop microextraction of chlorobenzenes prior to thermal desorption-gas chromatography-mass spectrometry, Analytical and Bioanalytical Chemistry 410 (2018) 4679–4687.

[120] T. Huang, X. Lei, S. Wang, C. Lin, X. Wu, Ionic liquid assisted in situ growth of nanoconfined ionic liquids/metal-organic frameworks nanocomposites for monolithic capillary microextraction of microcystins in environmental waters, Journal of Chromatography A 1692 (2023) 463849, https://doi.org/10.1016/j.chroma.2023.463849.

[121] J.I. Cacho, N. Campillo, P. Viñas, M. Hernández-Córdoba, In situ ionic liquid dispersive liquid-liquid microextraction coupled to gas chromatography-mass spectrometry for the determination of organophosphorus pesticides, Journal of Chromatography A 1559 (2018) 95–101, https://doi.org/10.1016/j.chroma.2017.12.059.

[122] A. Raoufi, A.M. Raoufi, A. Ismailzadeh, E. Soleimani Rad, A. Kiaeefar, Application of hollow fiber-protected liquid-phase microextraction combined with GC-MS in determining Endrin, Chlordane, and Dieldrin in rice samples, Environmental Geochemistry and Health 45 (2023) 5261–5277, https://doi.org/10.1007/s10653-023-01570-3.

[123] M.J. Trujillo-Rodríguez, H. Nan, J.L. Anderson, Expanding the use of polymeric ionic liquids in headspace solid-phase microextraction: determination of ultraviolet filters in water samples, Journal of Chromatography A 1540 (2018) 11–20, https://doi.org/10.1016/j.chroma.2018.01.048.

[124] R.R. Jha, C. Singh, A.B. Pant, D.K. Patel, Ionic liquid based ultrasound assisted dispersive liquid-liquid micro-extraction for simultaneous determination of 15 neurotransmitters in rat brain, plasma and cell samples, Analytica Chimica Acta 1005 (2018) 43–53, https://doi.org/10.1016/j.aca.2017.12.015.

[125] I.D. Souza, L.W. Hantao, M.E.C. Queiroz, Polymeric ionic liquid open tubular capillary column for on-line in-tube SPME coupled with UHPLC-MS/MS to determine endocannabinoids in plasma samples, Analytica Chimica Acta 1045 (2019) 108–116, https://doi.org/10.1016/j.aca.2018.08.062.

[126] T. Lu, D. Li, J. Feng, W. Zhang, Y. Kang, Efficient extraction performance and mechanisms of Cd2+ and Pb2+ in water by novel dicationic ionic liquids, Journal of Environmental Management 351 (2024) 119767, https://doi.org/10.1016/j.jenvman.2023.119767.

[127] S. Chen, Y. Sun, J. Chao, L. Cheng, Y. Chen, J. Liu, Dispersive liquid-liquid microextraction of silver nanoparticles in water using ionic liquid 1-octyl-3- methylimidazolium hexafluorophosphate, Journal of Environmental Sciences 41 (2016) 211–217, https://doi.org/10.1016/j.jes.2015.04.015.

[128] G.B. Grecco, K.F. Albini, L.S. Longo, M.A. Andreo, B.L. Batista, F.R. Lourenço, L.A. Calixto, Application of experimental design for dispersive liquid-liquid microextraction optimization for metallic impurities determination in arnica infusion employing green solvents, Journal of the Iranian Chemical Society 20 (2023) 371–380, https://doi.org/10.1007/s13738-022-02674-w.

[129] Y. Zeng, X. Zhu, J. Xie, L. Chen, Ionic liquid coated magnetic core/shell CoFe2O4@ SiO2 nanoparticles for the separation/analysis of trace gold in water sample, Advanced Nano Research 10 (2021) 295, https://doi.org/10.1016/j.saa.2014.08.113.

[130] A.P. Abbott, G. Capper, D.L. Davies, H.L. Munro, R.K. Rasheed, V. Tambyrajah, Preparation of novel, moisture-stable, Lewis-acidic ionic liquids containing quaternary ammonium salts with functional side chainsElectronic supplementary information (ESI) available: plot of conductivity etc. Temperature for the ionic liquid formed from zinc chloride and choline chloride (2:1). See, Chemical Communications (19) (2001) 2010−2011. http://www.rsc.org/suppdata/cc/b1/b106357j.

[131] A.P. Abbott, G. Capper, D.L. Davies, R.K. Rasheed, V. Tambyrajah, Novel solvent properties of choline chloride/urea mixtures, Chemical Communications (1) (2003) 70−71.

[132] A. Elik, H.U. Haq, G. Boczkaj, S. Fesliyan, Ö. Ablak, N. Altunay, Magnetic hydrophobic deep eutectic solvents for orbital shaker-assisted dispersive liquid-liquid microextraction (MAGDES-OS-DLLME)-Determination of nickel and copper in food and water samples by FAAS, Journal of Food Composition and Analysis 125 (2024) 105843.

[133] N. Faraz, H.U. Haq, M.B. Arain, R. Castro-Muñoz, G. Boczkaj, A. Khan, Deep eutectic solvent based method for analysis of Niclosamide in pharmaceutical and wastewater samples−A green analytical chemistry approach, Journal of Molecular Liquids 335 (2021) 116142−117134.

[134] R. Ahmadi, E.A. Azooz, Y. Yamini, A.M. Ramezani, Liquid-liquid microextraction techniques based on in-situ formation/decomposition of deep eutectic solvents, TrAC, Trends in Analytical Chemistry 161 (2023) 117019, https://doi.org/10.1016/j.trac.2023.117019.

[135] R. Hussein, M.S. Gburi, N.M. Muslim, E.A. Azooz, A greenness evaluation and environmental aspects of solidified floating organic drop microextraction for metals: a review, Trends in Environmental Analytical Chemistry 37 (2023) e00194, https://doi.org/10.1016/j.teac.2022.e00194.

[136] D. Snigur, E.A. Azooz, O. Zhukovetska, O. Guzenko, W. Mortada, Low-density solvent-based liquid-liquid microextraction for separation of trace concentrations of different analytes, TrAC, Trends in Analytical Chemistry 167 (2023) 117260, https://doi.org/10.1016/j.trac.2023.117260.

[137] E.A. Azooz, F.A. Wannas, R.K. Ridha, S.K. Jawad, E.A.J. Al-Mulla, A green approach for micro determination of silver(I) in water and soil samples using vitamin C, Analytical and Bioanalytical Chemistry Research 9 (2) (2022) 133−140, https://doi.org/10.22036/abcr.2021.277834.1609. http://www.analchemres.org/article_139800.html.

[138] S.C. Cunha, J.O. Fernandes, Extraction techniques with deep eutectic solvents, TrAC, Trends in Analytical Chemistry 105 (2018) 225−239.

[139] X. Li, K.H. Row, Development of deep eutectic solvents applied in extraction and separation, Journal of Separation Science 39 (18) (2016) 3505−3520.

[140] J. Chen, Y. Li, X. Wang, W. Liu, Application of deep eutectic solvents in food analysis: a review, Molecules 24 (24) (2019) 4594.

[141] Y.H. Choi, J. van Spronsen, Y. Dai, M. Verberne, F. Hollmann, I.W. Arends, R. Verpoorte, Are natural deep eutectic solvents the missing link in understanding cellular metabolism and physiology, Plant Physiology 156 (4) (2011) 1701−1705.

[142] H.U. Haq, A. Wali, F. Safi, M.B. Arain, L. Kong, G. Boczkaj, Natural deep eutectic solvent based ultrasound assisted liquid-liquid micro-extraction method for methyl violet dye determination in contaminated river water, Water Resources and Industry 29 (2023) 100210.

[143] F. Chemat, M. Abert Vian, H.K. Ravi, B. Khadhraoui, S. Hilali, S. Perino, A.S. Fabiano Tixier, Review of alternative solvents for green extraction of food and natural products: panorama, principles, applications and prospects, Molecules 24 (16) (2019) 3007.

[144] C. Florindo, F. Lima, B.D. Ribeiro, I.M. Marrucho, Deep eutectic solvents: overcoming 21st century challenges, Current Opinion in Green and Sustainable Chemistry 18 (2019) 31–36.

[145] P. Makoś-Chełstowska, M. Kaykhaii, J. Płotka-Wasylka, M. de la Guardia, Magnetic deep eutectic solvents—Fundamentals and applications, Journal of Molecular Liquids 365 (2022) 120158.

[146] F. Elahi, M.B. Arain, W.A. Khan, H.U. Haq, A. Khan, F. Jan, G. Boczkaj, Ultrasound-assisted deep eutectic solvent-based liquid–liquid microextraction for simultaneous determination of Ni (II) and Zn (II) in food samples, Food Chemistry 393 (2022) 133384.

[147] N. Altunay, A. Elik, R. Gürkan, Preparation and application of alcohol based deep eutectic solvents for extraction of curcumin in food samples prior to its spectrophotometric determination, Food Chemistry 310 (2020) 125933.

[148] H.U. Haq, M. Balal, R. Castro-Muñoz, Z. Hussain, F. Safi, S. Ullah, G. Boczkaj, Deep eutectic solvents based assay for extraction and determination of zinc in fish and eel samples using FAAS, Journal of Molecular Liquids 333 (2021) 115930.

[149] Q. Liu, X. Huang, P. Liang, Preconcentration of copper and lead using deep eutectic solvent modified magnetic nanoparticles and determination by inductively coupled plasma optical emission spectrometry, Horse Spectroscopy 41 (1) (2020) 36–42.

[150] Y. Dai, E. Rozema, R. Verpoorte, Y.H. Choi, Application of natural deep eutectic solvents to the extraction of anthocyanins from Catharanthus roseus with high extractability and stability replacing conventional organic solvents, Journal of Chromatography A 1434 (2016) 50–56.

[151] B. Ozturk, C. Parkinson, M. Gonzalez-Miquel, Extraction of polyphenolic antioxidants from orange peel waste using deep eutectic solvents, Separation and Purification Technology 206 (2018) 1–13.

[152] C. Ortega-Zamora, J. González-Sálamo, C. Hernández-Sánchez, J. Hernández-Borges, Menthol-based deep eutectic solvent dispersive liquid–liquid microextraction: a simple and quick approach for the analysis of phthalic acid esters from water and beverage samples, ACS Sustainable Chemistry & Engineering 8 (23) (2020) 8783–8794.

[153] N. Altunay, A. Elik, D. Bingöl, Simple and green heat-induced deep eutectic solvent microextraction for determination of lead and cadmium in vegetable samples by flame atomic absorption spectrometry: a multivariate study, Biological Trace Element Research 198 (2020) 324–331.

[154] H. Li, C. Zhao, H. Tian, Y. Yang, W. Li, Liquid–liquid microextraction based on acid–base-induced deep eutectic solvents for determination of β-carotene and lycopene in fruit juices, Food Analytical Methods 12 (2019) 2777–2784.

[155] H. Lores, V. Romero, I. Costas, C. Bendicho, I. Lavilla, Natural deep eutectic solvents in combination with ultrasonic energy as a green approach for solubilization of proteins: application to gluten determination by immunoassay, Talanta 162 (2017) 453–459.

[156] S.M. Yousefi, F. Shemirani, S.A. Ghorbanian, Improved headspace single drop microextraction method using deep eutectic solvent based magnetic bucky gels: application to the determination of volatile aromatic hydrocarbons in water and urine samples, Journal of Separation Science 41 (4) (2018) 966–974.

[157] A. Shishov, S. Savinov, N. Volodina, I. Gurev, A. Bulatov, Deep eutectic solvent-based extraction of metals from oil samples for elemental analysis by ICP-OES, Microchemical Journal 179 (2022) 107456.

[158] Y. Dai, R. Verpoorte, Y.H. Choi, Natural deep eutectic solvents providing enhanced stability of natural colorants from safflower (Carthamus tinctorius), Food Chemistry 159 (2014) 116−121.

[159] X. Li, H. Li, K. Lai, J. Miao, J. Zhao, New natural deep eutectic solvent for vortex-assisted liquid−liquid microextraction for the determination of bisphenols in beverages, Microchemical Journal 197 (February 2024) 109710, https://doi.org/10.1016/j.microc.2023.109710.

[160] N. Kizil, E. Basaran, M. Lutfi Yola, M. Soylak, Deep eutectic solvent dispersive liquid−liquid microextraction methods for the analysis of chlorophyll natural colorant (E140) via microwave assisted sample preparation, Microchemical Journal 196 (2024) 109577, https://doi.org/10.1016/j.microc.2023.109577.

[161] A. Hamedfar, A. Javadi, M.R. Afshar Mogaddam, H. Mirzaei, Deep eutectic solvent-based microwave-assisted extraction combined with in-syringe homogenous liquid−liquid microextraction of chloramphenicol and florfenicol from chicken meat, Microchemical Journal 196 (2024) 109636, https://doi.org/10.1016/j.microc.2023.109636.

[162] K. Abdi, M. Ezoddin, N. Pirooznia, Temperature-controlled Liquid−Liquid microextraction using a biocompatible hydrophobic deep eutectic solvent for microextraction of palladium from catalytic converter and road dust samples prior to ETAAS determination, Microchemical Journal 157 (2020) 104999, https://doi.org/10.1016/j.microc.2020.104999.

[163] M.K. Rajput, M. Konwar, D. Sarma, Hydrophobic natural deep eutectic solvent THY-DA as sole extracting agent for arsenic (III) removal from aqueous solutions, Environmental Technology & Innovation 24 (2021) 102017, https://doi.org/10.1016/j.eti.2021.102017.

[164] M. Mattonai, P. Massai, E. Ribechini, Sustainable microwave-assisted eutectic solvent extraction of polyphenols from vine pruning residues, Microchemical Journal 197 (2024) 109816, https://doi.org/10.1016/j.microc.2023.109816.

[165] Z. Triaux, H. Petitjean, E. Marchioni, M. Boltoeva, C. Marcic, Deep eutectic solvent-based headspace single-drop microextraction for the quantification of terpenes in spices, Analytical and Bioanalytical Chemistry 412 (2020) 933−948, https://doi.org/10.1007/s00216-019-02317-9.

[166] H. Qian, Q. Yang, Y. Qu, Z. Ju, W. Zhou, H. Gao, Hydrophobic deep eutectic solvents based membrane emulsification- assisted liquid-phase microextraction method for determination of pyrethroids in tea beverages, Journal of Chromatography A 1623 (2020) 461204, https://doi.org/10.1016/j.chroma.2020.461204.

[167] C. Dal Bosco, F. Mariani, A. Gentili, Hydrophobic eutectic solvent-based dispersive liquid-liquid microextraction applied to the analysis of pesticides in wine, Molecules 27 (2022) 908, https://doi.org/10.3390/molecules27030908.

[168] M. Khoubnasabjafari, A. Jouyban, M. Hosseini, M.A. Farajzadeh, R. Saboohi, M. Nemati, E. Marzi Khosrowshahi, M.R. Afshar Mogaddam, A mixed deep eutectic solvents-based air-assisted liquid-liquid microextraction of surfactants from exhaled breath condensate samples prior to HPLC-MS/MS analysis, Journal of Chromatography B 1204 (2022) 123289, https://doi.org/10.1016/j.jchromb.2022.123289.

[169] D. Baute-Pérez, Á. Santana-Mayor, A.V. Herrera-Herrera, B. Socas-Rodríguez, M.Á. Rodríguez-Delgado, Analysis of alkylphenols, bisphenols and alkylphenol ethoxylates in microbial-fermented functional beverages and bottled water: optimization of a

dispersive liquid-liquid microextraction protocol based on natural hydrophobic deep eutectic solvents, Food Chemistry 377 (2022) 131921, https://doi.org/10.1016/j.foodchem.2021.131921.

Further readings

[1] D. Snigur, E.A. Azooz, O. Zhukovetska, O. Guzenko, W. Mortada, Recent innovations in cloud point extraction towards a more efficient and environmentally friendly procedure, TrAC, Trends in Analytical Chemistry 164 (2023) 117113, https://doi.org/10.1016/j.trac.2023.117113.

[2] J.P. Hallett, T. Welton, Room-temperature ionic liquids: solvents for synthesis and catalysis. 2, Chemical Reviews 111 (5) (2011) 3508−3576.

Chapter 7

Advancements and innovations in solvent-based extraction techniques

Muhammad Farooque Lanjwani[1,2], Muhammad Yar Khuhawar[2], Mustafa Tuzen[1], Seçkin Fesliyan[3] and Nail Altunay[3]
[1]*Tokat Gaziosmanpaşa University, Faculty of Science and Arts, Chemistry Department, Tokat, Turkey;* [2]*Dr M. A. Kazi Institute of Chemistry, University of Sindh, Jamshoro, Sindh, Pakistan;* [3]*Sivas Cumhuriyet University, Faculty of Science, Department of Chemistry, Sivas, Turkey*

1. Introduction

The development of many new products, technologies and processes, involve significant attempts for ecological development. The compounds in chemical industry are usually observed to be the reason of several environmental difficulties. To confirm sustainability, new techniques must be considered from the life cycle point of view. Life cycle assessment (LCA) is a method in which whole life cycle of the yield or efficacy effects are investigated, like extraction, handling of raw materials, consumption, distribution, disposal, production, and possible environmental effects. Furthermore, energy alterations in a life cycle and resulting load on environment are evaluated. The health perspective and indoor environment are significant as outdoor environment. The chemicals must be employed in less amounts and must be less dangerous. A huge amounts of organic solvents are utilized in the extraction process. The traditional liquid liquid extraction (LLE), called as solvent extraction, and also liquid solid extraction (LSE) are methods in which many compounds can be isolated from individually and their comparative solubility. For the solid sample, separation of analyte from mixture through dissolving the desired constituent in suitable solvent. Extraction procedure generally needs many hours to complete, and is dependent on the temperature of extraction. These procedures are gradually being changed through more attractive replacements. The most broadly employed extraction methods are still Soxhlet and sonication extraction. The classical methods are commonly multi-step processes based on complete extractions from the sample matrix monitored by consecutive clean-up stages before examination. These sample preparation processes need huge amounts of sample, organic solvents and sorbents that are often dangerous and poisonous, resultant in higher costs of both disposal and purchase. These approaches also need wide physical handling that often produce work-related health issues [1].

The samples preparation for extraction is very important mainly if trace and ultra-trace level components are of interest. The separation of analytes from matrix is categorized via complex nonhomogeneous arrangement and can be treated as major analytical task. There are many variety of extraction approaches and techniques are reported. There are different parameters to be utilized for arrangement of extraction methods. The different extraction techniques are growing continually as extraction procedures based on liquid or solid phase, accelerated solvent extraction, Soxhlet extraction and multiplicity of applications like separations in chemistry, pharmaceutical, waste treatment, industrial practices in hydrometallurgy, and food engineering [2].

In these extraction techniques many solvents are used like volatile organic compounds (VOCs), including ketones, ethers, halogenated hydrocarbons, aromatic hydrocarbons alcohols, aldehydes, esters, and aliphatic and are mainly utilized as organic solvents [3]. The sample handling plays a basic part in analysis and huge effort are devoted to develop faster, safer and eco-friendly processes for sample extraction and cleanup of extract in past 2 decades. The desire to decrease the consumption of organic solvents are the main dynamic force for the progress of these methods. Some approaches comprising solutes adsorption on solids like miniaturization, solid phase extraction (SPE), headspace solid-phase microextraction (HS-SPME), and enhanced solvent-extraction, microwave assisted extraction (MAE), pressurized liquid extraction, matrix solid-phase dispersion (MSPD), stir bar sorptive extraction(SBSE) currently play significant role in handling of sample in chemistry laboratories. The attention was also paid to utilize alternative solvents, mostly supercritical fluids and more newly ionic liquids [4]. There are many extraction techniques used and various solvents employed in the extraction processes (Fig. 7.1). We discussed here different extraction techniques in which different solvents are used for samples analysis.

2. Dispersive liquid liquid microextraction

Dispersive liquid liquid microextraction (DLLME) is a sample presentation technique for the enrichment of analyte from low volume of water sample. It was initially proposed by Razaee et al. (2006) and is increasingly being used because of simplicity, ease in method development and low cost. Extraction of the analyte from water samples comprises two solvents, extracting solvent and dispersive solvent. The extracting solvent is generally immiscible in the water and dispersive solvent is miscible in the water. A mixture of dispersive and extracting is quickly introduced in the water sample. A rapid diffusion occurs,

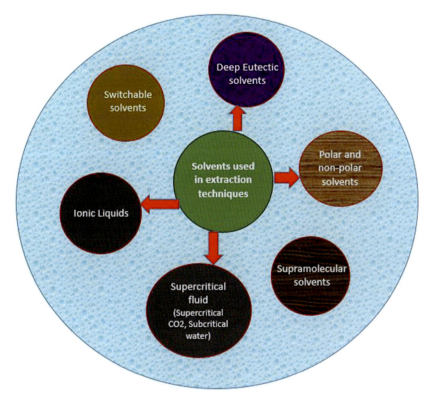

FIGURE 7.1 Different solvents used in extraction techniques.

which helps the transmission of the analyte from water to organic phase. The organic phase is collected with micro syringe after centrifugation, containing analyte. The extract is used for the analysis using different analytical techniques. Initially the extracting solvent were chlorinated solvents (carbon tetrachloride, chloroform, chlorobenzene, etc.) heavier than water and sedimented at bottom of test tube. The separation of the extracting solvent was easier with micro syringe. Dispersive solvents were acetonitrile, ethanol, acetone and methanol etc. The choice and volume of extracting and dispersive solvents are important and have to be optimized. The main factors for development of procedure are the addition of salt (salting effect) and pH of sample. These also require optimization [5–8].

During the years the use of DLLME has grown extensively for the analyses of organic, biological, pharmaceutical and inorganic compounds at trace levels from variety of different real samples, using different extracting solvents. A number of reviews have also appeared to cover the development in the field of DLLME. Zgoła-Grześkowiak & Grześkowiak, reviewed on the improvement made in DLLME since 2006. They described the derivatization of analytes simultaneously and joining the DLLME with other sample preparations. They also reviewed the use of extracting solvents less dense than water and ionic liquids. They included the applications of DLLME for the analyses of pesticides, pharmaceuticals, phenols and other compounds [9]. Saraji and Boroujeni examined the developments in DLLME during 2010—13. The review focused on the technological developments and applications of different extraction techniques. The main fields covered in the review were the nature of extracting and dispersive solvents, joining of DLLME with other extracting methods, automation, derivatization reactions and applications of DLLME for metal analyses [10]. Zuloaga et al. examined DLLME for the analyses of biological samples. The review covered mainly the analyses of plasma and urine samples. They discussed the use of high-density extracting solvents, low density extracting solvents and ionic liquids. Applications included were mostly organic compounds, but some bio analyses of inorganic substances were also discussed [11]. Geletu reviewed the applications and developments of DLLME. He examined the use of ionic liquids, solidification of extracting organic solvent moving on the surface of water, the solvents of mixed density and use of supra molecules. The review focused mainly on use of extracting solvents heavier and lighter than water for DLLME and their applications [12]. Campillo et al. reviewed the papers published on DLLME and allied techniques up to 2016. They also examined their applications [13]. Campillo et al. reviewed the significant advances during 2016—18, mainly to reduce the shortcomings in the use of DLLME for the reducing the extraction time, toxicity of solvents and improving the sensitivity and selectively of the methods for the suitable applications [14]. Lemos et al. reviewed in-syringe DLLME. They examined the methods using the techniques and discussed the procedures to improve the dispersion, analytical merits, advantages and disadvantage of the methods critically. They also described the trends in utilization of in-syringe microextraction [15]. Snigur et al. reviewed low-density solvents (LDSs)-based LLME for the separation and concentration of various analytes. They considered the advantages and disadvantage of LDSs mainly DESs, supramolecular solvents (SUPRAs) and switchable hydrophobicity solvents (SHSs) in different procedures of LLME during 2018—23 for the quantitation of analytes in different matrices [16].

2.1 Innovations in the use of extracting solvents

The introduction of the DLLME was with use of chlorinated solvents like carbon tetrachloride, chloroform, tetrachloroethane, and chlorobenzene. These solvents are immiscible with water and have higher density. These solvents are easily separated at the bottom of test tube after centrifugation. These solvents are still being used [17,18], because of immiscibility, better extraction abilities and ease in separation from aqueous phase. However, the chlorinated solvents are environmentally unfavorable and toxic. In order to decrease the drawback in DLLME for the use of chlorinated solvents, attempts were made to use less toxic brominated/iodinated solvents [19]. The results indicated that many brominated solvents had better extraction efficacy than chlorinated solvents. The solvents with lower density were examined, following the same procedure, to overcome necessity of use of chlorinated solvents. The less dense solvents have to be collected from surface of water, and therefore special devices were suggested [10]. The devices are generally homemade and are not available commercially. Farajzadeh et al. suggested a narrow tube as the extracting vessel for DLLME. The solvent from the surface was withdrawn by capillary tube. Good extraction efficiencies and recoveries were claimed. The common extracting solvents were n-hexane, toluene, xylene and n-octanol. If low density solvent has low melting point (10−30°C), the floating drop on the surface of water can be solidified through freezing in ice bath. The solidified drop is easily separated by spatula and the collected drop is reconverted to liquid and used for analysis using suitable analytical equipment [20]. Leong and Huang first used solidification of floating organic drop (SFOD) procedure. The same procedure of DLLME was used, where low density solvent (LDS) and dispenser solvents were injected together in water sample, and drop of LDS collected on the surface was frozen. The solvents used for the purpose were fatty alcohols: undecanol, decanol, and dodecanol. The solvents used are less toxic than chlorinated solvents, but the choice of the solvents is limited [21].

2.2 Ionic liquids (ILs)

The ILs are substances with melting point less than 100°C and are prepared through the combination of an organic or inorganic cation and organic anion. The ILs have been used in DLLME since 2008 [22], because of having higher thermal strength and low vapor pressure. Moreover, viscosity, solubility and density can be adjusted by proper arrangement of anions and cations [23]. Furthermore, they can possibly be modified to enhance extraction efficiency. Generally, ILs are denser than water and are collected on bottom of conical test tube after centrifugation. It makes easier than lower density organic solvents to collect the extract (Fig. 7.2). The ILs can be directly injected in HPLC after necessary dilution. Some of the applications of ILs for the preconcentration of organic and inorganic analyte are summarized in Table 7.1. The ILs used in DLLME are mostly immiscible with water and the condition has reduced the list of extracting solvents. The ILs commonly used in DLLME are: 1-butyl-3-methylimidazolium hexafluorophosphate, 1-decyl-3-methylimidazolium hexaflurophosphate, 1, 3-dibutylimidazolium hexafluorophosphate and 1-octyl-3-methylimidazolium bis(trifluoromethylsulfonyl) imide [9].

The DLLME using ILs is carried out by different procedures. In a classical procedure hydrophobic IL and dispenser solvent are injected in water sample, and the cloudy solution formed is centrifuged. The sediment phase is separated

194 Green Analytical Chemistry

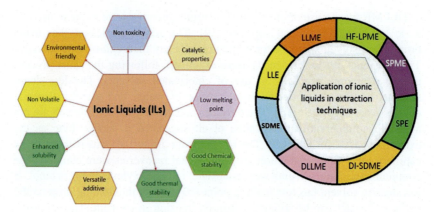

FIGURE 7.2 Application ionic liquids in extraction techniques.

and analyzed on instrument. In a typical procedure imidazolium based ILs with hexafluorophosphate dissolved in methanol was used for the extraction of organic compounds. The temperature based DLLME is carried out by the change in temperature (usually 50—90°C), which results in complete dispersion of the ILs in water. After thoroughly mixing, the contents are cooled in ice box. Cloudy solution formed is centrifuged and a drop sedimented at the bottom is separated. The dispersion solvent may not be necessary. The sedimented drop is generally dissolved in less volume of solvent and analyzed on suitable instrument. Imidazolium based ILs with hexafluorophosphate as anion are utilized for extraction of metal ions and organic compounds. In another procedure of DLLME, IL is added to water sample and agitated strongly by ultrasound, vortexing or microwaves to disperse the IL in solution. Sometimes water miscible organic solvent, hydrophilic ILs or surfactant is added to aid the dispersion. The procedure has been used for analyses of organic compounds and some metal ions. In order to further improve DLLME using ILs, in situ extraction solvent formation for micro extraction was introduced. The dissolved hydrophilic IL in water sample is added to interact analyte with IL. The addition of anion altercation reagent promoted to convert hydrophilic IL to hydrophobic IL, resulting in efficient extraction performance. The in-situ IL based DLLME has been employed for analyses of organic compounds and metal ions [10]. In spite of the extensive use of ILs in DLLME, the use of ILs has some limitations. They have some problems in coupling with GC, LC-MS and atomic absorption instruments, because of high viscosity and low volatility of ILs. It may result into matrix effects and drift in signals. The high thermal stability may also affect electrothermal atomic absorption spectroscopy, because high temperatures required for the pyrolysis of ILs may involve the loss of analyte [44].

TABLE 7.1 Application of ionic liquids in different microextraction techniques.

Method	Matrix	Analyte	LOD μg/L	Detection technique	References
DLLME	Beverages	Parabens	0.03–0.06	HPLC	[24]
IL-DLLME	Honey and milk	Quinolones	1.5	HPLC	[25]
SA-DLLME	Environmental water samples	Parabens	2.0	HPLC-UV	[26]
VA-IL-DLLME	Water samples	UV filters	0.81–9.6	HPLC	[27]
AA-LLME	Milk samples	Seven antibiotics	0.09–0.21	HPLC–DAD	[28].
DLLME	Water	Zinc	0.11	UV-visible	[29]
DLLME	Curcuma longa	Curcuminoids	3.0	HPLC	[30]
DLLME	Water	Pyrethroid pesticides	0.16–0.21	HPLC	[31]
UAE-IL-DLLME	M. cordata	Six alkaloids	0.08	UPLC–MS/MS	[32]
IL-UA-ME	Dairy products	Melamine residues	0.015	UV-visible	[33]
USA-IL-DLLME	Different rice samples	Se (IV)	0.018	FAAS	[34]
MRTILs-DLLME	Water	Au, Ag	0.0032 and 0.0073	ETAAS	[35]
SSIL-DLLME	Water	Pyrethroid insecticides	0.71–1.54	HPLC	[36]
Vortex-assisted IL-DLLME	Chocolates	Nickel and cobalt	0.3 and 0.2	FAAS	[37]

Continued

TABLE 7.1 Application of ionic liquids in different microextraction techniques.—cont'd

Method	Matrix	Analyte	LOD μg/L	Detection technique	References
In situ DLLME	Honey	Chlorophenol compounds	0.8–3.2		[38]
IL-DLLME	Beverage	Parabens	1.53	UPLC	[39]
CPE	Coarse cereals	Organophosphorus pesticides	2.0–4.2	HPLC	[40]
Air-assisted IL-DLLME	Juice	Fungicides	0.4–1.8	HPLC	[41]
MIL-UDSA DLLME	Rice	Inorganic selenium	0.018	GFAAS	[42]
IL-DLLME	Fish	BG and CV	2.7 and 1.4	UV-visible	[43]

AA-LLME, air–assisted liquid–liquid microextraction; *BG*, Brilliant Green; *CPE*, Cloud point extraction; *CV*, Crystal Violet; *SADLLME*, salting-out assisted dispersive liquid–liquid microextraction; *UHPLC-DAD*, ultra-high-performance liquid chromatography-diode array detector; *UPLC*, ultra-performance liquid chromatography; *VFAs*, Volatile fatty acids.

2.3 Supramolecular solvents (SUPRASs)

The SUPRASs are immiscible in water, can be generated by amphiphile aggregates, through the change of pH, temperature and salt addition. SUPRASs are planned by two procedures, (1) the amphiphilic molecules above the critical concentration are self-aggregated to develop reverse micelles or vesicles, and (2) the nano-sized groups change to bigger aggregates, which are immiscible in water (Fig. 7.3). The use of SUPRASs in micro extraction has widened enormously during last decade. Moradi et al. studied the utilization of SUPRASs in procedures based on liquid phase [45]. Jalili et al. reviewed the utilized of SUPRASs in procedures based on liquid phase for preconcentration and sampling of heavy metals [46]. SUPRASs are nonflammable and nonvolatile, and the characteristics of the solvents may be easily arranged by changing hydrophobic or hydrophilic groups of the amphiphile. These properties have made SAPRASs attractive to substitute for organic solvents [47].

FIGURE 7.3 Application of supramolecular solvents in extraction techniques.

A number of applications of SUPRASs for the preconcentration of different analytes are reported (Table 7.2). Ruiz et al. first time reported the capability of tetrabutylammonium (TBA) supported coacervation system using alkyl-carboxylic acid vesicles for the preconcentration of organic substances before HPLC determinations. The habitation of high combination of TBA and water (above critical micellar concentration (CMC)), the surfactant molecules produce three dimensional vesicles and these individual vesicles combine together, forming larger molecules called vascular SUPRAS [67]. These SUPRAS molecules can extract the analyte by analyte/vesicles binding interactions. The polar interactions of vesicles occur through carboxylic/carboxylate groups and ammonium groups by electrostatic and hydrogen bond formation, while hydrophobic reciprocal actions occur through hydrocarbon portion [68]. Moradi and Yamini examined the implementation of vesicular coacervate phase micro extraction on the basis of solidification of floating drop. In this procedure small mass of SUPRAS based on vesicles from coacervation of decanoic acid in existence of tetra butyl ammonium (melting point above 10°C) is hanging above the aqueous phase. The hovering drop after extraction can be collected after solidification by keeping in ice box. The solidified mass is separated easily and it is melted at the room conditions. The extract is then injected in HPLC-UV for determination [69].

2.4 Reverse micelles based SUPRAS

The Ruiz et al. introduced the reverse micelles of decanoic acid dispersed in the mixture of water/tetrahydrofuran (THF) for extraction of polar and nonpolar mixtures, based on hydrogen bonding and Van der Walls interactions [70]. The addition of the water to the mixture allows to dissolve the aggregates partially and encourages the formation of larger reverse micelles as immiscible liquid phase from THF/water total solution. The volume ratio of THF/water was approximately by the bulk of THF (4%—50%) and alkyl carboxylic acid (0.025%—4.0%), and alkyl carboxylic acid chain length (8—18 carbon atoms). The main factors responsible for the extraction of the solute in reverse micelles of alkyl carboxylic acid could be hydrophobic and hydrogen bond influences. The high enrichment factors are therefore reported [45]. Instead of alkyl carboxylic acid, aliphatic alcohols were tried for preparation of SUPRAS and resulted into the formation of rod like micelles. The equilibrium in the solution resulted in the dispersion of the analyte in sample and enhanced the extraction at lower SUPRS volume [71]. THF-based SUPRASs have been applied in DLLME mode. Hafez et al. proposed SUPRAS liquid phase microextraction method to preconcentrate and examine Al (III) in different samples. The process was based on use of undecanol/THF SUPRAS and thymol blue as chelating agent. The Al (III) complex created at pH 6 was extracted into undecanol/THF SUPRAS phase and centrifuged. The determination was carried out on spectrophotometer at 550 nm. LOD was reported 1.20 µg/L with

TABLE 7.2 Application of supramolecular solvents in different microextraction techniques.

Method	Matrix	Analyte	LOD µg/L	Detection technique	References
VA-SUPRAS-LPME	Meat	Nitrate	0.035	UV-visible	[48]
SsLLME	Foodstuff	Copper	0.2	GF-AAS	[49]
Ss-ME	Fruits and Vegetable	Carbaryl	30	LC-MS/MS	[50]
LPME	Water and rice	Phenoxy acid herbicides	0.001–0.002	HPLC	[51]
SUPR–FF—AA–DMSPE	Juices	Organophosphate diazinon and metalaxy	0.20, 0.80	GC-FID	[52]
DLLME	Environmental samples	Hg	0.561	AAS	[53]
	Vegetables	Manganese and zinc	0.06	FAAS	[54]
DLLME	Human urine	Steroid sex hormone	0.01–0.10	LC-MS	[55]
MSB-LPME	Human serum	Non-steroidal Anti-inflammatory drug	0.83–3.16	LC/MS–MS	[56]
DSPE	Food	Ionophore coccidiostats	0.004–0.07	LC/MS–MS	[57]
VA–SS–DLLME	Water samples	Orthophosphate	0.1	UV-visible	[58]
UA–SS–LLME	Water	Chromium	0.79	UV-visible	[59]

Continued

TABLE 7.2 Application of supramolecular solvents in different microextraction techniques.—cont'd

Method	Matrix	Analyte	LOD µg/L	Detection technique	References
SSME	Soil, water	Uranium	0.31	UV-visible	[60]
SS-LSME	Food	Lead	0.047	ETAAS	[61]
UA–SS–LLME	Water	Palladium	0.63	FAAS	[62]
SDME	Vegetable oils	Cadmium	0.002	ETAAS	[63]
SS-LLME	Human hair	Aluminum	0.16	UV-visible	[64]
VA-SMS-LLME	Biological samples	Mercury	0.30	UV-visible	[65]
UA-DS-LLME	Blood serum	Copper, Cobalt	2.9, 3.5	FAAS	[66]

DSPE, dispersive solid phase extraction; ETAAS, electrothermal atomic absorption spectrometry; FAAS, flame atomic absorption spectrophotometry; LLME, liquid-liquid microextraction; MSB-LPME, magnetic solvent bar-liquid-phase microextraction; SsLLME, supramolecular solvent-based liquid-liquid microextraction; SS-LSME, supramolecular aggregated liquid–solid microextraction; Ss–ME, Supramolecular solvent microextraction; SU/PR–FF–AA-DMSPE, supramolecular ferrofluid-air-assisted-dispersive micro solid phase extraction method; UA-DS-LLME, ultrasonic-assisted dispersion solidification–liquid–liquid microextraction; VA-SMS-LLME, vortex-assisted-supramolecular solvent-liquid–liquid microextraction; VA-SU/PRAS-LPME, vortex-assisted-supramolecular solvent-based liquid phase microextraction.

enrichment factor of 50 [72]. Yilmaz and Soylak, applied SUPRAS and prepared reverse Michelle of 1-decanol/THF in water for microextraction of Cu (II) and examined by FAAS. The Cu (II) was complexed with dimethyl dithiocarbamate and extracted hydrophobic complex in SUPRAS phase. The effects of different variables were optimized. The LOD and LOQ were 0.52 and 1.71 μg/L with preconcentration factor 60. External forced assisted SUPRAS micro extraction has been examined, particularly by using ultrasound waves and vortex for the improvement of partition coefficient among water and extracting solvent [73]. Maslov et al. developed methodology using vortex-assisted SUPRAS based DLLME with FAAS for analysis of Ni in plant samples. The SUPRAS was synthesized from tetra-n-butyl ammonium hydroxide and undecanol for the formation of nano structured vesicular SUPRAS. The mixture was vortexed and milky solution produced was centrifuged to isolate water phase from the SUPRAS. Crystal violet was used to complex Ni (II). The preconcentration factor obtained was 195 [74]. Similarly, Moradi et al. developed ultrasonically assisted SUPRAS micro extraction for separation and examination of phthalates with HPLC-UV. Reverse micelles comprising decanoic acid/THF based nano structural aggregates were developed. Sonication enhanced transfer of analytes into SUPRAS from aqueous sample, thus reduced the extraction time. The preconcentration factor at optimized conditions was reported 176−412 with calibration range 0.5−100 μg/L [75].

2.5 Deep eutectic solvents (DES) in DLLME

Deep eutectic solvents (DES) are comparable to ionic liquids, but are very easy to prepare from inexpensive chemicals. These are biodegradable and are classified as green solvents. The DES are generally prepared by mixing two materials: a hydrogen bond donor (HBD) and other hydrogen bond accepter (HBA) (Fig. 7.4). The mixture is then heated at 50−100°C and the product obtained, indicates melting point less than their starting materials. The hydrogen bonds formation may be responsible for their synthesis, but other forces like electrostatic and Van der Walls may also be involved [76]. DES may be classified as water soluble (hydrophilic) and water insoluble (hydrophobic). The hydrophilic DES are more reported, but hydrophobic DES have more importance in the micro extraction of the analytes from aqueous solutions [77].

Lu et al. summarized the literature to preconcentrate chemical contaminants at trace levels in water and food by DES for extraction in DLLME. They have reported the impacts of different variables extensively on the preconcentration efficacy. They have examined critically the use of DES in DLLME to enrich chemical contaminants [78]. Andruch et al. reviewed the uses of DES in liquid-liquid microextraction (LLME) (2017−22) for the analysis of organic and inorganic analytes in different matrixes [79]. Makos

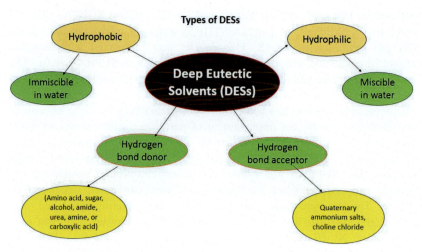

FIGURE 7.4 Types of DES solvents used in extraction process.

et al. examined available literature on the important physicochemical properties of hydrophobic deep eutectic solvents for their utility in aqueous sample preparation processes, its limitations and challenges in micro extraction [77]. Lanjwani et al. reviewed the applications of DES for the separation of inorganic and organic contaminants from food and water samples using different micro extraction techniques [76]. Li and Row, accessed the preparation, properties and classifications of DES for the use of DES in DLLME with different modes of modifications. The review also examined benefits and drawbacks of DES in the applications of DLLME and expected future trends [80]. The number of starting material for the preparation of DES is huge and the suitable physicochemical features of DES can be adjusted by selecting appropriate HBA and HBD [79]. The DES used in LLME are prepared from organic salts (hydrogen bond accepter, HBA) and hydrogen bond acceptors mixed in different ratios or by adding water. The common HBA investigated are choline chloride, N, N- + diethyl-2-hydroxyl ethanamide chloride, methyl triphenyl phosphonium bromide, tetra butyl ammonium chloride, tetramethylammonium chloride, N-benzyl-2-hydroxyl-N, N-dimethyl ethanamide chloride, tetra ethyl ammonium chloride, 2-acetyl-N, N, N-trimethyl ethanamide chloride. The use of hydrogen bond donors (HBD) includes urea, acetamide, glycerol, thiourea, imidazole, glucose, malonic acid and benzamide [80]. The hydrophobic DESs are also prepared through the grouping of metal chloride with HBD. The hydrogen bonding force among analyte and hydrophobic DES, and the polarity of DES have a significant impact on solubility of analyte in DES. The formation of hydrogen bond between analyte and DES- may reduce the solubility of analyte in water and favors the transfer of the analyte to hydrophobic DES [81]. The density of DES is also an important property and should be different from water, otherwise, it will take longer time to separate the phases. More DES are reported with density higher than water and density of DES depends upon the structure of HBSs. The molar ratios of the components of DES also have effect on the density of DES [82].

2.5.1 Applications of DES

The (Table 7.3) summaries some of the applications for the use of DES in DLLME for trace analyses. Tomai et al. developed method for low temperature mixture DES for DLLME for pesticides from surface water. The DES involved heating the mixture of choline chloride and acetylsalicylic acid in molar ratio 1:2. The obtained liquid was clear at room temperature and denser than water (1.2 g/mL). Different pesticides (24) were extracted from surface water (5 mL) and analyzed by HPLC-MS/MS. At the optimized situations the extraction of 18 polar compounds was reported 96% [116]. Li et al. developed sample preparation procedure for examination of fluoroquinolone antibiotics in surface water. The hydrophobic DES based on thymol and heptanoic acid indicated better extraction efficiency for a number of fluoroquinolones. At optimized conditions 10 μL of DES-was generated in 10 mL water sample with warming at 52°C for 5 min. The DES-was dilutee and was examined on HPLC-UV detection. The in-situ formation of the DES-was validated and indicated LOD 3.0 ng/L [117]. Sorouraddin et al. reported DLLME method of solidification of ternary DES for extraction of trace quantities of Cd (II), Cu (II) and Pb (II) from milk samples and analyzed by FAAS. Sorbitol-menthol-mandelic acid were mixed in appropriate ratio and heated to form low density DES. The DES was utilized as chelating agent and extracting solvent for the metal ions. At the optimized conditions LODs were reported within 0.38−0.42 μg/L [118]. Werner, reported sample preparation procedure with DES using ultrasound assisted DLLME based on solidification of DES, followed via HPLC-UV for the preconcentration and quantitation of aromatic amines in water samples. Prepared low density DES-was by mixing trihexyl(tetradecyl)phosphonium chloride and decanol at a molar ratio of 1:2. The extractant was distributed by ultrasound and separated by centrifugation. The solid DES obtained by cooling and dissolved in the methanol and then injected in the HPLC. At optimized conditions LODs obtained were 0.07 ng/mL for 2-chloroaniline, 0.11 ng/mL for 4-chloroaniline and 0.08 ng/mL 1-naphthylamine [119].

2.6 Switchable solvents for micro extraction

The switchable solvents are drawing attention gradually, because these are among the green solvents and are being employed for micro extraction of pollutants from different matrices. These solvents may be converted into two different forms with changed physical features and are used for micro

TABLE 7.3 Application of deep eutectic solvents (DESs) in different microextraction techniques.

Method	Matrix	Analyte	LOD µg/L	Detection technique	References
UV-DES-LPME	Water	Amaranth (E123)	23.0	UV-visible	[83]
TC-NADES-LLME	Food matrices	Chloramphenicol	0.03	UV-visible	[84]
USA-DLLME-SFO	Honey	β-Lactam	1.16–5.08	HPLC	[85]
AA-NADES–SH–LPME	Fruit juice	Patulin	3.5	UV-visible	[86]
DES-VA-DLLME	Food	Sudan dyes	0.5–1	HPLC	[87]
VA-NADES-LPME	Food	Indigo-carmine	3.3	UV-visible	[88]
UA-LLME-DES	Apple juice	Sulfonamides	0.02–0.05	HPLC	[89]
VALLME-DES	Chili oil	Sudan I dye	0.02	HPLC	[90]
USAEME-DES	Beverages	Quercetin	18.8	UV-visible	[91]
ESE-DES	Water	V	0.01	GF-AAS	[92]
SA-LPME-HDES	Water	Pb	0.24	FAAS	[93]
DES-LPME	Linden tea	Co	2.0	FAAS	[94]
NH-IS-SPE	Spinach	Cr	3.10	µS-FAAS	[95]
VAME-DES	Soil	As	0.10	GF-AAS	[96]
UA-DES-LPME	Tobacco	Pb	4.4	FAAS	[97]
NADES-UAME	Rice	Se	3.0	HG-AAS	[98]
DES-VA-DLLM	Mix fruit	Thiabendazole	0.1	UV-visible	[99]
TC-VA-LLME-SFDES	Water	Fe	1.5	UV-visible	[100]

DES-LPME	Coffee	Mn	0.52	FAAS	[101]
HDES-LPME	Wastewater	Cd	1.6	FAAS	[102]
DES-MCG-DSPE	Urine	Co	4.6	FAAS	[103]
ULLME-LV-HDES	Food	Se	0.25	HG-AAS	[104]
RP-DLLME	Fish oil	Zn	0.18	FAAS	[105]
DESMNF-LPME	Rosemary tea	Cd	0.25	SQT-FAAS	[106]
DES-UA-DLLME-SAP	Water	Pb, Cd, Co, Ni	0.05–0.13	HPLC	[107]
HDES-UA-DLLME	Milk and beverage	Phenols	0.25–1	HPLC	[108]
DES-AALLME	Water	Tetracyclines	1.2–8	HPLC	[109]
VA-DES-LPE	Wastewater	Methyl red dye	2.3	UV-visible	[110]
DES-UA-DLLME	Water and milk	Non-steroidal anti-inflammatory drugs	0.5–1	HPLC	[111]
DES-ME	Environmental samples	Tartrazine	0.084	UV-visible	[112]
DES-DLLME	Water	Neonicotinoid insecticide	0.001–0.003	HPLC	[113]
DES-DLLME	Water	Nitrophenol	0.2	HPLC	[114]
TDES-DLLME-SFOD	Water	Endocrine disrupting compounds	0.97–2.3	HPLC	[115]

COF, ciprofloxacin; *DES-LPME*, deep eutectic solvent based liquid phase microextraction; *DES-MCG-DSPE*, deep eutectic solvent Magnetic colloidal gel dispersive solid phase extraction; *DESMNF-LPME*, deep eutectic solvent-based magnetic nanofluid liquid phase microextraction method; *DLLME-SFOD*, dispersive liquid-liquid microextraction based on solidification floating organic droplet; *EA-DLLME*, effervescence-assisted dispersive liquid-liquid microextraction; *EDCs*, endocrine disrupting compounds; *EF*, Enrichment factor; *ER*, Extraction recovery time; *FAAS*, flame atomic absorption spectrophotometry; *FQ*, Fluoroquinolone; *GC–FID*, Gas chromatography flame ionization detector; *GF-AAS*, Graphite furnace atomic absorption spectrophotometry; *HG-AAS*, Hydrogen generation atomic absorption spectrophotometry; *HIDES-ME*, heat-induced deep eutectic solvent microextraction; *LOD*, Limit of detection; *LOF*, levofloxacin; *LOQ*, Limit of quantification; *LR*, Linear range; *NH-IS-SPE* μ*S-FAAS*, Needle hub in-syringe solid phase extraction coupled with micro-sampling flame atomic absorption spectrophotometry; *PF*, Pre-concentration factor; *RSD*, Relative standard deviation; *SQT-FAAS*, slotted quartz tube flame atomic absorption spectrophotometry; *TC-VA-LLME-SFDES*, temperature-controlled vortex-assisted liquid-liquid microextraction based on the solidification of a floating DES; *ULLME-LV-HDES*, ultrasound liquid-liquid microextraction method based on low viscous hydrophobic deep eutectic solvent; *VA-DES-LPE*, vortex assisted deep eutectic solvent based liquid phase extraction; *VM-NADES-ME*, Vortex-assisted natural deep eutectic solvent microextraction.

extraction. The switchable solvents involve in the change of polarity. The solvents might be converted from hydrophobic to hydrophilic form and vice versa. The suitable solvents are mainly based on secondary or tertiary amines. The change in the physical property of the solvent is carried out by passing CO_2 gas or through the adding of dry ice to the mixture of amine and water (1:1 v/v) [120]. The addition of CO_2 results into the protonation of amine and conversion of hydrophobic form to hydrophilic form. The removal of CO_2 results with deprotonation of amine and converts back amine to hydrophobic form. Other compounds used as switchable solvents are fatty acids [121]. The addition of NaOH to fatty acid converts the fatty acid to hydrophilic sodium salt. The addition of the acid to the salt reverts back to hydrophobic fatty acid. Thus, the conversion of hydrophilic form to hydrophilic state is controlled by pH [122]. Alternatively, the fatty acid is reacted with amine solution to develop fatty acid-amine complex, which is soluble in water (hydrophilic). The fatty acid-amine complex dissociates by bubbling CO_2 through the solution and the phase separation is induced [123]. Pochivalov et al. introduced di(2-ethylhexyl)phosphoric acid (BEHPA) as switchable solvent. The compound BEHPA was converted to hydrophilic form when treated with alkaline solution and reverted to hydrophobic form when reacted with acidic solution [124].

Amine based solvents can be converted quickly between hydrophobic to hydrophilic forms. Their polarity can be changed by bubbling CO_2 from their solution, which results into the formation of hydrophilic ammonium carbonate. The removal of CO_2 by passing nitrogen gas, the carbonate salt returns back to the hydrophobic amine form [125]. The use of switchable solvents for LLME have been reviewed. Musarurwa and Tavengwa, reviewed the use of amine based switchable solvents and medium chain fatty acids for preconcentration of pesticides contaminants from different samples. They also examined the combination of switchable solvents for micro extraction with other micro extraction processes for preconcentration of pesticides [47]. Alshana et al. reviewed the preconcentration of organic and inorganic compounds by using switchable solvents for LLME from environmental, pharmaceutical, food, biological and industrial materials. The synthesis of switchable solvents and their applications are reviewed. They also examined the combination of switchable solvents with other micro extraction techniques [126]. Herrero et al. assessed the characteristics and applications of Switchable solvents and gas expended liquids. Carbon dioxide was used in the methods to alter the physicochemical features of the solvents. The relevant examples are also included [127].

Some of the applications of switchable solvents are summarized in (Table 7.4). Chormey et al. reported switchable solvents for liquid phase micro extraction of selected pesticides, endocrine disrupters and hormones. GC-MS was used for the quantitative determinations. The method indicated LOD and LOQ of 0.2–13 and 0.9–46 ng/mL respectively. The method was employed

TABLE 7.4 Application of Switchable solvents in extraction techniques.

Method	Matrix	Analyte	LOD µg/L	Detection technique	References
SHS-UA-LLME	Food samples	Nitrite	0.1	UV-visible	[128]
TSSME	Water samples	PAEs	0.03–0.06	GC–MS	[129]
SHE-ME	Beverages	Polychlorinated biphenyls (PCBs)	0.007–0.02	GC-MS/MS	[130]
VA–SHS–LPME	Food samples	Vanillin	0.06	UV-visible	[131]
LPME-ETA-SS	Water	Diazinon	0.010	UV-visible	[132]
SHS-LLME	Fish and hair	Mercury	0.014	UV-visible	[133]
SPs-LPME	Cosmetics	Rhodamine B	2.96	UV-visible	[134]
SHS-LPME	Shellfish and fish sample	Brilliant green dye	1.4	HPLC	[135]
LPME	Soil	Manganese	0.70	FAAS	[136]
LPME	Milk	Lead (Pb)	0.01–0.02	GFAAS	[137]
LPME	Biological	Methionine	3.30	GC–MS	[138]
SHS-ME	Beverages	Bisphenols	0.27–0.40	HPLC-UV	[139]
SW-DLLME	Food and water	Cadmium	0.38	GFAAS	[140]
LPME	Urine sample	Ofloxacin	1.10	HPLC-FLD	[141]
LPME	Sediment	Uranium	0.30	UV-visible	[142]

Continued

TABLE 7.4 Application of Switchable solvents in extraction techniques.—cont'd

Method	Matrix	Analyte	LOD μg/L	Detection technique	References
LPME	Water	Triazine	1.80	HPLC-UV	[143]
ETA–SHS–ME	Road dust and water	Palladium	0.017	ETAAS	[144]
μS-SHS	Food and water	Vanadium	0.12	ETAAS	[145]
EA-DLLME-FA	Water and food	Silver and Cobalt	3.0	FAAS	[146]
SPS-LPME	Water and biological	Mercury	0.19	UV-visible	[147]

EA-DLLME-FA, effervescence-assisted dispersive liquid-liquid microextraction based on a medium-chain fatty acid; *ETA–SHS–ME*, effervescence tablet-assisted switchable solvent-based microextraction; *GC-MS/MS*, gas chromatography-tandem mass spectrometry; *HPLC-DAD*, high-performance liquid chromatography-diode array detection; *LPME-ETA-SS*, liquid-phase microextraction-based effervescent tablet-assisted switchable solvent; *PAEs*, phthalic acid esters; *SHS-LLME*, witchable hydrophilicity solvent-based liquid-liquid microextraction; *SHS-ME*, switchable hydrophilicity solvent-microextraction; *SHS-UA-LLME*, sustainable switchable hydrophilicity solvent-based ultrasound-assisted liquid-liquid microextraction; *SPS-LPME*, switchable solvent-based liquid phase microextraction; *SW-DLLME*, Switchable water dispersive liquid–liquid microextraction; *TSSME*, Temperature controlled switchable solvent based microextraction; *VA–SHS–LPME*, vortex-assisted switchable hydrophilicity solvent liquid phase microextraction.

for the analysis of tap water [148]. Di et al. examined N, N-dimethyl benzylamine as switchable solvent for LLME of pyrethroid insecticides using HPLC-UV detection. The reversible states between water insoluble and water soluble were triggered by CO_2. Phase separation was attained through adding of NaOH within 3 min. At optimized conditions the method indicated LOD of 0.2−0.5 ng/mL [149]. Gao et al. reported effervescence assisted micro extraction process, comprises switchable fatty acid with solidification of organic phase droplet for the analysis of antibiotics from seawater and seafood. Nonanoic acid was used as extracting solvent. The change between hydrophilic and hydrochloric forms was carried out by the adjustment of pH. At the optimized conditions LODs were 0.007−0.113 μg/L [150].

3. Pressurized fluid extraction (PFE)

The PFE is an extraction technique which uses solvents at subcritical pressure and temperature situations, mainly for recovery of semi volatile organic compounds from the solid constituents. The PFE technique has attained recognition as a standard extraction process and is usually employed in extraction of pollutants from water, soils and sediments samples [151]. Pressurized fluid extraction (PFE) is also known as pressurized liquid extraction (PLE), enhanced solvent extraction (ESE) and pressurized solvent extraction (PSE). In PFE extraction process liquid solvents are used at moderate pressure and temperature, less than those utilized in SFE. The main benefit of PFE is probability to work with liquid solvents beyond its boiling point, maintaining in liquid phase by a pressure rise. The two advantageous phenomena happen thus instantaneously: a decrease in viscosity of solvent and increase the solute solubility in solvent. Moreover, high temperatures help the breakdown of solute-matrix interactions, allowing the dispersion of solute on surface of matrix. In PFE method, extraction time is decreased as compared to conventional procedures like Soxhlet extraction. Furthermore, PFE may be done under dynamic or static conditions. In static type, the sample is kept in extractor, and filled with solvent. Afterward extraction, solvent is removed with nitrogen into vessel. But in dynamic type, the solvent is continuously pumped over the sample. It has drawback of needing larger volumes of solvent, which outcomes in dilution of the extract [152]. The PFE process employs the identical basic working conditions as traditional liquid solvent extractions, nevertheless the extractions are done at higher pressure and temperature. An increase in the temperature through the extraction provides more effective extraction with lowered solvent uses and time savings. Principle of PFE is alike to MAE, wherever solvent is heated through microwave energy, while in PFE method both higher pressure and temperature may be achieved, individually [1]. The PLE is extraction method where the required contains are removed from the target samples present in the solid matrix through using high pressure and temperatures, generally up to over 200 bar and 200°C, respectively, without achieving critical point by using liquid solvents. These situations increase mass transfer rates and solubility, resulting in raised diffusivity of solvent, and enhanced kinetics of matrix. The temperature, time, pressure, sample weight, many cycles, and solvent all effect on the extraction process. To increase the efficiency of PLE, these variables must be optimized through using an appropriate experimental design [153].

3.1 Instrumentation and principles of pressurized fluid extraction

The extractions by a pressurized liquid overhead its distinctive boiling point, need pressure restrictors or open/close regulators, in order to preserve pressure through the extraction process. The two types of instrumentation may be employed: static PFE that is a batch procedure with one or numerous extraction steps with solvent replacement in the process, and in dynamic PFE, extraction solvent is pumped constantly via extraction container comprising sample. The dynamic PFE is relatively comparable to static PFE, nonetheless needs sophisticated HPLC pump or high-pressure and also pressure restrictor somewhat than static close/open valve. However, column is changed with extraction cell and plumbing has somewhat broader inner-diameter. Presently, here is no any dynamic PFE tool existing on the local market. While there are some different kinds of marketable instruments available. Many of them perform both static and dynamic extractions at temperatures above than 200°C [1].

3.2 Solvent used in PFE

The solvent is most important parameter to optimize the extraction process. The solvent's help to solubilize analyte although reducing extraction of other constituents. Consequently, it is significant to select the solvent which has same polar characteristic of target samples. The water immiscible or Non-polar solvents and a mixture of medium-polarity solvent with non-polar solvents are utilized for extraction of lipophilic or non-polar compound. Therefore, higher polarity solvents are utilized to extract hydrophilic and polar compounds. Lastly, while extracting samples with dissimilar polarities, a combination of solvents with low and high polarity is usually used. Definitely, some researchers recommend two PLE extractions while target is for low and high polar analytes, therefore they may be extracted in two stages. Finally, miscibility and affinity are two parameters utilized in predictive methods to examine solubility of target samples in green solvents at changed temperature [154]. Extraction solvent in PFE must be very selective, with higher solvation capability for the target samples or compound and reduce co-extraction of other constituents. Polarity of solvent must be adjacent to that of target material. The Non-polar solvents like pentane and n-hexane or non-polar solvents with medium-polarity like cyclohexane/ethyl acetate, or pentane/dichloromethane have often been utilized in extraction of a lipophilic or polar compounds. The extra polar solvents, like ethyl acetate acetonitrile, water or methanol, were utilized in case of hydrophilic and polar compounds. Other main solvent distinctive of its capability is to support in discharge of compounds from matrix and in breakdown of interactions matrix-compounds [155]. The most frequently utilized extraction solvents for analysis of bioactive components from natural sources are methanol [156,157], ethanol [158,159], and n-hexane [160]. While water is applied as a solvent [161,162]. The high temperature, pressure and variations in pH, contribute to change the polarity of water, adjusting its dielectric constant (ε). At ambient temperature and pressure, water act as polar solvent with higher dielectric constant ($\varepsilon = 78$) however at and 23 MPa and 573 K this value drops to 21, that is comparable to value for the acetone ($\varepsilon = 20.7$ at 298 K) or ethanol ($\varepsilon = 24$ at 298 K). At moderate pressures and high temperature polarity of water could be decreased significantly; therefore, water may act as if acetone or ethanol were being utilized [163]. The water as a solvent is usual green method utilized for extraction of carbohydrates since at 100−150°C they are very soluble in water and dielectric constant of water is decreased at this temperature. Furthermore, subcritical water actions as alkali or acid, supporting the extraction of polysaccharides [154]. Other substitute eco-friendly solvent, like DES solvent is considered for extraction of carbohydrate. The DES are eutectic mixtures which is composed of hydrogen bonding donors (HBDs) and hydrogen bonding acceptors (HBAs). Because their stability, ease of synthesis and cost-competitiveness. The DES were proposed to dissolve many polysaccharides like chitin cellulose, lignin for biomass handling and starch. Furthermore, recently DES solvent was applied for extraction of alginates and fucoidans from the brown algae [164].

4. Dispersive solid phase extraction

Solid phase extraction (SPE), which is seen as an alternative to liquid phase extraction, has important advantages such as ease of use, requiring less solvent, saving time and offering good selectivity [165,166]. Anastassiades and his colleagues introduced a new method called dispersive solid phase extraction (DSPE) to researchers in 2003 [167]. DSPE is basically based on dipping the solid adsorbent into the sample matrix, ensuring its contact with the target analytes, and then centrifuging it. Although the method seems to be an adaptation of SPE, it is quite environmentally friendly compared to SPE because it works with lower amounts of adsorbent and solvent and is faster [168].

DSPE can basically be used for two different purposes. One of these is the extraction of the target analyte. At this stage, the adsorbent is added to the sample and dispersed in the sample, interacting with the analytes in various regions of the matrix through different forces and retaining the analytes. The adsorbent phase is then separated and removed from the matrix by a process such as centrifugation. By washing the adsorbent with a specific elution solvent, the analytes it holds are separated into different phase. Another procedure is method called clean-up. In this part, the analytes remain in the sample and absorb solid adsorbent matrix types, and different parts of the matrix are removed along with the adsorbent [169–171].

The clean-up procedure has become more widely known and widespread under the title QuEChERS. Various harmful effects of pesticides have led researchers to control their residues in foods [172,173]. However, the fact that pesticides are highly resistant to physical, biological and chemical degradation and matrices of the foods in which their residues are found, are quite crowded and have made the determination of various pesticide types difficult [174]. To eliminate these difficulties, Anastassiades and his team developed a new technique in 2003 and attributed the initials of the advantageous features of this technique to the name of the technique (Quick, Easy, Cheap, Effective, Rugged, and Safe (QuEChERS)). The method they developed was based on the application of liquid-liquid separation and subsequent dispersive solid phase extraction. 10 g of sample was sufficient for determination and utilize easily available chemicals such as acetonitrile, $MgSO_4$ and NaCl [167].

Like other analytical techniques, DSPE takes advantage of magnetic materials. In this context, more advanced DSPE procedures have been introduced [175]. The principle of the magnetic dispersive solid phase (MDSPE) procedure, which was first applied by Seferikova and her team in 1999 [176], and is hinged on the magnetic material distributed in the sample medium retaining the target analyte. The magnetic adsorbent, which is separated from the matrix by an external magnetic effect without the need of centrifugation step, is eluted with a specific solvent and the target analytes are extracted. The method attracts attention with its less energy consumption, environmental friendliness and simplicity [177,178].

An important type of dispersive solid phase extraction is undoubtedly dispersive micro solid phase extraction (DμSPE). A new trend emerged after Anastas introduced the 12 principles of analytical chemistry in 1998 [179]. Within the scope of this trend, which continues until today, the main point taken by researchers is green methodology. The DμSPE is the minimized version of SPE. The very small sizes and quantities of adsorbents used in this technique, add efficiency and selectivity to the method. In addition, when these adsorbents are dispersed in the sample, extraction can occur more simply and quickly [175]. Another advantage of the method being its environmentally friendly feature that meets the green trend. Solid adsorbent selection is a very important parameter in DμSPE. When choosing the adsorbent, the difference in polarity of the sample and the adsorbent should also be taken into consideration. In addition, the adsorbent present in small sizes in the sample tends to clump. For this reason, an externally applied energy source is required to prevent agglomeration and ensure a uniform adsorbent distribution. For this purpose, ultrasonic sound waves or vortex method are frequently used. In this way, uniform distribution of the solid adsorbent is ensured and the contact of the adsorbent with the target analytes is increased. Therefore, these external processes also help to achieve effective mass transfer [180–182].

In dispersive solid phase extraction, different types of solid sorbents such as molecularly imprinted polymers [183,184], carbonaceous sorbents [185,186], metal organic frameworks [187,188], layered double hydroxides [189,190], silica-based sorbents [191,192] can be widely used.

4.1 Applications of dispersive solid phase extraction (DLPE)

Many studies have been carried out on the application of DSPE and widely used in different field. Saylan et al. extracted copper ions from tap water samples before their determination by flame atomic absorption spectrophotometry (FAAS). For this purpose, a solid sorbent based on $Ni(OH)_2$ nanoflowers was developed and used in the dispersive solid phase extraction of copper ions. After optimizing the experimental parameters, the developed method was successfully applied under optimum conditions; Low detection limits and satisfactory percent recovery values were obtained. In a study aimed at the separation of trace perfluorinated compounds from environmental water and soil samples before their determination, a new magnetic, fluorinated carbon-based dispersive solid phase extraction sorbent with high pore volume and surface area was developed [193]. When analytes separated by dispersive solid phase extraction were determined by UHPLC-MS, wide linear range, low relative standard deviation and low detection limit were obtained. Additionally, the synthesized sorbent offered excellent adsorption capacity for perfluorinated compounds [194]. In a study aiming to control diseases by determining two biomarkers (8-hidroksi-2′-deoksiguanozin, 8-hidroksi-2′-deoksiadenozin), which are decisive in the early diagnosis of diseases, from saliva samples. Biomarkers were determined by LC-MS/MS after dispersive solid phase extraction and was used in the clean-up phase. A commercial polymer adsorbent was used as the solid phase. Experimental parameters affecting the efficiency of extraction were optimized with the Box-Behnken design and the optimum method was applied. As a result of the study, good sensitivity, low detection limits and good linearity were obtained [195]. In a different study non-steroidal anti-inflammatory drugs were determined from urine and water samples. For this purpose, polymer composite (Fe_3O_4@IL-HCP) containing magnetic ionic liquid was used as the extraction phase in dispersive solid phase extraction. Target analytes extracted from water and urine samples were determined by HPLC. As a result of the study, satisfactory recoveries were obtained with a wide linear range and low relative standard deviation [196].

5. Micellar-assisted extraction or microvawe assisted extraction

Micellar-assisted extraction (MAE) is an important technique that dates back to the 1970s and is frequently used by many researchers even today. This technique derives its main advantages from micelles, which is also included in its name. Micelles are different structures formed by the solvents used. To explain this extraction technique, it would be useful to first talk about micelles and the types of solvents that participate in them. Surfactants (surface active solvents) are the solvents used in this technique that form structures called micelles. Micelles are formed when surfactants are added to aqueous solution in amounts above the critical micelle concentration. This concentration is necessary for the self-assembly of surfactants [197]. Surfactants forming micelles consist of two groups; one of them is the head part, which is hydrophilic, and the other is the tail part, which is hydrophobic [198]. Surfactants are classified as anionic, cationic, zwitterionic (both anionic and cationic parts in the same molecule) and non-ionic surfactants according to the functional group their head parts contain [3]. While surfactants form micelles in water and reverse micelles in organic solvents, they can form cylindrical clusters and some hollow spheres depending on the increase in surfactant concentration [199]. The most commonly used surfactants are Triton-X 114, Triton-X 100, PONPE 7.5 (polyethyleneglycol mono-p-nonylphenylether), SDS (sodium dodecyl sulfate) and CTAB (cetyltrimethylammonium bromide). In the implementation of the micelle-assisted extraction process, in the first stage, the target analytes are bound to the micelles formed in the medium. Then, environmental factors such as temperature and pressure are adjusted and clouding occurs. At this moment, blurriness occurs. In fact, the surfactant phase that retains the target analyte is separated from the specimen matrix, but centrifugation is required to achieve clear separation. After the surfactant phase is separated, it is processed for subsequent analytical procedure [200].

Micelle-assisted extraction is an easy and inexpensive technique and is also very advantageous because non-toxic solvents are used. It is a very useful technique that promises a good level of separation and enrichment of many analyte groups, as the solvents used form both hydrophilic and hydrophobic aggregates [201]. With the development of techniques that utilize the energy of ultrasonic sound waves or the vortex effect, analyte-micelle interaction can occur more effectively and quickly. In this way, extraction efficiency will increase and the time and amount of solvent required for extraction may decrease [3]. A different combination that has received great attention recently, is with magnetic materials. Solvents used in this technique can be doped with magnetic materials. Therefore, the extraction phase can be easily collected with an externally applied magnetic field. This means saving energy and time [202,203]. Thanks to its wide range of advantages, MAE is utilized to separate and enrich various metals, organic compounds and biomolecules from different sample matrix.

5.1 Applications of micelle-assisted extractions

A new separation procedure was developed by utilizing the micelle formation of surfactants and was used to separate cadmium ions before flame atomic absorption spectrometry. Here, a mixture of Triton-X 114 and a nonionic IL (1-hexadecyl-3-methylimidazolium chloride) was used as the extraction solvent. When the method was applied to cadmium ions under optimum conditions, low relative standard deviation, low detection limits and a linear calibration range were obtained. It has also been noted by researchers that this method gave better results compared to methods applied without surfactant [204]. Another study investigated the use of trehalose lipid, a biosurfactant, for the extraction of antioxidant compounds from herbal tea. In the study, a method based on ultrasound-assisted micelle extraction was developed before ultra-high performance liquid chromatography (UHPLC) [205]. Experimental parameters were optimized with Box-Behnken design and the optimum method was applied. The new method developed was successfully applied to herbal tea leaves and high extraction recoveries were obtained. In a study aiming at the extraction of flavonoids from fruit juices and vegetable samples before their determination by HPLC-UV, dispersive magnetic solid phase extraction was used based on mixed micelle formation. As the solid phase in the extraction, the synthesized Fe_3O_4/SiO_2 nanoparticles were coated with cetyltrimethyl ammonium bromide, forming mixed micelles. Experimental parameters were optimized by the central composite design method and the optimum method was applied to the samples. As a result, satisfactory analytical parameters were obtained [206]. In a study examining the effectiveness of surface-active compounds Triton-X 100 and acetone-water mixture was used in the separation of phenol and antioxidants via micelle-assisted extraction [207]. In the study, dandelion leaves and flowers were chosen as the sample matrix. It was stated that when the method was combined with UHPLC and the analytes were determined, the relevant surfactants can be used effectively for the purpose of the study and that they are very advantageous, especially compared to classical organic solvents, in terms of their non-flammability and use in low amounts.

6. Supercritical fluid extraction (SFE)

The extraction of mixture or compounds from the different samples such as, food, environmental, biological, plant samples and other solid mediums may be performed through used different extraction processes. Conventional methods like Soxhlet extraction, steam distillation, hydro-distillation, persistent method and solvent extraction suffer from many limitations like being too much time-consuming, using extreme organic solvents, consuming of volatile compounds, thermal decomposition of compounds because of high working temperature, occurrence of toxic solvent deposits in extract, and low income and extraction efficacy [208–211]. These disadvantages have generated efforts to progress green solvents [212]. Among these advanced methods, SFE is considered as one of greatest methods for improving chemical compounds from different samples where supercritical fluid (SCF) is utilized as a solvent in the place of organic solvents. Supercritical fluid is generally defined as a fluid with pressure and temperature above a critical point where there are no distinct liquid and gas phases. The supercritical portion is just the liquid state, which has properties similar to liquids and gases (see Fig. 7.5).

The SFE is an analytical method introduced in 1980s and was established as best candidate for observing good analytical methods to prepare various samples, such as it is rapid, economical, simple and does not include the use of toxic solvents [213]. The SFE not only decreases the uses of hazardous solvents, also likely to integrate it into automatic analysis schemes. It is proper for the preparation of heat-sensitive materials for use as high-purity extracts [214]. The SFE has various advantages over common extraction processes due to usages of supercritical solvents with dissimilar physicochemical properties like viscosity, density, dielectric constant and diffusion coefficient. The supercritical fluids increase diffusion features in comparison to liquids and is capable to definitely extent in solid materials, therefore facilities more fast extraction. The most interesting feature of supercritical fluids is that they allow adjustment of density by changing the temperature and/or pressure of the fluid. Since density affects dissolubility, the effectiveness of the solvent can be adjusted by varying the extraction pressure. In SFE, usage of solvents which are normally considered safe and efficacy of separation procedure to growth the production and reduce time of extraction, and probability of direct link to chromatographic analysis procedure such as supercritical fluid chromatography or gas chromatography (GC) [215]. There are many applications reported for the uses of Supercritical solvents (Table 7.5).

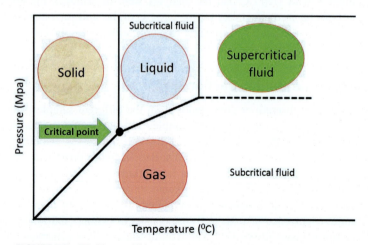

FIGURE 7.5 Working conditions of supercritical fluid in extraction technique.

TABLE 7.5 Application of Supercritical solvents in different microextraction techniques.

Method	Matrix	Analyte	LOD μg/L	Detection technique	References
SFE	Soil and sediment	PAHs	3.9–31.3	GC-MS	[216]
SPE	Ginger	Carbamate Residues	7.01	HPLC	[217]
PTT-SSFE-SFC	Cosmetics	Prohibited substances	0.001–3.09	MS/MS	[218]
SFE-SFC	Uncaria rhynchophylla	Indole alkaloids	0.17	MS/MS	[219]
HS-SPME	Gurum seed oil	Volatile flavor compounds	0.23	GC-MS	[220]
SFE	Organic honey	Pesticides pollutants	0.005–0.009	HPLC-DAD	[221]
SFE-SSME	Apple peels	PAHs	0.34–1.27	HPLC-FLD	[222]
SFE	Wine-making by-products	Polyphenols and vitamins	0.02–1.23	GC–MS	[223]
SFE	Microalgae	Bioactive compounds	0.05	HPLC	[224]
SFE	Capsicum annuum industrial waste	Oleoresin	0.12	HPLC	[225]
SFE	Vegetable waste	Carotenoids	0.73	HPLC	[226]
SFE-SFC	Garlic	Phenolic compounds	0.6–12	MS/MS	[227]
SFE-UA-DLLME	Soil samples	Alkylphenols	10.0	GC–MS	[228]

HS-SPME, headspace solid phase microextraction; *PAHs,* Polycyclic aromatic hydrocarbons; *PTT-SSFE-SFC,* phase transition trapping-selective supercritical fluid extraction-supercritical fluid chromatography; *SFE,* Supercritical fluid extraction; *SFE-SFC,* supercritical fluid extraction-supercritical fluid chromatography; *SFE-SSME,* Supercritical fluid extraction followed supramolecular solvent microextraction; *SFE-UA-DLME,* Supercritical fluid extraction combined with ultrasound-assisted dispersive liquid liquid microextraction; *SPE,* Solid phase extraction.

6.1 Properties of supercritical fluid

The SCF is a substance at pressure and temperature above their consistent critical values [229]. The reputation of SCFs' use is attributed to its solvation capacity, proper density and transport characteristics [230]. Nowadays, the SFE that utilized SCF solvent has develop a sustainable substitute to traditional extraction process in several industries like food, cosmetic, pharmaceuticals and agriculture. The used SCF must have good selectivity, efficiency and diffusivity [231,232]. Currently several research works are reported on uses of SCFs as substitute solvents for extraction. The SFE is normally done in packed bed (PB) columns because of low cost and high specific surface. In this column, extraction is achieved through constantly connection with solid feedstock (substrate) with the supercritical solvent to separate the desired compound or solute. To prepare like extraction process, it is essential to have a detailed knowledge about the fundamental process of the packed bed column. The hydrodynamic (such as drag coefficient, velocity profile, pressure drop and flow pattern), heat transfer (such as heat transfer coefficient, temperature distribution, and heat rate) and mass transfer (such as contour, mass transfer coefficient and concentration spreading), and are many of vital procedures fundamental to completely understand the fundamental sciences of SFE procedure in the packed bed column. Though, difficulty of SFE procedure, it is very difficult to completely know the passage phenomena in SFE packed bed column and depends on experimental works. Obtaining details of temperature, velocity, concentration and pressure in packed bed column by experimental work may be challenging [233]. The SCF is substance at a pressure and temperature is beyond its critical point, where separate liquid and gas phases did not occur. The SCFs may diffuse by solid matrices alike to a gas and then dissolve constituents like a normal liquid. Meanwhile many SFEs are led nearby to fluid's critical point, while small variations in temperature or pressure result in huge variations of density of the fluid's [234]. This distinctive property indicates to density-depended selectivity of SFE. Through the past few decades there are considerable explores on extraction of seed oil from the various biological sources have been reported [235]. The isolation, extraction and fractionation of high value-constituents from seeds are commonly adjusted to the produce of definite bioactive compounds, thus avoiding its deterioration and damage of physiological functionality throughout processing. Among the utilized methods, SFE is the environmental and consumer-friendly substitute for extraction of bioactive mixtures [234].

6.2 Solvent used in SFE

The carbon dioxide (CO_2) is frequently used as a solvent in SFE because it is particularly non-toxic. It is also non-sticky, non-corrosive and labor less to procure in large amounts. It is low cost and has good purity and also it may change the features of solvency through varying the temperature or pressure of the liquid. The organic carbon dioxide is usually considered safe through the EFSA (European Food Safety Authority) and by FDA (Food and Drug Administration). In comparison to traditional process, the chief benefits of SFE are low energy prices, equable temperature and making of great purity extracts [236]. Nevertheless, these features are compromised by various factors, such as (1) high pressure related with them, that needs suitable protection equipment, comprising safety systems, (2) higher costs as compared to traditional methods. Another drawback of SFE procedures based on CO_2 has less polarity of extraction solvent, which may come through by polar changers of supercritical fluid and rise in solvation strength of investigated analyte. High percentage of SFE analytical methods are carried out with CO_2. It is comparatively lower temperature (32°C) and lower critical pressure (74 bar). The supercritical CO_2 is preferred due to its higher diffusivity and simply adaptable solvent resistance. Additional advantage is that at ambient temperature and pressure CO_2 is gaseous state, which creates recovery exploration easy and delivers solvent-free analysis. Furthermore, the capability of SFE to employ CO_2 at lower temperature by nonoxidizing media, and was able to extract easily oxidizable and thermally unstable mixtures [213]. Lately some other solvents are used for extraction of seeds oils instead of CO_2 such as dimethyl ether (DME) by Ref. [237]. The DME is non-toxic, easy to transport, and cheap than other similar fuels. The DME has capability to extract constituents to examine and to isolate bioactive composites from green algae [238]. Subratti et al. effectively extracted oil 31% from hemp seeds by using DME within 50 min, while yields achieved by using n-hexane and petroleum ether were 26% and 23% respectively [237]. Another current study indicated that, the rubber seed oil was extracted by supercritical DME and 41% yield was obtained [239].

6.3 Instrumentation used in SFE

Based on a simple foundation, the SFE method has similar contents to liquid chromatography. The pump (reciprocator or syringe) which cools the head to confirm that liquid pumping is useful to transport the extract in high-pressure extracting chamber comprising sample of interest. The battery is kept at temperature greater than critical temperature of liquid through placing battery in the oven. The extraction units of different geometries and sizes are applied, commonly in-built stainless steel capable to survive the high utilized pressure. The cell geometry may effect on extraction; however, it is less significant if cells are fully packed. The SFE may be executed in dynamic or static mode. Pressure in system is sustained through a fixed or adjustable flow limiter. The fixed limiter tube with a small diameter decreases the flow of fluid. The corrugated stainless-steel tube or a piece of fused silica is used. The problems occur with fixed flow limiters, comprising blockage of matrix or blocked analyte and unequal flow. The restriction block reduces the efficiency at temperature different from that of the flow restrictor. The flexible restrictor is a controller that is used to turn on or off the regulator fluid flow. In addition to being cheap, it controls the flow and pressure of the supercritical fluid. The different collection process is to directly deposit the sample without the need for interphase collection by directly combining the SFE to the system [213].

6.4 Applications of supercritical fluid

The SFE is used for extraction of environmental samples like water, soil, biological samples and plants samples. The SPE is employed to identify removal of contamination from different types of samples. Meskar et al. reported the effects of fluid pressure, temperature, duration of time and also impact of extraction mode (dynamic and static) on extraction of analyte. The SFE procedure was widely utilized to extract various metal ions from many substrates. The metal ions extraction by CO_2-based SFE may be done by changing the charged metal in neutral metal solvate by using appropriate complexing agent [240]. Wicker et al. used online supercritical SFE liquid mass spectrometry to identify 16 polycyclic aromatic hydrocarbons (PAHs) in soil sample. The link of SFE to SFC delivers least sample preparation, negligible sample reduced overall analysis time. This method assesses 10−1500 ng/g of concentration of PAH in soil sample in deposits of standard reference material (CRM), sand or clay [241]. The SFE is well utilized for analysis of food and vegetables samples. The essential oils extract from represents a slight portion of herbs ad plants, composed of lipophilic constituents, which comprise volatile aromatic compounds specific oxygenated derivatives like ethers, ketones, alcohols, acids, esters, aldehydes and phenols. The necessary oils are extracted from bark, leaves, branches, roots, flowers, rhizomes and fruits also extract from herbaceous seeds. Bagheri et al. presented an SFE to extract the three pyrethroid residues, fenvalerate, fenpropathrin, and cialotrine from vegetable and fruit samples. They produced a novel magnetic adsorbent modified with ionic liquid (IL). The pyrethroid residues were analyzed by high-performance liquid chromatography-ultraviolet for ultra trace analysis [242]. Bayrak et al. reported the extraction of colchicine from colchicum speciosum L. by SFE method. First seeds were degreased from extraction by 50 mL of petroleum ether nearly achieved 7.92% oil. Seeds were then utilized for routine extraction and SFE. The SFE process extracted 10 g seeds by SFE-CO_2 system. Under CO_2-SFE study situations, the maximum yield was achieved by methanol modifier 0.5 mL/min [243]. Liu et al. reported offline SFE−SFC−MS/MS technology for extraction, analysis and identify the 15 phenolic compounds in the garlic. Phenolic compound was firstly examined through an automatic SFE, and extracted constituents have been then studied through supercritical liquid chromatography-mass spectrometry. Under the optimum conditions, linearity was 0.02−8 μg/g, and correlation coefficient was more than 0.933 [227]. Deniz et al. evaluated the impact of temperature, time duration, solvent ratio, pressure on SCF of phycocyanin (PC) of Spirulina platensis. Afterward the experimental circumstances were chosen as 60°C, 250 bar and also used 10% ethanol as a cosolvent. The dynamic time was 45 min, quantity of PC was 90.74%, and purity was 75.12% [244].

7. Liquid liquid extraction (LLE)

A number of methods are developed for sample pre-treatment for the extraction of different contamination in different samples. Currently chief pre-treatment methods are applied to analyze different samples include liquid liquid extraction (LLE), supercritical fluid extraction, simultaneous distillation extraction, solvent-assisted flavor evaporation, solid phase extraction and stirring rod adsorption extraction [227]. As compared with other preconcentration and extraction approaches, LLE is very fit for liquid sample investigation and is very easy to operate. Furthermore, LLE overcomes several limitations of other extraction process for samples analysis. Currently LLE method is widely employed to recognize the characteristics of different compounds in different samples (such as agricultural products, beverages and many other samples).

7.1 Principle of LLE

The implementation of LLE dates back to the early 19th century and it was Bucholz who realized it. LLE can be briefly defined as the process of taking a component dissolved in a liquid phase into a different phase. One of the liquid phases may be an aqueous solution (e.g., buffer medium) and the other may be a nonpolar phase [245]. The LLE is a liquid phase sample preparation technique generally applied to environmental, biological and different food samples. It is a simple method and requires easily accessible equipment [246]. In the general and classical method, a separating funnel is used. The phases in the separatory funnel are shaken and the component passes from one phase to the other. As a result of the process, the two phases are separated from each other [247]. Ismaili et al. conducted a study aiming to examine the amount of oleuropein in the fruit and leaf parts of olives and used LLE as the extraction procedure. LLE is combined with ultrasound and supplemented with salt medium. In this method, there was no need for mixing and the use of large amounts of organic solvents [248].

7.2 The history of LLE

Although the classical LLE procedure was previously a frequently preferred procedure, the interest of researchers has shifted to different techniques, especially since it requires the use of high volumes of toxic organic solvents.

Researchers are now motivated to use more minimal methods, use less solvent, and process fewer samples. The main purposes are to improve fast, globally friendly and effective separation approaches, and to increase the sensitivity and selectivity of extraction procedures through decreasing volume percentage of the relevant solvent to the water phase [14]. The liquid-liquid microextraction (LLME), was publicized in 1996. It is micro level pre-treatment process based on LLE. As compared with traditional LLE process, the LLME procedures have main advantages utilized decreased specimen and solvent consumption, well extraction efficacy and higher enrichment [249]. The LLME was widely utilized for extraction of pharmaceutical, organic, and inorganic mixtures from different biological matrices and environmental samples. The enhancement of LLME, a new LLME procedures were reported, like solidified floating organic drop DLLME, vortex-assisted DLLME, dispersive liquid liquid microextraction (DLLME) [250], air-assisted DLLME [251,252] and effervescent tablet consisted DLLME.

7.3 Solvents used in LLE

The ionic liquids (ILs) were first used into LLE in 2005 to support organic solvent environment. The purpose was to improve the separation and enrichment of metals or other substances. Ionic liquids are organic or inorganic salts containing various anions and cations. ILs frequently used in LLE are notable for their hydrophobic character and thermal stability with very low vapor pressure. The ILs are considered lower polluting friendly than classical solvents [253]. Lately, ionic liquids with magnetic properties, that combine the benefits and magnetic features of ionic liquids, were extensively utilized in LLE to separate heavy metals, drugs and pesticides [254]. The DESs, as alternate solvents, are more low-polluting than ILs and have concerned wide research consideration in the area of chemistry [255]. Mostly DES are binary or ternary mixtures achieved by hydrogen bonding to create eutectic mixtures. Usually, components have a hydrogen bond donor and hydrogen bond acceptor which form solvents with low melting points, normally, higher densities as compared to those of single components. The mixtures of these components, commonly at a 1:2 or 1:1 M ratio, create liquids at room temperature which show higher thermal stability, high viscosity and low volatility. Furthermore, these solvents are more friendly and have other interesting features, comprising simple preparation procedures and chemical arrangement tunability allowing for organized design [256]. Latest studies have revealed the prospective of DES in various applications and different areas. For example, Liu et al. applied an alcohol amino DES to get the carbon dioxide (CO_2) [257]. Cherkashina et al. utilized a hydrophobic-DES to extract tetracycline from milk [258]. Khosrowshahi et al. reported the extraction of antiepileptic drugs from milk and used a DES solvent [259]. Switchable solvents have been firstly utilized in the LLE in 2015 [260]. The individuality of switchable solvents hinge on its capability to change polarity from polar to nonpolar and contrary. The switchable solvents have possessed significant consideration as an proper extractants because of their time-saving and simplicity in consequence of decreased number of the extraction steps; furthermore, its low hazard and moderate to environment. There are two main forms of switchable solvents: one amine-based and the other fatty acid-based. The amine-based switchable solvents may potentially switch among hydrophilic and hydrophobic forms reversibly and suddenly. Its polarity switches could be activated simply through addition of CO_2 into solution or eliminating it. While amine reacts with CO_2, formed create hydrophilic ammonium carbonate. After CO_2 is detached with N_2, carbonate is reformed. The fatty acid-based switchable solvents could react with sodium hydroxide and change it into hydrophilic sodium salt to form a homogeneous solution. The adding of acids into the homogeneous solutions outcomes in creation of hydrophobic forms of the fatty acids, which are then separated from aqueous solution [121]. There is increasing tendency in literature of joining switchable solvents and LLME. The main advantages of such a method comprise the fast procedure, high enrichment coefficient, low extraction time, simple, good extraction effect, low cost and lower detection limit [261,262].

In current years, ionic liquids, DES and switchable solvents were developed as novel green extractants and are usually joined with DLLME to obtain fast green and effective extraction. At present-day, LLE method is mostly combined with acetone, acetonitrile, methanol and many other toxic solvents for food samples analysis. Though, there are few published literature to apply the green extractants joined with LLE for samples analysis. The future of LLE process focused could be combined with emergent extraction solvent to improve the selectivity and sensitivity of the LLE method. In adding, further studies may further discover the usage of LLE, extraction column technology, counter-current extraction and some other extraction technologies [250].

8. Liquid phase microextraction (LPME)

In early 1990s, presented the SPME method to address the requirement for fast and preparation of solvent-free sample that provides instantaneous preconcentration and separation for volatile and non-volatile parts in composite samples. The fiber-coated SPME is traditional method, and is made of thin fused-silica fiber coated with an adsorbent layer as extraction phase. In SPME procedure, the extracting phase is uncovered to sample matrix for definite time and after successful equilibrium, the adsorbed mixtures are analyzed through placing fiber into injection port of the gas chromatography (GC). Normally, the time required to reach equilibrium depending on physicochemical characteristics of the analytes. The equilibration times may be decreased by several strategies, comprising shaking the sample, exploiting the sample-headspace interface, warming the sample or cooling the fiber [263]. The overall indications of SPME techniques and modern developments were provided in through a wide array of review articles. Carabajal et al. presents applications of LPME for extraction of analytes in complex samples. This procedure was introduced to simplify the extraction approaches, and improve their selectivity. Sample cleaning and efficacy, is permitting the extraction of an extensive variety of analytes [264]. The development of the techniques, concepts, and devices related with SPME has been discussed in a response to growth of understanding of the important principles behind DPME technique. Fernández-Amado et al. observed the strengths and flaws of DPME technique during 2009–15 to recognize the research gaps which may be discussed in the future. In-tube SPME, strong point is in coupling liquid chromatography (LC), and techniques like electrochemically-controlled in tube SPME and magnetic SPME to show potentials in the development of capillaries and coatings. The in-tube SPME research were performed for biomedical, chemical and environmental analyses [265]. Queiroz et al. discussed the fundamental theory and advances related to in-tube SPME schemes, experimental parameters, new selective capillary coatings, extraction configurations, and highlighted its main applications for food, environmental and clinical analyses by using SPME technique [266].

8.1 Specific extraction media in SPME

The selection of suitable adsorbent is a critical factor to achieve high recovery and good precision in SPME process. It is well-known that efficacy of microextraction process is dependent on the extraction phase [267]. An extensive range of marketable materials (polydimethylsiloxane, carboxen, Tenax TA) are utilized as adsorbent in SPME methods [268]. In order to increase the efficiency of extraction and to reduce the limitations, such as limited vapor capability of adsorbents, new materials synthesis are presently under progress [269]. Alili et al. discussed the airgel-based adsorbents utilized in microextraction techniques, like SPME, in-tube SPME, and needle trap method. The airgel adsorbents have distinctive properties and applications such as carbon airgel, hybrid airgel, silica airgel and graphene airgel in microextraction methods and were studied with a discussion concerning their future scenarios [270]. The SPME are widely utilized in analytical chemistry for greater extraction efficiency in an appropriate processes. The different constituents give different extraction outcomes. Among current resources, carbon-based materials are attracting considerations from scientists because of their excellent chemical and physical properties as well as their adjustable surfaces, which might enhance the adsorption impacts of SPME [271]. The applications include various types of carbon-based material coatings on fibers. Furthermore, future research guidelines for carbon material compounds were described. The synthesis and applications are focused on the carbon fiber coating, porous carbon coating, carbon nanotube composite, graphene coating, coating and fullerene coating in SPME [263]. Lashgari, & Yamini discussed five well-established collections of lab-made SPME coatings materials comprising ionic liquids (ILs), metal organic frameworks (MOFs), conductive polymers (CPs), molecularly imprinted polymers (MIPs), layered double hydroxides (LDHs), and their derivations were evaluated, with their unique structures, and well-proven important extraction presentations as well as disadvantages were also discussed. The molecularly imprinted polymers (MIPs), ionic liquids (ILs), polymeric ILs, carbon nanotubes, and metal organic frameworks (MOFs) and graphene were deliberated, based on current trends in the SPME-fiber coatings. Furthermore, the usage of conventional constituents in SPME coatings with comparatively new ideas were mentioned and described for the potentials of developing SPME fibers by these materials [272].

8.2 Solvent used in LPME

The selection of suitable extractant phase is important in LPME. The final choice must be based on comparison of extraction efficiency, selectivity, incidence of drop loss, rate of drop dissolution (particularly for extended extraction times and faster stirring rates) or evaporation and toxicity level. Therefore, the information of the physical properties of both extractant phase and analyte is advisable. Therefore, extractant solvent must be friendly with analytical technique eventually employed. The ionic liquids or Water-immiscible organic solvents are mostly used for two-phase LPME methods, while low-volatility extractant phases (including ionic liquids, organic solvents or even water) might be utilized in three-phase LPME methods such as HF-LLLME, HS-SDME or LLLME. The density of extractant solvent plays a main role when methods such as DSDME, DLLME or SFOME are applied. Therefore, extractant solvent must have higher density than water in the DLLME method, while it must display lower density than water in SFOME and DSDME. Furthermore, organic solvents show a melting point near to room temperature are essential to carry out extraction of target analytes through SFOME [273].

In liquid phase microextraction (LPME) techniques, the choice of the suitable extraction solvent is important for the efficiency in the extraction process. The usually utilized substances for extraction process are immiscible organic solvents in aqueous phase, like acetonitrile, methanol, carbon tetrachloride, chloroform, trichlorethylene among others. Though, most solvents are detrimental to the humans, animals and environment. Therefore, little toxicity solvents are considered better options and are generally used in various analytical sections [274,275]. In this situation, DESs are very feasible substitute solvent in the LPME processes [276,277]. Many studies comprising LPME apply DESs as a receptor phase. The various combinations of HBD with HBA are studied, as several interactions may occur among the DES and extracted constituent. Some examples of these interactions are such as van der Waals dispersion, ion-dipole electrostatic, hydrogen bond, π-cation, hydrophobic, dipole-dipole and π-π [278,279].

9. Conclusion

The increasing demand for analyzing compounds at lower concentrations in complex matrices need a primary step of analytes isolation and enrichment by the use of an extraction technique with high sensitivity and low LOQ level with environmentally friendly solvent and reduction of the extracting solvent. The "greenness" is very significant as well selectivity to avoid consuming harmful organic solvents in maintainable extraction methods. These solvents may produce hazardous, and toxic wastes while utilizing large resources, quantity and volume. The developing of new green solvents are key objects in Green Chemistry to reduce the strength of anthropogenic activities associated to analytical laboratories. The lot of eco-friendly solvents have been utilized as extractant phases. These solvents are more eco-friendly, low cost, provide shorter extraction times, simplicity and good selectivity. There are various solvents developed and introduced in analytical processes, and used in the sample preparation procedures. These solvents are cleaner and greener substitutes to conventional molecular solvents. These solvents include ionic liquids, deep eutectic solvents, superheated water, surfactants, water, supramolecular solvents, switchable solvents, bio-derived solvents, methanol, ethanol, supercritical fluids etc and are utilized in the extraction processes.

10. Future recommendations

It seems that because of wide range of polarity of analytes, no universal solvent occurs which can be utilized in all the extraction practices. To substitute the conventional solvents such as (chloroform, hexane, Benzene, toluene etc.) with an appropriate green solvent then we may succeed a significant decrease the cost of analysis and also reduce the negative effects on people, plants and for environment. The water is most important and freely available good solvent, must be considered as a solvent. The water is very polar and did not seem a suitable extracting solvent for analysis of organic samples; however water can be used as a solvent in extraction process in future to avoid the negative impacts of solvent for human and environment.

References

[1] C. Turner, M. Waldebäck, Principles of pressurized fluid extraction and environmental, food and agricultural applications, in: *Separation*, Extraction and Concentration Processes in the Food, Beverage and Nutraceutical Industries, 2013, pp. 39–70.

[2] Z. Li, K.H. Smith, G.W. Stevens, The use of environmentally sustainable bio-derived solvents in solvent extraction applications—a review, Chinese Journal of Chemical Engineering 24 (2) (2016) 215–220.

[3] J. Płotka-Wasylka, M. Rutkowska, K. Owczarek, M. Tobiszewski, J. Namieśnik, Extraction with environmentally friendly solvents, TrAC, Trends in Analytical Chemistry 91 (2017) 12–25.

[4] A. Ballesteros-Gómez, M.D. Sicilia, S. Rubio, Supramolecular solvents in the extraction of organic compounds. A review, Analytica Chimica Acta 677 (2) (2010) 108–130.

[5] M. Rezaee, Y. Assadi, M.R.M. Hosseini, E. Aghaee, F. Ahmadi, S. Berijani, Determination of organic compounds in water using dispersive liquid–liquid microextraction, Journal of Chromatography 1116 (1–2) (2006) 1–9.

[6] A. Quigley, W. Cummins, D. Connolly, Dispersive liquid-liquid microextraction in the analysis of milk and dairy products: a review, Journal of Chemistry 2016 (2016).

[7] T.S. Rao, M. Sridevi, C.G. Naidu, B. Nagaraju, Ionic liquid-based vortex-assisted DLLME followed by RP-LC-PDA method for bioassay of daclatasvir in rat serum: application to pharmacokinetics, Journal of Analytical Science and Technology 10 (1) (2019) 1–11.

[8] Y. Li, W. Zhang, R.G. Wang, P.L. Wang, X.O. Su, Development of an efficient and sensitive dispersive liquid–liquid microextraction technique for extraction and pre-concentration of 10 β2-agonists in animal urine, PLoS One 10 (9) (2015) e0137194.

[9] A. Zgoła-Grześkowiak, T. Grześkowiak, Dispersive liquid-liquid microextraction, TrAC, Trends in Analytical Chemistry 30 (9) (2011) 1382–1399.

[10] M. Saraji, M.K. Boroujeni, Recent developments in dispersive liquid–liquid microextraction, Analytical and Bioanalytical Chemistry 406 (2014) 2027–2066.

[11] O. Zuloaga, M. Olivares, P. Navarro, A. Vallejo, A. Prieto, Dispersive liquid–liquid microextraction: trends in the analysis of biological samples, Bioanalysis 7 (17) (2015) 2211–2225.

[12] A.K. Geletu, Recent developments and appl, Journal of Current Research 9 (08) (2017) 55524–55531.

[13] N. Campillo, P. Vinas, J. Šandrejová, V. Andruch, Ten years of dispersive liquid–liquid microextraction and derived techniques, Applied Spectroscopy Reviews 52 (4) (2017) 267–415.

[14] N. Campillo, K. Gavazov, P. Viñas, I. Hagarova, V. Andruch, Liquid-phase microextraction: update May 2016 to December 2018, Applied Spectroscopy Reviews 55 (4) (2020) 307–326.

[15] V.A. Lemos, J.A. Barreto, L.B. Santos, R.D.S. de Assis, C.G. Novaes, R.J. Cassella, In-syringe dispersive liquid-liquid microextraction, Talanta 238 (2022) 123002.

[16] D. Snigur, E.A. Azooz, O. Zhukovetska, O. Guzenko, W. Mortada, Low-density solvent-based liquid-liquid microextraction for separation of trace concentrations of different analytes, Trends in Analytical Chemistry (2023) 117260.

[17] Y. Çağlar, Dispersive liquid–liquid microextraction for the spectrophotometric determination of Fe^{3+} with a water soluble Cu (II) phthalocyanine compound, Turkish Journal of Analytical Chemistry 5 (1) (2023) 70–76.

[18] W.A. Soomro, M.Y. Khuhawar, T.M. Jahangir, M.F. Lanjwani, I.K. Rind, Determination of copper from environmental samples by solvent microextraction method using AAS. Multivariate modelling and factorial design, International Journal of Environmental Analytical Chemistry (2023) 1–15.

[19] M.I. Leong, C.C. Chang, M.R. Fuh, S.D. Huang, Low toxic dispersive liquid–liquid microextraction using halosolvents for extraction of polycyclic aromatic hydrocarbons in water samples, Journal of Chromatography A 1217 (34) (2010) 5455–5461.

[20] M.A. Farajzadeh, D. Djozan, P. Khorram, Development of a new dispersive liquid–liquid microextraction method in a narrow-bore tube for preconcentration of triazole pesticides from aqueous samples, Analytica Chimica Acta 713 (2012) 70–78.

[21] M.I. Leong, S.D. Huang, Dispersive liquid–liquid microextraction method based on solidification of floating organic drop combined with gas chromatography with

electron-capture or mass spectrometry detection, Journal of Chromatography A 1211 (1−2) (2008) 8−12.
[22] P. Zhang, L. Hu, R. Lu, W. Zhou, H. Gao, Application of ionic liquids for liquid−liquid microextraction, Analytical Methods 5 (20) (2013) 5376−5385.
[23] K. Yavir, K. Konieczna, Ł. Marcinkowski, A. Kloskowski, Ionic liquids in the microextraction techniques: the influence of ILs structure and properties, Trends in Analytical Chemistry 130 (2020) 115994.
[24] L. Qiao, Y. Tao, H. Qin, R. Niu, Multi-magnetic center ionic liquids for dispersive liquid-liquid microextraction coupled with in-situ decomposition based back-extraction for the enrichment of parabens in beverage samples, Journal of Chromatography A 1689 (2023) 463771.
[25] J. Hong, X. Liu, X. Yang, Y. Wang, L. Zhao, Ionic liquid-based dispersive liquid−liquid microextraction followed by magnetic solid-phase extraction for determination of quinolones, Microchimica Acta 189 (2022) 1−10.
[26] Y. Tao, L. Jia, H. Qin, R. Niu, L. Qiao, A new magnetic ionic liquid based salting-out assisted dispersive liquid−liquid microextraction for the determination of parabens in environmental water samples, Analytical Methods 14 (46) (2022) 4775−4783.
[27] Y.M. Pestana, É.M. Sousa, D.L. Lima, L.K. Silva, J.F. Pinheiro, E.R. Sousa, G.S. Silva, Multivariate optimization of dispersive liquid-liquid microextraction using ionic liquid for the analysis of ultraviolet filters in natural waters, Talanta 259 (2023) 124469.
[28] R. Ghasemi, H. Mirzaei, M.R.A. Mogaddam, J. Khandaghi, A. Javadi, Application of magnetic ionic liquid-based air−assisted liquid−liquid microextraction followed by back-extraction optimized with centroid composite design for the extraction of antibiotics from milk samples prior to their determination by HPLC−DAD, Microchemical Journal 181 (2022) 107764.
[29] A. Preethi, R. Vijayalakshmi, A.B. Lakshmi, Dispersive liquid-liquid microextraction of zinc from environmental water samples using ionic liquid, Chemical Engineering Communications 208 (2021) 914−923.
[30] Y. Shu, M. Gao, X. Wang, R. Song, J. Lu, X. Chen, Separation of curcuminoids using ionic liquid based aqueous two-phase system coupled with in situ dispersive liquid−liquid microextraction, Talanta 149 (2016) 6−12.
[31] C. Fan, Y. Liang, H. Dong, G. Ding, W. Zhang, G. Tang, Y. Cao, In-situ ionic liquid dispersive liquid-liquid microextraction using a new anion-exchange reagent combined Fe$_3$O$_4$ magnetic nanoparticles for determination of pyrethroid pesticides in water samples, Analytica Chimica Acta 975 (2017) 20−29.
[32] L. Li, M. Huang, J. Shao, B. Lin, Q. Shen, Rapid determination of alkaloids in *Macleaya cordata* using ionic liquid extraction followed by multiple reaction monitoring UPLC−MS/MS analysis, Journal of Pharmaceutical and Biomedical Analysis 135 (2017) 61−66.
[33] N. Altunay, A. Elik, S. Kaya, A simple and quick ionic liquid-based ultrasonic-assisted microextraction for determination of melamine residues in dairy products: theoretical and experimental approaches, Food Chemistry 326 (2020) 126988.
[34] M. Tuzen, O.Z. Pekiner, Ultrasound-assisted ionic liquid dispersive liquid−liquid microextraction combined with graphite furnace atomic absorption spectrometric for selenium speciation in foods and beverages, Food Chemistry 188 (2015) 619−624.
[35] A. Beiraghi, M. Shokri, S. Seidi, B.M. Godajdar, Magnetomotive room temperature dicationic ionic liquid: a new concept toward centrifuge-less dispersive liquid−liquid microextraction, Journal of Chromatography A 1376 (2015) 1−8.
[36] L. Hu, P. Zhang, W. Shan, X. Wang, S. Li, W. Zhou, H. Gao, In situ metathesis reaction combined with liquid-phase microextraction based on the solidification of sedimentary

ionic liquids for the determination of pyrethroid insecticides in water samples, Talanta 144 (2015) 98−104.
[37] N. Altunay, A. Elik, R. Gürkan, Vortex assisted-ionic liquid based dispersive liquid liquid microextraction of low levels of nickel and cobalt in chocolate-based samples and their determination by FAAS, Microchemical Journal 147 (2019) 277−285.
[38] C. Fan, N. Li, X. Cao, Determination of chlorophenols in honey samples using in-situ ionic liquid-dispersive liquid−liquid microextraction as a pretreatment method followed by high-performance liquid chromatography, Food Chemistry 174 (2015) 446−451.
[39] Q. Yin, Y. Zhu, Y. Yang, Dispersive liquid−liquid microextraction followed by magnetic solid-phase extraction for determination of four parabens in beverage samples by ultra-performance liquid chromatography tandem mass spectrometry, Food Analytical Methods 11 (2018) 797−807.
[40] L. Zhang, R. Yu, M. Zhou, C. Wang, D. Zhang, W. Ren, Y. Shao, Ionic liquid-based cloud point extraction five organophosphorus pesticides in coarse cereals, Food Chemistry 379 (2022) 132161.
[41] X. You, X. Chen, F. Liu, F. Hou, Y. Li, Ionic liquid-based air-assisted liquid−liquid microextraction followed by high performance liquid chromatography for the determination of five fungicides in juice samples, Food Chemistry 239 (2018) 354−359.
[42] X. Wang, P. Chen, L. Cao, G. Xu, S. Yang, Y. Fang, X. Hong, Selenium speciation in rice samples by magnetic ionic liquid-based up-and-down-shaker-assisted dispersive liquid-liquid microextraction coupled to graphite furnace atomic absorption spectrometry, Food Analytical Methods 10 (2017) 1653−1660.
[43] S. Sadeghi, Z. Nasehi, Simultaneous determination of *Brilliant Green* and *Crystal Violet* dyes in fish and water samples with dispersive liquid-liquid micro-extraction using ionic liquid followed by zero crossing first derivative spectrophotometric analysis method, Spectrochimica Acta Part A: Molecular and Biomolecular Spectroscopy 201 (2018) 134−142.
[44] J. Grau, C. Azorín, J.L. Benedé, A. Chisvert, A. Salvador, Use of green alternative solvents in dispersive liquid-liquid microextraction: a review, Journal of Separation Science 45 (1) (2022) 210−222.
[45] M. Moradi, Y. Yamini, N. Feizi, Development and challenges of supramolecular solvents in liquid-based microextraction methods, Trends in Analytical Chemistry 138 (2021) 116231.
[46] V. Jalili, R. Zendehdel, A. Barkhordari, Supramolecular solvent-based microextraction techniques for sampling and preconcentration of heavy metals: a review, Reviews in Analytical Chemistry 40 (1) (2021) 93−107.
[47] H. Musarurwa, N.T. Tavengwa, Emerging green solvents and their applications during pesticide analysis in food and environmental samples, Talanta 223 (2021) 121507.
[48] N. Altunay, A. Elik, A green and efficient vortex-assisted liquid-phase microextraction based on supramolecular solvent for UV−VIS determination of nitrite in processed meat and chicken products, Food Chemistry 332 (2020) 127395.
[49] A.A. Gouda, M.S. Elmasry, H. Hashem, H.M. El-Sayed, Eco-friendly environmental trace analysis of thorium using a new supramolecular solvent-based liquid-liquid microextraction combined with spectrophotometry, Microchemical Journal 142 (2018) 102−107.
[50] Z.A. Alothman, E. Yilmaz, M.A. Habila, B. Alhenaki, M. Soylak, A.B.H. Ahmed, E.A. Alabdullkarem, Development of combined-supramolecular microextraction with ultra-performance liquid chromatography-tandem mass spectrometry procedures for ultra-trace analysis of carbaryl in water, fruits and vegetables, International Journal of Environmental Analytical Chemistry 102 (7) (2022) 1491−1501.

[51] K. Seebunrueng, P. Phosiri, R. Apitanagotinon, S. Srijaranai, A new environment-friendly supramolecular solvent-based liquid phase microextraction coupled to high performance liquid chromatography for simultaneous determination of six phenoxy acid herbicides in water and rice samples, Microchemical Journal 152 (2020) 104418.

[52] L. Adlnasab, M. Ezoddin, M. Shabanian, B. Mahjoob, Development of ferrofluid mediated CLDH@ Fe$_3$O$_4$@ tanic acid-based supramolecular solvent: application in air-assisted dispersive micro solid phase extraction for preconcentration of diazinon and metalaxyl from various fruit juice samples, Microchemical Journal 146 (2019) 1−11.

[53] J. Ali, M. Tuzen, T.G. Kazi, Evaluation of mercury in environmental samples by a supramolecular solvent−based dispersive liquid−liquid microextraction method before analysis by a cold vapor generation technique, Journal of AOAC International 100 (3) (2017) 782−788.

[54] N. Altunay, K.P. Katin, Ultrasonic-assisted supramolecular solvent liquid-liquid microextraction for determination of manganese and zinc at trace levels in vegetables: experimental and theoretical studies, Journal of Molecular Liquids 310 (2020) 113192.

[55] Y. Zong, J. Chen, J. Hou, W. Deng, X. Liao, Y. Xiao, Hexafluoroisopropanol-alkyl carboxylic acid high-density supramolecular solvent based dispersive liquid-liquid microextraction of steroid sex hormones in human urine, Journal of Chromatography A 1580 (2018) 12−21.

[56] X. Li, A. Huang, X. Liao, J. Chen, Y. Xiao, Restricted access supramolecular solvent based magnetic solvent bar liquid-phase microextraction for determination of non-steroidal anti-inflammatory drugs in human serum coupled with high performance liquid chromatography-tandem mass spectrometry, Journal of Chromatography A 1634 (2020) 461700.

[57] S. González-Rubio, D. García-Gómez, A. Ballesteros-Gómez, S. Rubio, A new sample treatment strategy based on simultaneous supramolecular solvent and dispersive solid-phase extraction for the determination of ionophore coccidiostats in all legislated foodstuffs, Food Chemistry 326 (2020) 126987.

[58] A. Najafi, M. Hashemi, Feasibility of liquid phase microextraction based on a new supramolecular solvent for spectrophotometric determination of orthophosphate using response surface methodology optimization, Journal of Molecular Liquids 297 (2020) 111768.

[59] N. Ozkantar, M. Soylak, M. Tuzen, Ultrasonic-assisted supramolecular solvent liquid-liquid microextraction for inorganic chromium speciation in water samples and determination by UV-vis spectrophotometry, Atomic Spectroscopy 41 (1) (2020) 43−50.

[60] M. Khan, E. Yilmaz, M. Soylak, Supramolecular solvent microextraction of uranium at trace levels from water and soil samples, Turkish Journal of Chemistry 41 (1) (2017) 61−69.

[61] H. Kahe, M. Chamsaz, M.A. Zavar, A novel supramolecular aggregated liquid−solid microextraction method for the preconcentration and determination of trace amounts of lead in saline solutions and food samples using electrothermal atomic absorption spectrometry, RSC Advances 6 (54) (2016) 49076−49082.

[62] M. Ezoddin, B. Majidi, K. Abdi, Ultrasound-assisted supramolecular dispersive liquid−liquid microextraction based on solidification of floating organic drops for preconcentration of palladium in water and road dust samples, Journal of Molecular Liquids 209 (2015) 515−519.

[63] J.S. Almeida, T.A. Anunciação, G.C. Brandão, A.F. Dantas, V.A. Lemos, L.S. Teixeira, Ultrasound-assisted single-drop microextraction for the determination of cadmium in

vegetable oils using high-resolution continuum source electrothermal atomic absorption spectrometry, Spectrochimica Acta Part B: Atomic Spectroscopy 107 (2015) 159—163.

[64] M. Khan, M. Soylak, Supramolecular solvent based liquid—liquid microextraction of aluminum from water and hair samples prior to UV-visible spectrophotometric detection, RSC Advances 5 (77) (2015) 62433—62438.

[65] A.A. Gouda, A.M. Alshehri, R. El Sheikh, W.S. Hassan, S.H. Ibrahim, Development of green vortex-assisted supramolecular solvent-based liquid—liquid microextraction for preconcentration of mercury in environmental and biological samples prior to spectrophotometric determination, Microchemical Journal 157 (2020) 105108.

[66] A. Shokrollahi, F. Ebrahimi, Supramolecular-based ultrasonic-assisted dispersion solidification liquid—liquid microextraction of copper and cobalt prior to their flame atomic absorption spectrometry determination, Journal of AOAC International 100 (6) (2017) 1861—1868.

[67] F.J. Ruiz, S. Rubio, D. Pérez-Bendito, Tetrabutylammonium-induced coacervation in vesicular solutions of alkyl carboxylic acids for the extraction of organic compounds, Analytical Chemistry 78 (20) (2006) 7229—7239.

[68] F.J. López-Jiménez, S. Rubio, D. Pérez-Bendito, Supramolecular solvent-based microextraction of Sudan dyes in chilli-containing foodstuffs prior to their liquid chromatography-photodiode array determination, Food Chemistry 121 (3) (2010) 763—769.

[69] M. Moradi, Y. Yamini, Application of vesicular coacervate phase for microextraction based on solidification of floating drop, Journal of Chromatography A 1229 (2012) 30—37.

[70] F.J. Ruiz, S. Rubio, D. Perez-Bendito, Water-induced coacervation of alkyl carboxylic acid reverse micelles: phenomenon description and potential for the extraction of organic compounds, Analytical Chemistry 79 (19) (2007) 7473—7484.

[71] M.J. Dueñas-Mas, A. Ballesteros-Gómez, S. Rubio, Supramolecular solvent-based microextraction of emerging bisphenol A replacements (colour developers) in indoor dust from public environments, Chemosphere 222 (2019) 22—28.

[72] E.M. Hafez, R. El Sheikh, M. Fathallah, A.A. Sayqal, A.A. Gouda, An environment-friendly supramolecular solvent-based liquid—phase microextraction method for determination of aluminum in water and acid digested food samples prior to spectrophotometry, Microchemical Journal 150 (2019) 104100.

[73] E. Yilmaz, M. Soylak, Development a novel supramolecular solvent microextraction procedure for copper in environmental samples and its determination by microsampling flame atomic absorption spectrometry, Talanta 126 (2014) 191—195.

[74] M.M. Maslov, A. Elik, A. Demirbaş, K.P. Katin, N. Altunay, Theoretical and experimental studies aimed at the development of vortex-assisted supramolecular solvent microextraction for determination of nickel in plant samples by FAAS, Microchemical Journal 159 (2020) 105491.

[75] M. Moradi, Y. Yamini, M. Tayyebi, H. Asiabi, Ultrasound-assisted liquid-phase microextraction based on a nanostructured supramolecular solvent, Analytical and Bioanalytical Chemistry 405 (2013) 4235—4243.

[76] M.F. Lanjwani, M. Tuzen, M.Y. Khuhawar, M.R. Afshar Mogaddam, M.A. Farajzadeh, Deep eutectic solvents for extraction and preconcentration of organic and inorganic species in water and food samples: a review, Critical Reviews in Analytical Chemistry (2022) 1—14.

[77] P. Makoś, E. Słupek, J. Geogonek bicki, Hydrophobic deep eutectic solvents in microextraction techniques—A review, Microchemical journal 152 (104384.10) (2020) 1016.

[78] W. Lu, S. Liu, Z. Wu, Recent application of deep eutectic solvents as green solvent in dispersive liquid−liquid microextraction of trace level chemical contaminants in food and water, Critical Reviews in Analytical Chemistry 52 (3) (2022) 504−518.

[79] V. Andruch, A. Kalyniukova, J. Płotka-Wasylka, N. Jatkowska, D. Snigur, S. Zaruba, J. Werner, Application of deep eutectic solvents in analytical sample pretreatment (update 2017−2022). Part A: liquid phase microextraction, Microchemical Journal 189 (2023) 108509.

[80] G. Li, K.H. Row, Utilization of deep eutectic solvents in dispersive liquid-liquid microextraction, Trends in Analytical Chemistry 120 (2019) 115651.

[81] Y. Shi, X. Li, Y. Shang, T. Li, K. Zhang, J. Fan, Effective extraction of fluorescent brightener 52 from foods by in situ formation of hydrophobic deep eutectic solvent, Food Chemistry 311 (2020) 125870.

[82] K. Shahbaz, F.G. Bagh, F.S. Mjalli, I.M. AlNashef, M.A. Hashim, Prediction of refractive index and density of deep eutectic solvents using atomic contributions, Fluid Phase Equilibria 354 (2013) 304−311.

[83] M. Soylak, S. Baran, F. Uzcan, Ultrasound assisted deep eutectic solvent based liquid phase microextraction for the preconcentration and spectrophotometric determination of amaranth (E123) in water and food samples, Instrumentation Science and Technology 50 (2) (2022) 203−218.

[84] A. Elik, N. Altunay, Chemometric approach for the spectrophotometric determination of chloramphenicol in various food matrices: using natural deep eutectic solvents, Spectrochimica Acta Part A: Molecular and Biomolecular Spectroscopy 276 (2022) 121198.

[85] M. Shirani, B. Akbari-adergani, F. Shahdadi, M. Faraji, A. Akbari, A hydrophobic deep eutectic solvent-based ultrasound-assisted dispersive liquid−liquid microextraction for determination of β-lactam antibiotics residues in food samples, Food Analytical Methods (2021) 1−10.

[86] H. Taşpınar, A. Elik, S. Kaya, N. Altunay, Optimization of green and rapid analytical procedure for the extraction of patulin in fruit juice and dried fruit samples by air-assisted natural deep eutectic solvent-based solidified homogeneous liquid phase microextraction using experimental design and computational chemistry approach, Food Chemistry 358 (2021).

[87] D. Ge, Z. Shan, T. Pang, X. Lu, B. Wang, Preparation of new hydrophobic deep eutectic solvents and their application in dispersive liquid−liquid microextraction of Sudan dyes from food samples, Analytical and Bioanalytical Chemistry 413 (15) (2021) 3873−3880.

[88] N. Altunay, An optimization approach for fast, simple and accurate determination of indigo-carmine in food samples, Spectrochimica Acta Part A: Molecular and Biomolecular Spectroscopy 257 (2021) 119791.

[89] Y. Ji, Z. Meng, J. Zhao, H. Zhao, L. Zhao, Eco-friendly ultrasonic assisted liquid−liquid microextraction method based on hydrophobic deep eutectic solvent for the determination of sulfonamides in fruit juices, Journal of Chromatography A 1609 (2020) 460520.

[90] W. Liu, B. Zong, X. Wang, J. Cai, J. Yu, A highly efficient vortex-assisted liquid−liquid microextraction based on natural deep eutectic solvent for the determination of Sudan I in food samples, RSC Advances 9 (30) (2019) 17432−17439.

[91] G.S. Kanberoglu, E. Yilmaz, M. Soylak, Application of deep eutectic solvent in ultrasound-assisted emulsification microextraction of quercetin from some fruits and vegetables, Journal of Molecular Liquids 279 (2019) 571−577.

[92] J. Ali, M. Tuzen, T.G. Kazi, Green and innovative technique develop for the determination of vanadium in different types of water and food samples by eutectic solvent extraction method, Food Chemistry 306 (2020) 125638.

[93] A. Elik, A. Demirbaş, N. Altunay, Experimental design of ligandless sonication-assisted liquid-phases microextraction based on hydrophobic deep eutectic solvents for accurate determination of Pb (II) and Cd (II) from waters and food samples at trace levels, Food Chemistry 371 (2022) 131138.

[94] Z. Tekin, T. Unutkan, F. Erulaş, E.G. Bakırdere, S. Bakırdere, A green, accurate and sensitive analytical method based on vortex assisted deep eutectic solvent-liquid phase microextraction for the determination of cobalt by slotted quartz tube flame atomic absorption spectrometry, Food Chemistry 310 (2020) 125825.

[95] M. Shirani, F. Salari, S. Habibollahi, A. Akbari, Needle hub in-syringe solid phase extraction based a novel functionalized biopolyamide for simultaneous green separation/preconcentration and determination of cobalt, nickel, and chromium (III) in food and environmental samples with micro sampling flame atomic absorption spectrometry, Microchemical Journal 152 (2020) 104340.

[96] M. Ataee, T. Ahmadi-Jouibari, N. Noori, N. Fattahi, The speciation of inorganic arsenic in soil and vegetables irrigated with treated municipal wastewater, RSC Advances 10 (3) (2020) 1514−1521.

[97] Z.M. Memon, E. Yilmaz, A.M. Shah, T.G. Kazi, B.R. Devrajani, M. Soylak, A green ultrasonic-assisted liquid−liquid microextraction technique based on deep eutectic solvents for flame atomic absorption spectrometer determination of trace level of lead in tobacco and food samples, Journal of the Iranian Chemical Society 16 (4) (2019) 687−694.

[98] A. Elik, A. Demirbas, N. Altunay, Developing a new and simple natural deep eutectic solvent based ultrasonic-assisted microextraction procedure for determination and preconcentration of as and Se from rice samples, Analytical Methods 11 (27) (2019) 3429−3438.

[99] M. Tuzen, N. Altunay, A. Elik, M.R.A. Mogaddam, K. Katin, Experimental and theoretical investigation for the spectrophotometric determination of thiabendazole in fruit samples, Microchemical Journal 168 (2021) 106488.

[100] K. Zhang, R. Guo, Y. Wang, Q. Nie, G. Zhu, One-step derivatization and temperature-controlled vortex-assisted liquid-liquid microextraction based on the solidification of floating deep eutectic solvents coupled to UV−Vis spectrophotometry for the rapid determination of total iron in water and food samples, Food Chemistry 384 (2022b) 132414.

[101] B. Tışlı, T.U. Gösterişli, B.T. Zaman, E.G. Bakırdere, S. Bakırdere, Determination of manganese in coffee and wastewater using deep eutectic solvent based extraction and flame atomic absorption spectrometry, Analytical Letters 54 (6) (2021) 979−989.

[102] D. Çıtak, D. Sabancı, Response surface methodology and hydrophobic deep eutectic solvent based liquid phase microextraction combination for determination of cadmium in food and water samples, Journal of Food Measurement and Characterization 15 (2) (2021) 1843−1850.

[103] T. Borahan, B.T. Zaman, G. Özzeybek, S. Bakırdere, Accurate and sensitive determination of cobalt in urine samples using deep eutectic solvent-assisted magnetic colloidal gel-based dispersive solid phase extraction prior to slotted quartz tube equipped flame atomic absorption spectrometry, Chemical Papers 75 (2021) 2937−2944.

[104] N. Altunay, M. Tuzen, A simple and green ultrasound liquid−liquid microextraction method based on low viscous hydrophobic deep eutectic solvent for the preconcentration and separation of selenium in water and food samples prior to HG-AAS detection, Food Chemistry 364 (2021) 130371.

[105] S.M. Sorouraddin, M.A. Farajzadeh, T. Okhravi, Application of deep eutectic solvent as a disperser in reversed-phase dispersive liquid-liquid microextraction for the extraction of Cd (II) and Zn (II) ions from oil samples, Journal of Food Composition and Analysis 93 (2020) 103590.

[106] N.A. Kasa, B.T. Zaman, S. Bakırdere, Ultra-trace cadmium determination in eucalyptus and rosemary tea samples using a novel method: deep eutectic solvent based magnetic nanofluid liquid phase microextraction-slotted quartz tube-flame atomic absorption spectrometry, Journal of Analytical Atomic Spectrometry 35 (11) (2020) 2565−2572.

[107] J. Werner, Ligandless, deep eutectic solvent-based ultrasound-assisted dispersive liquid-liquid microextraction with solidification of the aqueous phase for preconcentration of lead, cadmium, cobalt and nickel in water samples, Journal of Separation Science 43 (7) (2020) 1297−1305.

[108] L. Qiao, R. Sun, Y. Tao, Y. Yan, New low viscous hydrophobic deep eutectic solvents for the ultrasound-assisted dispersive liquid-liquid microextraction of endocrine-disrupting phenols in water, milk and beverage, Journal of Chromatography A 1662 (2022) 462728.

[109] H. Sereshti, G. Abdolhosseini, S. Soltani, F. Jamshidi, N. Nouri, Natural thymol-based ternary deep eutectic solvents: application in air-bubble assisted-dispersive liquid-liquid microextraction for the analysis of tetracyclines in water, Journal of Separation Science 44 (19) (2021) 3626−3635.

[110] N. Atsever, T. Borahan, A. Girgin, D.S. Chormey, S. Bakırdere, A simple and effective determination of methyl red in wastewater samples by UV−Vis spectrophotometer with matrix matching calibration strategy after vortex assisted deep eutectic solvent based liquid phase extraction and evaluation of green profile, Microchemical Journal 162 (2021) 105850.

[111] L. Qiao, R. Sun, C. Yu, Y. Tao, Y. Yan, Novel hydrophobic deep eutectic solvents for ultrasound-assisted dispersive liquid-liquid microextraction of trace non-steroidal anti-inflammatory drugs in water and milk samples, Microchemical Journal 170 (2021) 106686.

[112] M. Soylak, F. Uzcan, A novel ultrasonication-assisted deep eutectic solvent microextraction procedure for tartrazine at trace levels from environmental samples, Journal of the Iranian Chemical Society 17 (2020) 461−467.

[113] R. Kachangoon, J. Vichapong, Y. Santaladchaiyakit, R. Burakham, S. Srijaranai, An eco-friendly hydrophobic deep eutectic solvent-based dispersive liquid−liquid microextraction for the determination of neonicotinoid insecticide residues in water, soil and egg yolk samples, Molecules 25 (12) (2020) 2785.

[114] M. Faraji, F. Noormohammadi, M. Adeli, Preparation of a ternary deep eutectic solvent as extraction solvent for dispersive liquid-liquid microextraction of nitrophenols in water samples, Journal of Environmental Chemical Engineering 8 (4) (2020b) 103948.

[115] A.K. El-Deen, K. Shimizu, A green air assisted-dispersive liquid-liquid microextraction based on solidification of a novel low viscous ternary deep eutectic solvent for the enrichment of endocrine disrupting compounds from water, Journal of Chromatography A 1629 (2020) 461498.

[116] P. Tomai, A. Lippiello, P. D'Angelo, I. Persson, A. Martinelli, V. Di Lisio, A. Gentili, A low transition temperature mixture for the dispersive liquid-liquid microextraction of pesticides from surface waters, Journal of Chromatography A 1605 (2019) 360329.

[117] K. Li, Y. Jin, D. Jung, K. Park, H. Kim, J. Lee, In situ formation of thymol-based hydrophobic deep eutectic solvents: application to antibiotics analysis in surface water based on liquid-liquid microextraction followed by liquid chromatography, Journal of Chromatography A 1614 (2020) 460730.

[118] S.M. Sorouraddin, M.A. Farajzadeh, H. Dastoori, Development of a dispersive liquid-liquid microextraction method based on a ternary deep eutectic solvent as chelating agent and extraction solvent for preconcentration of heavy metals from milk samples, Talanta 208 (2020) 120485.

[119] J. Werner, Novel deep eutectic solvent-based ultrasounds-assisted dispersive liquid-liquid microextraction with solidification of the aqueous phase for HPLC-UV determination of aromatic amines in environmental samples, Microchemical Journal 153 (2020) 104405.

[120] G.D. Bozyiğit, M.F. Ayyıldız, D.S. Chormey, G.O. Engin, S. Bakırdere, Development of a sensitive and accurate method for the simultaneous determination of selected insecticides and herbicide in tap water and wastewater samples using vortex-assisted switchable solvent-based liquid-phase microextraction prior to determination by gas chromatography-mass spectrometry, Environmental Monitoring and Assessment 192 (2020) 1−8.

[121] X. Jing, X. Cheng, W. Zhao, H. Wang, X. Wang, Magnetic effervescence tablet-assisted switchable hydrophilicity solvent-based liquid phase microextraction of triazine herbicides in water samples, Journal of Molecular Liquids 306 (2020) 112934.

[122] Q. Chen, L. Wang, G. Ren, Q. Liu, Z. Xu, D. Sun, A fatty acid solvent of switchable miscibility, Journal of Colloid and Interface Science 504 (2017) 645−651.

[123] Y. Zhang, R. Fu, Q. Lu, T. Ren, X. Guo, X. Di, Switchable hydrophilicity solvent for extraction of pollutants in food and environmental samples: a review, Microchemical Journal (2023) 108566.

[124] A. Pochivalov, C. Vakh, S. Garmonov, L. Moskvin, A. Bulatov, An automated in-syringe switchable hydrophilicity solvent-based microextraction, Talanta 209 (2020) 120587.

[125] P. Pollet, C.A. Eckert, C.L. Liotta, Switchable solvents, Chemical Science 2 (4) (2011) 609−614.

[126] U. Alshana, M. Hassan, M. Al-Nidawi, E. Yilmaz, M. Soylak, Switchable-hydrophilicity solvent liquid-liquid microextraction, Trends in Analytical Chemistry 131 (2020) 116025.

[127] M. Herrero, J.A. Mendiola, E. Ibanez, Gas expanded liquids and switchable solvents, Current Opinion in Green and Sustainable Chemistry 5 (2017) 24−30.

[128] A. Elik, A.Ö. Altunay, M.F. Lanjwani, M. Tuzen, A new ultrasound-assisted liquid-liquid microextraction method utilizing a switchable hydrophilicity solvent for spectrophotometric determination of nitrite in food samples, Journal of Food Composition and Analysis 119 (2023) 105267.

[129] S.K. Shahvandi, M.H. Banitaba, H. Ahmar, P. Karimi, A novel temperature controlled switchable solvent based microextraction method: application for the determination of phthalic acid esters in water samples, Microchemical Journal 152 (2020) 104300.

[130] S. Li, J. Qi, B. Zhou, J. Guo, Y. Tong, Q. Zhou, S. Yuan, Sensitive determination of polychlorinated biphenyls from beverages based on switchable solvent microextraction: a robust methodology, Chemosphere 297 (2022) 134185.

[131] A. Elik, N. Altunay, Optimization of vortex-assisted switchable hydrophilicity solvent liquid phase microextraction for the selective extraction of vanillin in different matrices prior to spectrophotometric analysis, Food Chemistry 399 (2023) 133929.

[132] H. Kakaei, S.J. Shahtaheri, K. Abdi, N. Rahimi Kakavandi, Separation and quantification of diazinon in water samples using liquid-phase microextraction-based effervescent tablet-assisted switchable solvent method coupled to gas chromatography-flame ionization detection, Biomedical Chromatography 37 (6) (2023) e5624.

[133] M. Soylak, H.E.H. Ahmed, M. Khan, Switchable hydrophilicity solvent based microextraction of mercury from water, fish and hair samples before its spectrophotometric detection, Sustainable Chemistry and Pharmacy 32 (2023) 101006.

[134] Z. Erbas, M. Soylak, Determination of rhodamine B in water and cosmetics by switchable solvent-based liquid phase microextraction with spectrophotometric determination, Instrumentation Science and Technology 51 (3) (2023) 290−302.

[135] J.J. Coutinho, L.B. Santos, A.S. Melo, J.A. Barreto, V.A. Lemos, Switchable-hydrophilicity solvent-based liquid phase microextraction for determination of brilliant green in water, shellfish, and fish samples, Microchemical Journal (2023) 109617.

[136] B.T. Zaman, N.B. Turan, E.G. Bakirdere, S. Topal, O. Sağsöz, S. Bakirdere, Determination of trace manganese in soil samples by using eco-friendly switchable solvent based liquid phase microextraction-3 holes cut slotted quartz tube-flame atomic absorption spectrometry, Microchemical Journal 157 (2020) 104981.

[137] H. Wang, M. Xia, Y. Ling, F. Qian, J. Xu, Z. Wang, X. Wang, Switchable hydrophilicity solvent based and solidification-assisted liquid-phase microextraction combined with GFAAS for quantification of trace soluble lead in raw bovine and derivative milk products, Food Additives and Contaminants: Part A 36 (11) (2019) 1654−1666.

[138] S. Erarpat, S. Bodur, E. Öztürk Er, S. Bakırdere, Combination of ultrasound-assisted ethyl chloroformate derivatization and switchable solvent liquid-phase microextraction for the sensitive determination of l-methionine in human plasma by GC−MS, Journal of Separation Science 43 (6) (2020) 1100−1106.

[139] X. Wang, M. Gao, Z. Zhang, H. Gu, T. Liu, N. Yu, H. Wang, Development of CO_2-mediated switchable hydrophilicity solvent-based microextraction combined with HPLC-UV for the determination of bisphenols in foods and drinks, Food Analytical Methods 11 (2018) 2093−2104.

[140] M. Tuzen, T.G. Kazi, Simple and green switchable dispersive liquid−liquid microextraction of cadmium in water and food samples, RSC Advances 6 (34) (2016) 28767−28773.

[141] C. Vakh, A. Pochivalov, V. Andruch, L. Moskvin, A. Bulatov, A fully automated effervescence-assisted switchable solvent-based liquid phase microextraction procedure: liquid chromatographic determination of ofloxacin in human urine samples, Analytica Chimica Acta 907 (2016) 54−59.

[142] M. Soylak, M. Khan, E. Yilmaz, Switchable solvent based liquid phase microextraction of uranium in environmental samples: a green approach, Analytical Methods 8 (5) (2016) 979−986.

[143] E. Yilmaz, M. Soylak, Switchable solvent-based liquid phase microextraction of copper (II): optimization and application to environmental samples, Journal of Analytical Atomic Spectrometry 30 (7) (2015) 1629−1635.

[144] A. Shishov, I. Sviridov, I. Timofeeva, N. Chibisova, L. Moskvin, A. Bulatov, An effervescence tablet-assisted switchable solvent-based microextraction: on-site preconcentration of steroid hormones in water samples followed by HPLC-UV determination, Journal of Molecular Liquids 247 (2017) 246−253.

[145] Z. Zhou, B. Guo, B. Lv, H. Guo, G. Jing, Performance and reaction kinetics of CO_2 absorption into AMP solution with [Hmim][Gly] activator, International Journal of Greenhouse Gas Control 44 (2016) 115−123.

[146] A.G. Moghadam, M. Rajabi, M. Hemmati, A. Asghari, Development of effervescence-assisted liquid phase microextraction based on fatty acid for determination of silver and cobalt ions using micro-sampling flame atomic absorption spectrometry, Journal of Molecular Liquids 242 (2017) 1176−1183.

[147] M. Khan, M. Soylak, Switchable solvent based liquid phase microextraction of mercury from environmental samples: a green aspect, RSC Advances 6 (30) (2016) 24968−24975.

[148] D.S. Chormey, S. Bodur, D. Baskın, M. Fırat, S. Bakırdere, Accurate and sensitive determination of selected hormones, endocrine disruptors, and pesticides by gas chromatography—mass spectrometry after the multivariate optimization of switchable solvent liquid-phase microextraction, Journal of Separation Science 41 (14) (2018) 2895—2902.

[149] X. Di, Z. Zhang, Y. Yang, X. Guo, Switchable hydrophilicity solvent based homogeneous liquid-liquid microextraction for enrichment of pyrethroid insecticides in wolfberry, Microchemical Journal 171 (2021) 106868.

[150] M. Gao, J. Wang, X. Song, X. He, R.A. Dahlgren, Z. Zhang, X. Wang, An effervescence-assisted switchable fatty acid-based microextraction with solidification of floating organic droplet for determination of fluoroquinolones and tetracyclines in seawater, sediment, and seafood, Analytical and Bioanalytical Chemistry 410 (2018) 2671—2687.

[151] S. Fuller, A. Gautam, A procedure for measuring microplastics using pressurized fluid extraction, Environmental Science and Technology 50 (11) (2016) 5774—5780.

[152] V. Belandria, P.M.A. de Oliveira, A. Chartier, J.A. Rabi, A.L. de Oliveira, S. Bostyn, Pressurized-fluid extraction of cafestol and kahweol diterpenes from green coffee, Innovative Food Science and Emerging Technologies 37 (2016) 145—152.

[153] A. Mena-García, A.I. Ruiz-Matute, A.C. Soria, M.L. Sanz, Green techniques for extraction of bioactive carbohydrates, Trends in Analytical Chemistry 119 (2019) 115612.

[154] S. Sarkar, S. Sarkar, M.S. Manna, K. Gayen, T.K. Bhowmick, Extraction of carbohydrates and proteins from algal resources using supercritical and subcritical fluids for high-quality products, in: Innovative and Emerging Technologies in the Bio-Marine Food Sector, Academic Press, 2022, pp. 249—275.

[155] J.F. Osorio-Tobón, M.A.A. Meireles, J.F. Osorio-Tobón, M.A.A. Meireles, Recent applications of pressurized fluid extraction: curcuminoids extraction with pressurized liquids, Food and Public Health 3 (6) (2013) 289—303.

[156] W. Li, Z. Wang, Y.P. Wang, C. Jiang, Q. Liu, Y.S. Sun, Y.N. Zheng, Pressurised liquid extraction combining LC—DAD—ESI/MS analysis as an alternative method to extract three major flavones in Citrus reticulata 'Chachi'(Guangchenpi), Food Chemistry 130 (4) (2012) 1044—1049.

[157] K. Skalicka-Woźniak, K. Głowniak, Pressurized liquid extraction of coumarins from fruits of Heracleum leskowii with application of solvents with different polarity under increasing temperature, Molecules 17 (4) (2012) 4133—4141.

[158] A. Mustafa, L.M. Trevino, C. Turner, Pressurized hot ethanol extraction of carotenoids from carrot by-products, Molecules 17 (2) (2012) 1809—1818.

[159] F. Pouralinazar, M.A.C. Yunus, G. Zahedi, Pressurized liquid extraction of Orthosiphon stamineus oil: experimental and modeling studies, The Journal of Supercritical Fluids 62 (2012) 88—95.

[160] S.Y. Koo, K.H. Cha, D.G. Song, D. Chung, C.H. Pan, Optimization of pressurized liquid extraction of zeaxanthin from *Chlorella ellipsoidea*, Journal of Applied Phycology 24 (2012) 725—730.

[161] G. Alvarez-Rivera, M. Bueno, D. Ballesteros-Vivas, J.A. Mendiola, E. Ibanez, Pressurized liquid extraction, in: Liquid-phase Extraction, Elsevier, 2020, pp. 375—398.

[162] N. Katsinas, A. Bento da Silva, A. Enríquez-de-Salamanca, N. Fernández, M.R. Bronze, S. Rodríguez-Rojo, Pressurized liquid extraction optimization from supercritical defatted olive pomace: a green and selective phenolic extraction process, ACS Sustainable Chemistry and Engineering 9 (16) (2021) 5590—5602.

[163] M. Plaza, M.L. Marina, Pressurized hot water extraction of bioactives, Trends in Analytical Chemistry (2023) 117201.
[164] M.S. Jagirani, M. Soylak, Deep eutectic solvents-based adsorbents in environmental analysis, Trends in Analytical Chemistry (2022) 116762.
[165] M. Faraji, Y. Yamini, M. Gholami, Recent advances and trends in applications of solid-phase extraction techniques in food and environmental analysis, Chromatographia 82 (8) (2019) 1207–1249.
[166] P. Ścigalski, P. Kosobucki, Recent materials developed for dispersive solid phase extraction, Molecules 25 (21) (2020) 4869.
[167] M. Anastassiades, S.J. Lehotay, D. Štajnbaher, F.J. Schenck, Fast and easy multiresidue method employing acetonitrile extraction/partitioning and "dispersive solid-phase extraction" for the determination of pesticide residues in produce, Journal of AOAC International 86 (2) (2003) 412–431.
[168] Q. Wu, C. Wang, Z. Liu, C. Wu, X. Zeng, J. Wen, Z. Wang, Dispersive solid-phase extraction followed by dispersive liquid–liquid microextraction for the determination of some sulfonylurea herbicides in soil by high-performance liquid chromatography, Journal of Chromatography A 1216 (29) (2009) 5504–5510.
[169] A. Wilkowska, M. Biziuk, Determination of pesticide residues in food matrices using the quechers methodology, Food Chemistry 125 (3) (2011) 803–812.
[170] S. Walorczyk, Development of a multi-residue method for the determination of pesticides in cereals and dry animal feed using gas chromatography–tandem quadrupole mass spectrometry: II. Improvement and extension to new analytes, Journal of Chromatography A 1208 (1–2) (2008) 202–214.
[171] B. Socas-Rodríguez, A.V. Herrera-Herrera, M. Asensio-Ramos, J. Hernández-Borges, Dispersive solid-phase extraction, Analytical Separation Science (2015) 1525–1570.
[172] H. Musarurwa, L. Chimuka, V.E. Pakade, N.T. Tavengwa, Recent developments and applications of quechers based techniques on food samples during pesticide analysis, Journal of Food Composition and Analysis 84 (2019) 103314.
[173] W. Wittayanan, T. Chaimongkol, W. Jongmevasna, Multiresidue method for determination of 20 organochlorine pesticide residues in fruits and vegetables using modified quechers and GC-ECD/GC-MSD, International Food Research Journal 24 (6) (2017) 2340–2346.
[174] A.B. Bordin, L. Minetto, I. do Nascimento Filho, L.L. Beal, S. Moura, Determination of pesticide residues in whole wheat flour using modified quechers and LC–MS/MS, Food Analytical Methods 10 (2017) 1–9.
[175] M. Ghorbani, M. Aghamohammadhassan, M. Chamsaz, H. Akhlaghi, T. Pedramrad, Dispersive solid phase microextraction, Trends in Analytical Chemistry 118 (2019) 793–809.
[176] M. Šafaříková, I. Šafařík, Magnetic solid-phase extraction, Journal of Magnetism and Magnetic Materials 194 (1–3) (1999) 108–112.
[177] R. Chen, X. Qiao, F. Liu, Ionic liquid-based magnetic nanoparticles for magnetic dispersive solid-phase extraction: a review, Analytica Chimica Acta 1201 (2022) 339632.
[178] C. Liu, S. Wu, Y. Yan, Y. Dong, X. Shen, C. Huang, Application of magnetic particles in forensic science, Trends in Analytical Chemistry 121 (2019b) 115674.
[179] P.T. Anastas, Green chemistry and the role of analytical methodology development, Critical Reviews in Analytical Chemistry 29 (3) (1999) 167–175.
[180] S. Zarabi, R. Heydari, S.Z. Mohammadi, Dispersive micro-solid phase extraction in microchannel, Microchemical Journal 170 (2021) 106676.
[181] A. Chisvert, S. Cárdenas, R. Lucena, Dispersive micro-solid phase extraction, Trends in Analytical Chemistry 112 (2019) 226–233.

[182] G. Lasarte-Aragonés, R. Lucena, S. Cárdenas, M. Valcárcel, Effervescence-assisted dispersive micro-solid phase extraction, Journal of Chromatography A 1218 (51) (2011) 9128–9134.

[183] W. Lu, W. Ming, X. Zhang, L. Chen, Molecularly imprinted polymers for dispersive solid-phase extraction of phenolic compounds in aqueous samples coupled with capillary electrophoresis, Electrophoresis 37 (19) (2016) 2487–2495.

[184] J. Qiao, M. Wang, H. Yan, G. Yang, Dispersive solid-phase extraction based on magnetic dummy molecularly imprinted microspheres for selective screening of phthalates in plastic bottled beverages, Journal of Agricultural and Food Chemistry 62 (13) (2014) 2782–2789.

[185] J.M. Jiménez-Soto, S. Cárdenas, M. Valcárcel, Dispersive micro solid-phase extraction of triazines from waters using oxidized single-walled carbon nanohorns as sorbent, Journal of Chromatography A 1245 (2012) 17–23.

[186] J. Nasiri, M.R. Naghavi, E. Motamedi, H. Alizadeh, M.R.F. Moghadam, M. Nabizadeh, A. Mashouf, Carbonaceous sorbents alongside an optimized magnetic solid phase extraction (MSPE) towards enrichment of crude Paclitaxel extracts from callus cultures of Taxus baccata, Journal of Chromatography B 1043 (2017) 96–106.

[187] X. Liu, C. Wang, Z. Wang, Q. Wu, Z. Wang, Nanoporous carbon derived from a metal organic framework as a new kind of adsorbent for dispersive solid phase extraction of benzoylurea insecticides, Microchimica Acta 182 (2015) 1903–1910.

[188] N. Li, Z. Wang, L. Zhang, L. Nian, L. Lei, X. Yang, A. Yu, Liquid-phase extraction coupled with metal−organic frameworks-based dispersive solid phase extraction of herbicides in peanuts, Talanta 128 (2014) 345–353.

[189] S. Tang, H.K. Lee, Application of dissolvable layered double hydroxides as sorbent in dispersive solid-phase extraction and extraction by co-precipitation for the determination of aromatic acid anions, Analytical Chemistry 85 (15) (2013) 7426–7433.

[190] X. Guo, Y. Li, B. Zhang, L. Yang, X. Di, Development of dispersive solid phase extraction based on dissolvable Fe_3O_4-layered double hydroxide for high-performance liquid chromatographic determination of phenoxy acid herbicides in water samples, Microchemical Journal 152 (2020) 104443.

[191] I. Sowa, M. Wójciak-Kosior, M. Strzemski, J. Sawicki, M. Staniak, S. Dresler, M. Latalski, Silica modified with polyaniline as a potential sorbent for matrix solid phase dispersion (MSPD) and dispersive solid phase extraction (d-SPE) of plant samples, Materials 11 (4) (2018) 467.

[192] S. Zhang, F. Lu, X. Ma, M. Yue, Y. Li, J. Liu, J. You, Quaternary ammonium-functionalized MCM-48 mesoporous silica as a sorbent for the dispersive solid-phase extraction of endocrine disrupting compounds in water, Journal of Chromatography A 1557 (2018) 1–8.

[193] M. Saylan, R. Demirel, M.F. Ayyıldız, D.S. Chormey, G. Çetin, S. Bakırdere, Nickel hydroxide nanoflower−based dispersive solid-phase extraction of copper from water matrix, Environmental Monitoring and Assessment 195 (1) (2023) 133.

[194] Y.M. Lin, J.N. Sun, X.W. Yang, R.Y. Qin, Z.Q. Zhang, Fluorinated magnetic porous carbons for dispersive solid-phase extraction of perfluorinated compounds, Talanta 252 (2023) 123860.

[195] G. Peris-Pastor, S. Alonso-Rodríguez, J.L. Benedé, A. Chisvert, High-throughput determination of oxidative stress biomarkers in saliva by solvent-assisted dispersive solid-phase extraction for clinical analysis, Advances in Sample Preparation 6 (2023) 100067.

[196] P. Han, K. Hu, Y. Wang, L. Li, P. Wang, W. Zhu, S. Zhang, Dispersive solid-phase extraction of non-steroidal anti-inflammatory drugs in water and urine samples using a

magnetic ionic liquid hypercrosslinked polymer composite, Journal of Chromatography A 1689 (2023) 463745.
[197] I. Pacheco-Fernández, V. Pino, Green solvents in analytical chemistry, Current Opinion in Green and Sustainable Chemistry 18 (2019) 42−50.
[198] W.N. Maclay, Factors affecting the solubility of nonionic emulsifiers, Journal of Colloid Science 11 (3) (1956) 272−285.
[199] M. Moradi, Y. Yamini, Surfactant roles in modern sample preparation techniques: a review, Journal of Separation Science 35 (18) (2012) 2319−2340.
[200] C. Padron Sanz, R. Halko, Z. Sosa Ferrera, J.J. Santana Rodriguez, Micellar extraction of organophosphorus pesticides and their determination by liquid chromatography, Analytica Chimica Acta 524 (2004) 265e270.
[201] E.K. Paleologos, D.L. Giokas, M.I. Karayannis, Micelle-mediated separation and cloud-point extraction, Trends in Analytical Chemistry 24 (5) (2005) 426−436.
[202] Y. Liu, L. Jia, Analysis of estrogens in water by magnetic octadecylsilane particles extraction and sweeping micellar electrokinetic chromatography, Microchemical Journal 89 (1) (2008) 72−76.
[203] M.Q. Farooq, I. Ocaña-Rios, J.L. Anderson, Analysis of persistent contaminants and personal care products by dispersive liquid-liquid microextraction using hydrophobic magnetic deep eutectic solvents, Journal of Chromatography A 1681 (2022) 463429.
[204] F. Bamdad, A. Raziani, Application of surface-active ionic liquids in micelle-mediated extraction methods: pre-concentration of cadmium ions by surface-active ionic liquid-assisted cloud point extraction, Journal of the Iranian Chemical Society 17 (2) (2020) 327−332.
[205] K.Z. Du, J. Li, L. Wang, J. Hao, X.J. Yang, X.M. Gao, Y.X. Chang, Biosurfactant trehalose lipid-enhanced ultrasound-assisted micellar extraction and determination of the main antioxidant compounds from functional plant tea, Journal of Separation Science 43 (4) (2020) 799−807.
[206] T.H. Sani, M. Hadjmohammadi, M.H. Fatemi, Extraction and determination of flavonoids in fruit juices and vegetables using Fe_3O_4/SiO_2 magnetic nanoparticles modified with mixed hemi/ad-micelle cetyltrimethylammonium bromide and high performance liquid chromatography, Journal of Separation Science 43 (7) (2020) 1224−1231.
[207] M. Miłek, D. Marcinčáková, J. Legáth, Polyphenols content, antioxidant activity, and cytotoxicity assessment of *Taraxacum officinale* extracts prepared through the micelle-mediated extraction method, Molecules 24 (6) (2019) 1025.
[208] P.A. Uwineza, A. Waśkiewicz, Recent advances in supercritical fluid extraction of natural bioactive compounds from natural plant materials, Molecules 25 (17) (2020) 3847.
[209] B. Pavlić, L. Pezo, B. Marić, L.P. Tukuljac, Z. Zeković, M.B. Solarov, N. Teslić, Supercritical fluid extraction of raspberry seed oil: experiments and modelling, The Journal of Supercritical Fluids 157 (2020) 104687.
[210] T. Arumugham, K. Rambabu, S.W. Hasan, P.L. Show, J. Rinklebe, F. Banat, Supercritical carbon dioxide extraction of plant phytochemicals for biological and environmental applications−A review, Chemosphere 271 (2021) 129525.
[211] M. Zorić, M. Banožić, K. Aladić, S. Vladimir-Knežević, S. Jokić, Supercritical CO_2 extracts in cosmetic industry: current status and future perspectives, Sustainable Chemistry and Pharmacy 27 (2022) 100688.
[212] S. Singh, D.K. Verma, M. Thakur, S. Tripathy, A.R. Patel, N. Shah, C.N. Aguilar, Supercritical fluid extraction (SCFE) as green extraction technology for high-value

metabolites of algae, its potential trends in food and human health, Food Research International 150 (2021) 110746.

[213] T. Khezeli, M. Ghaedi, S. Bahrani, A. Daneshfar, Supercritical fluid extraction in separation and preconcentration of organic and inorganic species, in: New Generation Green Solvents for Separation and Preconcentration of Organic and Inorganic Species, Elsevier, 2020, pp. 425–451.

[214] Z. Huang, X.H. Shi, W.J. Jiang, Theoretical models for supercritical fluid extraction, Journal of Chromatography A 1250 (2012) 2–26.

[215] R.P. Da Silva, T.A. Rocha-Santos, A.C. Duarte, Supercritical fluid extraction of bioactive compounds, Trends in Analytical Chemistry 76 (2016) 40–51.

[216] T. Kariyawasam, G.S. Doran, J.A. Howitt, P.D. Prenzler, Optimization and comparison of microwave-assisted extraction, supercritical fluid extraction, and Eucalyptus oil–assisted extraction of polycyclic aromatic hydrocarbons from soil and sediment, Environmental Toxicology and Chemistry 42 (5) (2023) 982–994.

[217] Y. Shi, H.F. Jin, M.Z. Shi, J. Cao, L.H. Ye, Carbon black-assisted miniaturized solid-phase extraction of carbamate residues from ginger by supercritical fluid chromatography combined with ion mobility quadrupole time-of-flight mass spectrometry, Microchemical Journal 194 (2023) 109335.

[218] B. He, J. Feng, J. Liu, Q. Zhong, T. Zhou, Inline phase transition trapping-selective supercritical fluid extraction-supercritical fluid chromatography: a green and efficient integrated method for determining prohibited substances in cosmetics, Analytica Chimica Acta 1279 (2023) 341831.

[219] Q. Fu, W. Dong, D. Ge, Y. Ke, Y. Jin, Supercritical fluid-based method for selective extraction and analysis of indole alkaloids from *Uncaria rhynchophylla*, Journal of Chromatography A 1710 (2023) 464410.

[220] E. Karrar, I.A.M. Ahmed, W. Wei, F. Sarpong, C. Proestos, R. Amarowicz, X. Wang, Characterization of volatile flavor compounds in supercritical fluid separated and identified in gurum (*Citrulluslanatus* var. *colocynthoide*) seed oil using HSME and GC–MS, Molecules 27 (12) (2022) 3905.

[221] T. Tolcha, T. Gemechu, S. Al-Hamimi, N. Megersa, C. Turner, Multivariate optimization of a combined static and dynamic supercritical fluid extraction method for trace analysis of pesticides pollutants in organic honey, Journal of Separation Science 44 (8) (2021) 1716–1726.

[222] Z. Falsafi, F. Raofie, P.A. Ariya, Supercritical fluid extraction followed by supramolecular solvent microextraction as a fast and efficient preconcentration method for determination of polycyclic aromatic hydrocarbons in apple peels, Journal of Separation Science 43 (6) (2020) 1154–1163.

[223] A. Aresta, P. Cotugno, N. De Vietro, F. Massari, C. Zambonin, Determination of polyphenols and vitamins in wine-making by-products by supercritical fluid extraction (SFE), Analytical Letters 53 (16) (2020) 2585–2595.

[224] A. Molino, S. Mehariya, G. Di Sanzo, V. Larocca, M. Martino, G.P. Leone, D. Musmarra, Recent developments in supercritical fluid extraction of bioactive compounds from microalgae: role of key parameters, technological achievements and challenges, Journal of CO_2 Utilization 36 (2020) 196–209.

[225] A.C.F. Soldan, S. Arvelos, E.O. Watanabe, C.E. Hori, Supercritical fluid extraction of oleoresin from Capsicum annuum industrial waste, Journal of Cleaner Production 297 (2021) 126593.

[226] M. de Andrade Lima, I. Kestekoglou, D. Charalampopoulos, A. Chatzifragkou, Supercritical fluid extraction of carotenoids from vegetable waste matrices, Molecules 24 (3) (2019) 466.
[227] J. Liu, F. Ji, F. Chen, W. Guo, M. Yang, S. Huang, Y. Liu, Determination of garlic phenolic compounds using supercritical fluid extraction coupled to supercritical fluid chromatography/tandem mass spectrometry, Journal of Pharmaceutical and Biomedical Analysis 159 (2018) 513−523.
[228] B. Daneshvand, F. Raofie, Supercritical fluid extraction combined with ultrasound-assisted dispersive liquid−liquid microextraction for analyzing alkylphenols in soil samples, Journal of the Iranian Chemical Society 12 (2015) 1287−1292.
[229] J. Wang, J. Yang, B. Sundén, Q. Wang, Assessment of flow pattern and temperature profiles by residence time distribution in typical structured packed beds, Numerical Heat Transfer, Part A: Applications 77 (6) (2020) 559−578.
[230] F.J.G. Ortiz, A. Kruse, The use of process simulation in supercritical fluids applications, Reaction Chemistry and Engineering 5 (3) (2020) 424−451.
[231] M. Chaturvedi, R. Rani, D. Sharma, J.P. Yadav, Effect of temperature and pressure on antimycobacterial activity of *Curcuma caesia* extract by supercritical fluid extraction method, The International Journal of Mycobacteriology 9 (3) (2020) 296−302.
[232] D. Panadare, G. Dialani, V. Rathod, Extraction of volatile and non-volatile components from custard apple seed powder using supercritical CO_2 extraction system and its inventory analysis, Process Biochemistry 100 (2021) 224−230.
[233] A. Banafi, S.K. Wee, A.N.T. Tiong, Z.Y. Kong, A. Saptoro, J. Sunarso, Modeling of supercritical fluid extraction bed: a critical review, Chemical Engineering Research and Design 2 (193) (2023) 685−712.
[234] H. Ahangari, J.W. King, A. Ehsani, M. Yousefi, Supercritical fluid extraction of seed oils−A short review of current trends, Trends in Food Science and Technology 111 (2021) 249−260.
[235] K. Srinivas, J.W. King, Developments in the processing of foods and natural products using pressurized fluids, in: Alternatives to Conventional Food Processing, 2018, pp. 196−250.
[236] A.L.B. Dias, C.S.A. Sergio, P. Santos, G.F. Barbero, C.A. Rezende, J. Martínez, Effect of ultrasound on the supercritical CO_2 extraction of bioactive compounds from dedo de moça pepper (*Capsicum baccatum* L. var. pendulum), Ultrasonics Sonochemistry 31 (2016) 284−294.
[237] A. Subratti, L.J. Lalgee, N.K. Jalsa, Liquified dimethyl ether (DME): a green solvent for the extraction of hemp (*Cannabis sativa* L.) seed oil, Sustainable Chemistry and Pharmacy 12 (2019) 100144.
[238] F.E. Babadi, P. Boonnoun, K. Nootong, S. Powtongsook, M. Goto, A. Shotipruk, Identification of carotenoids and chlorophylls from green algae *Chlorococcum humicola* and extraction by liquefied dimethyl ether, Food and Bioproducts Processing 123 (2020) 296−303.
[239] P. Boonnoun, A. Shotipruk, H. Kanda, M. Goto, Optimization of rubber seed oil extraction using liquefied dimethyl ether, Chemical Engineering Communications 206 (6) (2019) 746−753.
[240] M. Meskar, M. Sartaj, J.A.I. Sedano, Optimization of operational parameters of supercritical fluid extraction for PHCs removal from a contaminated sand using response surface methodology, Journal of Environmental Chemical Engineering 6 (2) (2018) 3083−3094.
[241] A.P. Wicker, J.,D.D. Carlton, K. Tanaka, M. Nishimura, V. Chen, T. Ogura, K.A. Schug, On-line supercritical fluid extraction−supercritical fluid chromatography-mass

spectrometry of polycyclic aromatic hydrocarbons in soil, Journal of Chromatography B 1086 (2018) 82−88.
[242] H. Bagheri, Y. Yamini, M. Safari, H. Asiabi, M. Karimi, A. Heydari, Simultaneous determination of pyrethroids residues in fruit and vegetable samples via supercritical fluid extraction coupled with magnetic solid phase extraction followed by HPLC-UV, The Journal of Supercritical Fluids 107 (2016) 571−580.
[243] S. Bayrak, M. Sökmen, E. Aytaç, A. Sökmen, Conventional and supercritical fluid extraction (SFE) of colchicine from *Colchicum speciosum*, Industrial Crops and Products 128 (2019) 80−84.
[244] I. Deniz, M.O. Ozen, O. Yesil-Celiktas, Supercritical fluid extraction of phycocyanin and investigation of cytotoxicity on human lung cancer cells, The Journal of Supercritical Fluids 108 (2016) 13−18.
[245] M. Heydarzadeh, R. Heydari, Determination of 2,4-dichlorophenoxyacetic acid in environmental and food samples using salt-assisted liquid-liquid extraction coupled with micro-channel and high-performance liquid chromatography, Separation Science 5 (7) (2022) 305−313.
[246] S.A. Khatibi, S. Hamidi, M.R. Siahi-Shadbad, Application of liquid-liquid extraction for the determination of antibiotics in the foodstuff: recent trends and developments, Critical Reviews in Analytical Chemistry 52 (2) (2022) 327−342.
[247] F.M. Antony, D. Pal, K. Wasewar, Separation of bio-products by liquid−liquid extraction, Physical Sciences Reviews 6 (4) (2021) 20180065.
[248] A. Ismaili, R. Heydari, R. Rezaeepour, Monitoring the oleuropein content of olive leaves and fruits using ultrasound- and salt-assisted liquid-liquid extraction optimized by response surface methodology and high-performance liquid chromatography, Journal of Separation Science 39 (2) (2016) 405−411.
[249] V. Jalili, A. Barkhordari, A. Ghiasvand, Liquid-phase microextraction of polycyclic aromatic hydrocarbons: a review, Reviews in Analytical Chemistry 39 (1) (2020) 1−19.
[250] H. Jiang, S. Yang, H. Tian, B. Sun, Research progress in the use of liquid-liquid extraction for food flavour analysis, Trends in Food Science and Technology 132 (2023) 138−149.
[251] R. Heydari, M. Mousavi, Simultaneous determination of saccharine, caffeine, salicylic acid and benzoic acid in different matrixes by salt and air-assisted homogeneous liquid-liquid extraction and high-performance liquid chromatography, Journal of the Chilean Chemical Society 61 (3) (2016) 3090−3094.
[252] M. Moradi, S. Zarabi, R. Heydari, Spectrophotometric determination of trace amounts of Sb(III) and Sb(V) in water and biological samples by in-tube dispersive liquid−liquid microextraction and air-assisted liquid−liquid microextraction, Chemical Papers 75 (12) (2021) 6499−6650.
[253] B. Herce-Sesa, J.A. Lopez-Lopez, C. Moreno, Advances in ionic liquids and deep eutectic solvents-based liquid phase microextraction of metals for sample preparation in Environmental Analytical Chemistry, Trends in Analytical Chemistry 143 (2021). Article 116398.
[254] M.S. Alves, L.C. Ferreira, C. Scheid, J. Merib, An overview of magnetic ionic liquids: from synthetic strategies to applications in microextraction techniques, Journal of Separation Science 45 (1) (2022) 258−281.
[255] X. Li, K.H. Row, Development of deep eutectic solvents applied in extraction and separation, Journal of Separation Science 39 (18) (2016) 3505−3520.

[256] D.C. de Andrade, S.A. Monteiro, J. Merib, A review on recent applications of deep eutectic solvents in microextraction techniques for the analysis of biological matrices, Advances in Sample Preparation 1 (2022) 100007.

[257] X. Liu, Q. Ao, S. Shi, S. Li, CO_2 capture by alcohol ammonia based deep eutectic solvents with different water content, Materials Research Express 9 (1) (2022) 015504.

[258] K. Cherkashina, A. Pochivalov, V. Simonova, F. Shakirova, A. Shishov, A. Bulatov, A synergistic effect of hydrophobic deep eutectic solvents based on terpenoids and carboxylic acids for tetracycline microextraction, Analyst 146 (11) (2021) 3449–3453.

[259] E.M. Khosrowshahi, A. Jouyban, M.A. Farajzadeh, M. Tuzen, M.R.A. Mogaddam, M. Nemati, pH-induced homogeneous liquid-liquid microextraction method based on new switchable deep eutectic solvent for the extraction of three antiepileptic drugs from breast milk, Bioanalysis 13 (14) (2021) 1087–1099.

[260] G. Lasarte-Aragones, R. Lucena, S. Cardenas, M. Valcarcel, Use of switchable solvents in the microextraction context, Talanta 131 (2015) 645–649.

[261] G.D. Bozyigit, M.F. Ayyildiz, D.S. Chormey, G.O. Engin, S. Bakirdere, Trace level determination of eleven nervous system-active pharmaceutical ingredients by switchable solvent-based liquid-phase microextraction and gas chromatography-mass spectrometry with matrix matching calibration strategy, Environmental Monitoring and Assessment 194 (2) (2022) 1–11.

[262] P. Chaikhan, Y. Udnan, R.J. Ampiah-Bonney, W.C. Chaiyasith, Fast sequential multi element analysis of lead and cadmium in canned food samples using effervescent tablet-assisted switchable solvent based liquid phase microextraction (EA-SS-LPME) coupled with high-resolution continuum source flame atomic absorption spectrometry (HR-CS-FAAS), Food Chemistry 375 (2022). Article 131857.

[263] V. Jalili, A. Barkhordari, A. Ghiasvand, A comprehensive look at solid-phase microextraction technique: a review of reviews, Microchemical Journal 152 (2020) 104319.

[264] M. Carabajal, C.M. Teglia, S. Cerutti, M.J. Culzoni, H.C. Goicoechea, Applications of liquid-phase microextraction procedures to complex samples assisted by response surface methodology for optimization, Microchemical Journal 152 (2020) 104436.

[265] M. Fernández-Amado, M.C. Prieto-Blanco, P. López-Mahía, S. Muniategui-Lorenzo, D. Prada-Rodríguez, Strengths and weaknesses of in-tube solid-phase microextraction: a scoping review, Analytica Chimica Acta 906 (2016) 41–57.

[266] M.E.C. Queiroz, I.D. de Souza, C. Marchioni, Current advances and applications of in-tube solid-phase microextraction, Trends in Analytical Chemistry 111 (2019) 261–278.

[267] K. Kedziora, W. Wasiak, Extraction media used in needle trap devices—progress in development and application, Journal of Chromatography 1505 (2017) 1–17.

[268] H. Piri-Moghadam, M.N. Alam, J. Pawliszyn, Review of geometries and coating materials in solid phase microextraction: opportunities, limitations, and future perspectives, Analytica Chimica Acta 984 (2017) 42–65.

[269] M. Saraji, A. Shahvar, Metal-organic aerogel as a coating for solid-phase microextraction, Analytica Chimica Acta 973 (2017) 51–58.

[270] V. Jalili, A. Barkhordari, M. Heidari, The role of aerogel-based sorbents in microextraction techniques, Microchemical Journal 147 (2019) 948–954.

[271] J. Guo, S.J. Park, L.Y. Meng, X. Jin, Applications of carbon-based materials in solid phase micro-extraction: a review, Carbon letters 24 (2017) 10–17.

[272] M. Lashgari, Y. Yamini, An overview of the most common lab-made coating materials in solid phase microextraction, Talanta 191 (2019) 283–306.

[273] F. Pena-Pereira, I. Lavilla, C. Bendicho, Liquid-phase microextraction approaches combined with atomic detection: a critical review, Analytica Chimica Acta 669 (1–2) (2010) 1–16.

[274] V.A. Lemos, R.V. Oliveira, W.N.L. dos Santos, R.M. Menezes, L.B. Santos, S.L.C. Ferreira, Liquid phase microextraction associated with flow injection systems for the spectrometric determination of trace elements, Trends in Analytical Chemistry 110 (2019) 357–366.

[275] M. Sajid, Dispersive liquid-liquid microextraction coupled with derivatization: a review of different modes, applications, and green aspects, Trends in Analytical Chemistry 106 (2018) 169–182.

[276] T. Cai, H. Qiu, Application of deep eutectic solvents in chromatography: a review, Trends in Analytical Chemistry 120 (2019) 115623.

[277] J. Chen, Y. Li, X. Wang, W. Liu, Application of deep eutectic solvents in food analysis: a review, Molecules 24 (24) (2019) 4594.

[278] M. Faraji, M. Mahmoodi-Maymand, F. Dastmalchi, Green, fast and simple dispersive liquid-liquid microextraction method by using hydrophobic deep eutectic solvent for analysis of folic acid in fortified flour samples before liquid chromatography determination, Food Chemistry 320 (2020b) 126486.

[279] M. Faraji, Determination of some red dyes in food samples using a hydrophobic deep eutectic solvent-based vortex assisted dispersive liquid-liquid microextraction coupled with high performance liquid chromatography, Journal of Chromatography A 1591 (2019) 15–23.

Chapter 8

Solvent free extraction procedures

Moumita Saha[1,2], Rahul Makhija[1] and Vivek Asati[3]
[1]*Department of Pharmaceutical Analysis, ISF College of Pharmacy, Moga, Punjab, India;*
[2]*Department of Pharmaceutical Quality Assurance, Manipal College of Pharmaceutical Sciences MAHE, Manipal, Karnataka, India;* [3]*Department of Pharmaceutical Chemistry, ISF College of Pharmacy, Moga, Punjab, India*

1. Introduction

A scientific and philosophical trend known as "green chemistry" aims to create materials and chemical processes that are safer, more environmentally friendly, and perform on par with or better than current ones while having the least number of negative effects on the environment. Green chemistry has many uses, from energy production and agriculture to materials research and pharmaceuticals [1,2]. Solvent-free extraction, sometimes referred to as solventless extraction, is a core concept in green chemistry that involves doing away with the requirement for solvents during the extraction process. This tactic can lessen the harm that conventional extraction methods employing organic solvents due to the environment and the health risks they represent [3,4]. The main goal of solvent-free extraction is to minimize or completely avoid the requirement for volatile organic solvents during the extraction procedure. This lowers the risk of exposure to hazardous chemicals for industrial and laboratory workers as well as the release of harmful emissions into the atmosphere. Solvent-free extraction promotes the employment of alternative methods including heat, mechanical force, or supercritical fluids, which helps reduce waste, save energy, and save precious resources. Solvent-free extraction is being explored widely and novel approaches are being developed by enterprises and researchers, for making a viable strategy for a safer, more sustainable, and greener future [5–7].

1.1 Principles of solvent-free extraction [3,8−10]

No use of hazardous organic solvents: Solvent-free extraction seeks to avoid or reduce the use of organic solvents that pose a risk to the environment and human health.

Lowered waste generation: By minimizing the use of solvents, hazardous waste related to solvent disposal is produced at a much lower rate, making the process cleaner and more environmentally friendly.

Energy sustainability: By using various mechanisms release and recovery of target molecules can be done employing energy-efficient solvent-free extraction techniques.

2. Methods of solvent-free extraction

2.1 Grinding

It is a mechanical process of breaking down solid materials into smaller particles without the use of traditional organic solvents. This method is employed in various industries, including pharmaceuticals, food, and natural product extraction, to liberate target compounds from raw materials. An example of grinding in solvent-free extraction is the preparation of herbal extracts from plant material [11−13].

2.2 Pressing

This process involves applying mechanical force to squeeze or press solid materials to release valuable compounds without the use of traditional organic solvents. This method is particularly relevant in the extraction of oils or juices from seeds, fruits, or other plant materials [14−16].

2.3 Gas-assisted mechanical methods

Some innovative techniques involve using gases, such as carbon dioxide or nitrogen, to facilitate mechanical extraction methods like grinding or pressing. These gases can improve the release of target compounds without the need for additional solvents. For example, Gas-assisted grinding (GAG) where nitrogen gas helps in the dispersion of the sample particles, enhancing the contact between the solid material and the mechanical force [17]. This increased surface area facilitates the extraction of compounds without the need for additional solvents. Another technique is Gas-Assisted Jetting Extraction (GJET), that breaks up the sample matrix mechanically and extracts chemicals by using a high-velocity gas jet. The gas jet is directed at the solid sample, promoting the release of analytes [18].

2.4 Subcritical water extraction

Subcritical water extraction, alternatively referred in as pressurized hot water extraction and subcritical water chromatography, is an environment friendly technique that uses water at temperature above the boiling point (100°C) and below the critical temperature (374°C), along with elevated pressure to keep the water liquid throughout the extraction process, to extract compounds from a variety of materials (as shown in Fig. 8.1). One could classify this method as a type of pressurized liquid extraction (PLE) [19,20].

A wide range of analytes can be effectively extracted using the subcritical water extraction technique due to its unique properties, which include increased solubility (dissolving polar as well as non-polar compounds), rapid penetrability, minimized thermal degradation, specific extraction, and efficient recovery [21,22]. This technique is very useful for removing chemicals from natural sources that are susceptible to heat. It works especially well for removing heat-sensitive materials from botanical sources, such as essential oils and bioactive chemicals [23,24].

2.5 Pressurized liquid extraction (PLE)

PLE, sometimes referred by the term accelerated solvent extraction, is a contemporary extraction method that increases the effectiveness of analyte extraction from solid or semi-solid samples by using high temperature and pressure. Instead of using conventional organic solvents, it frequently uses water or other ecologically friendly solvents. Toxins, pharmaceuticals, natural

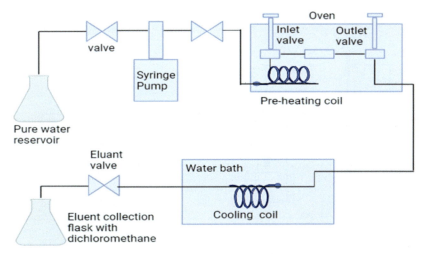

FIGURE 8.1 Schematic representation of subcritical water extraction technique.

products, and environmental pollutants are just a few of the materials that PLE is particularly useful in extracting. Improved analyte recovery, faster extraction times, and increased selectivity are just a few advantages of this approach [25–27].

Two varieties of PLE exist. There are two types of PLE: dynamic and static. Analytes are effectively extracted in dynamic PLE because pressurized solvent is constantly pumped into the sample matrix throughout the extraction procedure (as shown in Fig. 8.2). In static PLE, the pressurized solvent is introduced into the sample matrix, and the system is then sealed to maintain the pressure allowing solvent to interact with sample over a set period of time [28,29]. Dynamic PLE involves continuous solvent circulation, offering faster extraction, while static PLE operates with a non-flowing solvent, providing simplicity and ease of operation. The choice between the two depends on the specific requirements of the analytical method and the nature of the sample being analyzed [26,27,30,31].

2.6 Supercritical fluid extraction

Supercritical fluid extraction (SFE) is a separation technique that uses supercritical fluids as the extracting solvent to efficiently extract and separate target compounds from solid or liquid samples. Supercritical carbon dioxide (CO_2) is commonly used as a green solvent replacement. The gas is pressurized and heated to bring it in a supercritical state (critical temperature is 31.1°C, and the critical pressure is 73.8 atm), where it exhibits the properties of both gases and liquids [32,33]. In this state, it can effectively extract desired compounds from

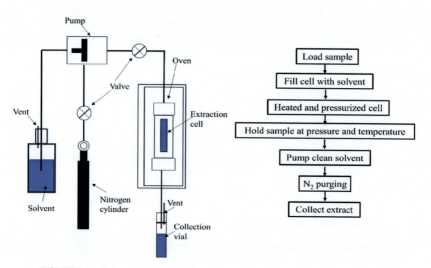

FIGURE 8.2 Schematic representation of pressurized liquid extraction technique.

the sample. Once the extraction is complete, the pressure is reduced, and the gas returns to its gaseous state, leaving behind the extracted compounds (Fig. 8.3) [34−36]. It provides simple CO_2 separation and recycling while extracting a broad variety of components, such as bioactive compounds, flavors, and essential oils. In mild extraction settings, this method's selective solvency allows for selective extraction while maintaining environmental sustainability [37,38]. Fig. 8.4 represents schematic comparision between PLE and SFE.

2.7 Microwave-assisted extraction (MAE)

Using microwave energy to improve the extraction of target molecules from a variety of matrices, including plant materials, food, and environmental samples, is known as microwave-assisted extraction (MAE) [39,40]. Utilizing the power of microwaves to quickly and evenly heat substances with high dielectric constants, including water and some polar organic solvents, this technique makes use of this property to help release analytes into a solvent [41−43]. This technique can be used to extract chemicals from solid or semisolid substances. The technique is useful for several tasks, such as sample preparation and natural product extraction, because of its rapid and accurate heating qualities [39,41,44,45].

2.8 Ultrasound-assisted extraction

Ultrasonic waves are used in the process of Ultrasound-Assisted Extraction (UAE), which improves the extraction of bioactive chemicals from liquid or solid materials. In order to produce strong local heating and pressure changes that result in microscale shockwaves that help release target compounds from the sample matrix, this technique uses sonic cavitation, which is the production and collapse

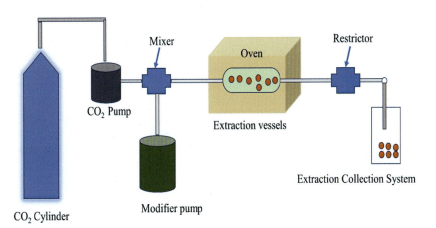

FIGURE 8.3 Schematic representation of supercritical fluid extraction technique.

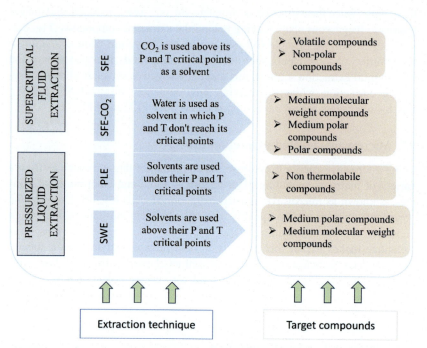

FIGURE 8.4 Schematic comparison between pressurized liquid extraction and supercritical fluid extraction technique (P denotes pressure, T denotes tempareture).

of small bubbles in a solvent [3,46,47]. It is possible to conduct UAE with little or no solvent use. Because of its short extraction time, increased efficiency, and selectivity, it is beneficial in solvent-free extraction method [48–50].

2.9 Ohmic heating extraction

Applying an electric current through a resistive material causes particles to heat up due to the resistance the current encounters. This phenomenon is known as ohmic heating [51,52]. Ohmic heating extraction is a technique which combines ohmic heating with processes of extraction to enhance the extraction of bioactive compounds from solid matrices, such as plant materials or food products. Ohmic heating heats the sample matrix selectively to promote the release of particular compounds into a solvent. Because of its advantages in selective and rapid heating, it is a potential method for food processing and the extraction of natural products [53].

2.10 Ionic liquids

Ionic liquid extraction comes under the category of solvent-free extraction, in which the materials are extracted using ionic liquids rather than conventional solvents. Salts that are in a liquid condition at low temperatures typically below 100°C are known as ionic liquids. Their distinct qualities render them appealing for solvent-free extraction methods [54–56]. Despite being solvents in theory, they are better for the environment because of their special qualities. Benefits of this approach include solvent recyclability, low volatility, decreased environmental impact, and adaptability [57].

2.11 Biocompatible extractions

The application of biocompatible extraction techniques, which release chemicals from source materials using enzymes or other biological agents, is still being studied. These methods have the potential to be both environmentally benign and extraction specific.

2.12 Solid phase microextraction (SPME)

Another sample preparation method called solid phase microextraction (SPME) is used to extract and concentrate volatile and semi-volatile substances from a variety of sample matrices. Analytes can be directly captured and concentrated from the sample matrix using the SPME method, which uses a solid phase fiber coated with an extracting phase (such as polydimethylsiloxane (PDMS), divinylbenzene (DVB), or a mixture of both) [58,59]. It frequently replaces liquid solvents when used for headspace analysis or volatile component extraction [60].

2.12.1 Headspace solid phase microextraction (HS-SPME)

A specialized version of the SPME method used in green chemistry is called Headspace solid phase microextraction (HS-SPME). It works especially well for removing volatile and semi-volatile substances from the headspace (i.e. gas phase) [61,62]. The headspace is the area above a liquid or solid sample that contains gaseous volatile chemicals. With this method, volatile chemicals that have partitioned into the gas phase (headspace) above a sealed sample are the goal of the extraction process, which exposes the SPME fiber to this headspace [59,63,64]. It is particularly useful for the analysis of liquid or solid samples with volatile and semi-volatile compounds such as beverages, perfumes, and samples with headspace components [64]. Fig. 8.5 represents difference between Solid phase extraction and Headspace solid phase extaction technique.
  Advantages of HS-SPME [65,66]:

- Solvent-free: HS-SPME is a solvent-free extraction technique, making it environmentally friendly and safer to use in comparison to traditional liquid–liquid extraction methods.
- Concentration and preconcentration: It provides concentration and preconcentration of analytes, enhancing sensitivity and detection limits in analytical techniques.

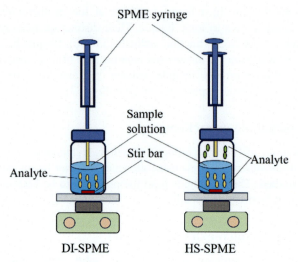

FIGURE 8.5 Schematic comparison representing solid phase extraction and headspace solid phase extaction technique.

- Minimal sample handling: Sample handling is simplified because there is no need for extensive sample preparation or solvent evaporation.
- Rapid analysis: HS-SPME allows for rapid and efficient extraction and analysis of volatile compounds.
- Selectivity: The choice of fiber coating allows for selectivity, enabling the extraction of specific compounds of interest.

HS-SPME is widely used in various fields, including [64,67,68]:

- Food analysis: It is employed for the analysis of volatile compounds in foods, beverages, and flavorings.
- Environmental monitoring: HS-SPME is used to assess air quality, analyze soil and water samples, and detect volatile pollutants.
- Pharmaceuticals: It is applied to analyze pharmaceutical products, including characterization of drug formulations.
- Chemical and petrochemical industry: HS-SPME is used in process monitoring and quality control.
- Forensic analysis: It is employed in the analysis of volatile compounds in forensic science.

Researchers and analysts continue to refine and expand its applications in diverse fields.

3. Advantages

- <u>Environmental benefits:</u> As solvent-free extraction does not require huge amounts of organic solvents, it significantly decreases the environmental

impact of chemical processes. As a result of this, the amount of volatile organic compounds (VOCs) released into the atmosphere is reduced [9].
- Health and safety: The safety of industrial operators and laboratory personnel can be improved by the lack of hazardous organic solvents. It makes the workplace safer by lowering the possibility of being exposed to dangerous chemicals [63,69].
- Resource efficiency: Solvent-free extraction methods can be more resource-efficient because they use less energy and produce fewer greenhouse gas emissions.
- Sustainability: Solvent-free extraction helps in making chemical processes more sustainable and compatible with international initiatives to save resources and lessen environmental effect. It reduces reliance on fossil fuels, minimizes waste output, and encourages the use of recycled materials by emphasizing renewable raw materials [9].
- Quality of extracts: Higher-quality extracts can be produced using solvent-free extraction techniques since there is no chance of contamination or solvent residues. It enables the minimally altered or degraded extraction of active chemicals from natural sources. Consequently, the extracted chemicals' purity can be guaranteed [70].
- Cost savings: Solvent-free extraction equipment may have a greater initial cost, but over time, these costs are typically offset by lower costs associated with purchasing, storing, and disposing of waste [70].

4. Considerations

- Regulatory Compliance: Implementing solvent-free extraction techniques requires strict adherence to safety standards and regulatory requirements, particularly in businesses with stringent quality and safety laws [9].
- Method Selection: Depending on the particular molecules being targeted and the characteristics of the source material, a solvent-free extraction method might be selected. Numerous methods are available, including thermal, supercritical fluid, and mechanical approaches [71].
- Equipment: For solvent-free extraction, specialized tools and technologies may be needed to efficiently use mechanical force, heat, or supercritical fluids. These could need a one time financial outlay.
- Optimization: To get the highest yield and efficiency, the extraction parameters and circumstances must be optimized.
- Scale-Up Challenges: Solvent-free extraction may present difficulties when scaling up the methodology between laboratory to manufacturing level, so it should be carefully evaluated.
- Waste Management: Although waste from solvents is kept to a minimum, waste from the extraction process which could include plant residues or other byproducts should also be taken into account [3,9].

- Quality Control: Strict quality assurance and analytical methods are required to guarantee the uniformity and pure nature of the extracted substances.

5. Applications

- Natural Product Extraction: To extract natural products from plants, such as essential oils, flavors, perfumes, and bioactive components, solvent-free extraction is frequently utilized [9].
- Pharmaceuticals: The pharmaceutical industry can also benefit from this method for obtaining active medicinal components from botanical sources [72].
- Food Processing: To extract flavors, antioxidants, and other food ingredients, the food business uses solvent-free extraction techniques [73,74].
- Environmental rehabilitation: To remove pollutants from soil or water, several solvent-free extraction processes are employed in environmental rehabilitation projects.
- Material Science: In material science, this approach is used to extract compounds or polymers from various materials.

6. Conclusion

All things considered, solvent free extraction is consistent with green chemistry's tenets, which prioritize sustainability, safety, and a minimal negative influence on the environment. It lessens waste, lessens the harmful impacts of dangerous solvents, and frequently results in more economical and effective procedures. The use of solvent free extraction techniques has changed sample preparation procedures significantly and has several benefits for the environment, human safety, and productivity [75]. These methods are flexible and suitable to a wide range of sample matrices, including liquids, solids, and complicated mixtures. They also address the limitations of conventional solvent based approaches. Because of their versatility, they can be used in forensic science, food and beverage testing, environmental analysis, and medicines, and other domains [76]. The ease of this approach makes it possible to prepare samples quickly, which cuts down on the amount of time needed for analysis overall and minimizes the matrix effect—especially for complicated sample matrices. They work well for trace-level analyte detection and quantification [77]. The subject of green chemistry is being advanced by industries and researchers who are constantly investigating and creating novel solvent free extraction procedures. The chemical and natural product sectors may have a more ecologically friendly and greener future if these techniques are used. These methods are expected to play a major role in developing analytical chemistry as they become more sophisticated and widely used, meeting the increasing demand for ecologically friendly research and industrial processes.

References

[1] Us epa o, Basics of Green Chemistry, 2013. Available from: https://www.epa.gov/greenchemistry/basics-green-chemistry.
[2] P.T. Anastas, E.S. Beach, Green chemistry: the emergence of a transformative framework, Green Chemistry Letters and Reviews 1 (1) (2007) 9−24.
[3] F. Chemat, M.A. Vian, G. Cravotto, Green extraction of natural products: concept and principles, International Journal of Molecular Sciences 13 (7) (2012) 8615−8627.
[4] T. Welton, Solvents and sustainable chemistry, Proceedings of the Royal Society A: Mathematical, Physical and Engineering Sciences 471 (2183) (2015) 20150502.
[5] N. Ullah, M. Tuzen, A comprehensive review on recent developments and future perspectives of switchable solvents and their applications in sample preparation techniques, Green Chemistry 25 (5) (2023) 1729−1748.
[6] P. Kapadia, A.S. Newell, J. Cunningham, M.R. Roberts, J.G. Hardy, Extraction of high-value chemicals from plants for technical and medical applications, International Journal of Molecular Sciences 23 (18) (2022) 10334.
[7] F.P. Byrne, S. Jin, G. Paggiola, T.H.M. Petchey, J.H. Clark, T.J. Farmer, et al., Tools and techniques for solvent selection: green solvent selection guides, Sustainable Chemical Processes 4 (1) (2016) 7.
[8] Q.W. Zhang, L.G. Lin, W.C. Ye, Techniques for extraction and isolation of natural products: a comprehensive review, Chinese Medicine 13 (1) (2018) 20.
[9] F. Chemat, A.S. Fabiano-Tixier, M. Vian, T. Allaf, E. Vorobiev, Solvent-free extraction of food and natural products, TrAC, Trends in Analytical Chemistry 71 (2015).
[10] M.A. Vian, T. Allaf, E. Vorobiev, F. Chemat, Solvent-free extraction: myth or reality? in: F. Chemat, M.A. Vian (Eds.), Alternative Solvents for Natural Products Extraction Springer, Berlin, Heidelberg, 2014, pp. 25−38.
[11] P. Wang, J. Tang, Solvent-free mechanochemical extraction of chondroitin sulfate from shark cartilage, Chemical Engineering and Processing: Process Intensification 48 (6) (2009) 1187−1191.
[12] L. Chen, X. Hu, M. Gong, Z. Chen, Y. Li, L. Zhou, et al., Study on a novel enzymatic colon-targeted particle of total saponins of Pulsatilla by mechanical grinding technology in a solvent free system, Biomedicine & Pharmacotherapy 155 (2022) 113645.
[13] M. Jug, P.A. Mura, Grinding as solvent-free green chemistry approach for cyclodextrin inclusion complex preparation in the solid state, Pharmaceutics 10 (4) (2018) 189.
[14] D.K. Bredeson, Mechanical extraction, Journal of the American Oil Chemists' Society 55 (11) (1978) 762−764.
[15] D.M. Kasote, Y.S. Badhe, M.V. Hegde, Effect of mechanical press oil extraction processing on quality of linseed oil, Industrial Crops and Products 42 (2013) 10−13.
[16] I. Reyes-Ocampo, M.S. Córdova-Aguilar, G. Guzmán, A. Blancas-Cabrera, G. Ascanio, Solvent-free mechanical extraction of *Opuntia ficus-indica* mucilage, Journal of Food Process Engineering 42 (1) (2019) e12954.
[17] M. Koubaa, F.J. Barba, H. Mhemdi, N. Grimi, W. Koubaa, E. Vorobiev, Gas assisted mechanical expression (GAME) as a promising technology for oil and phenolic compound

recovery from tiger nuts, Innovative Food Science & Emerging Technologies 32 (2015) 172−180.
[18] N. Lebovka, E. Vorobiev, F. Chemat, Enhancing Extraction Processes in the Food Industry, CRC Press, 2016, p. 558.
[19] Y. Cheng, F. Xue, S. Yu, S. Du, Y. Yang, Subcritical water extraction of natural products, Molecules 26 (13) (2021) 4004.
[20] A.G. Carr, R. Mammucari, N.R. Foster, A review of subcritical water as a solvent and its utilisation for the processing of hydrophobic organic compounds, Chemical Engineering Journal 172 (1) (2011) 1−17.
[21] L.G.G. Rodrigues, S. Mazzutti, I. Siddique, M. da Silva, L. Vitali, S.R.S. Ferreira, Subcritical water extraction and microwave-assisted extraction applied for the recovery of bioactive components from Chaya (*Cnidoscolus aconitifolius* Mill.), The Journal of Supercritical Fluids 165 (2020) 104976.
[22] M. Castro-Puyana, M. Herrero, J.A. Mendiola, E. Ibáñez, Subcritical water extraction of bioactive components from algae, in: Functional Ingredients from Algae for Foods and Nutraceuticals, Elsevier, 2013, pp. 534−560.
[23] M. Sarfarazi, S.M. Jafari, G. Rajabzadeh, J. Feizi, Development of an environmentally-friendly solvent-free extraction of saffron bioactives using subcritical water, LWT 114 (2019) 108428.
[24] S. Gbashi, O.A. Adebo, L. Piater, N.E. Madala, P.B. Njobeh, Subcritical water extraction of biological materials, Separation and Purification Reviews 46 (1) (2017) 21−34.
[25] A. Perez-Vazquez, M. Carpena, P. Barciela, L. Cassani, J. Simal-Gandara, M. Prieto, Pressurized liquid extraction for the recovery of bioactive compounds from seaweeds for food industry application: a review, Antioxidants 12 (3) (2023) 612. https://doi.org/10.3390/antiox12030612.
[26] G. Alvarez-Rivera, M. Bueno, D. Ballesteros-Vivas, J.A. Mendiola, E. Ibañez, Chapter 13 − Pressurized liquid extraction, in: C.F. Poole (Ed.), Liquid-Phase Extraction, Elsevier, 2020, pp. 375−398.
[27] F.P. Cardenas-Toro, S.C. Alcázar-Alay, J.P. Coutinho, H.T. Godoy, T. Forster-Carneiro, M.A.A. Meireles, Pressurized liquid extraction and low-pressure solvent extraction of carotenoids from pressed palm fiber: experimental and economical evaluation, Food and Bioproducts Processing 94 (2015) 90−100.
[28] Pressurized hot water extraction with on-line fluorescence monitoring: a comparison of the static, dynamic, and static−dynamic modes for the removal of polycyclic aromatic hydrocarbons from environmental solid samples, Analytical Chemistry 16 (2002) 4213−4219.
[29] S. Morales-Muñoz, J.L. Luque-Garcia, M.D. Luque de Castro, Static extraction with modified pressurized liquid and on-line fluorescence monitoring: independent matrix approach for the removal of polycyclic aromatic hydrocarbons from environmental solid samples, Journal of Chromatography A 978 (1) (2002) 49−57.
[30] K.B. Mustapha, M.T. Bakare-Odunola, G. Magaji, O.O. Obodozie-Ofoegbu, D.D. Akumka, Pharmacokinetics of chloroquine and metronidazole in rats, Journal of Applied Pharmaceutical Science 5 (8) (2015) 090−094.
[31] J. Felipe Osorio-Tobón, A. Angela, M. Meireles, Recent applications of pressurized fluid extraction: curcuminoids extraction with pressurized liquids, Fph 3 (6) (2013) 289−303.
[32] G.N. Sapkale, S.M. Patil, U.S. Surwase, P.K. Bhatbhage, Supercritical Fluid Extraction, 2010.
[33] R. Mohamed, G.A. Mansoori, The use of supercritical fluid extraction technology in food processing, Food Technology, WMRC (2002).

[34] J.L. Martinez, Supercritical Fluid Extraction of Nutraceuticals and Bioactive Compounds, CRC Press, 2007, p. 420.
[35] S.M. Pourmortazavi, S.S. Hajimirsadeghi, Supercritical fluid extraction in plant essential and volatile oil analysis, Journal of Chromatography A 1163 (1) (2007) 2−24.
[36] U.R. Askin, M. Sasaki, M. Goto, Sub- and supercritical fluid extraction of bioactive compounds from *Ganoderma lucidum*, in: AIChE Annual Meeting, Conference Proceedings, 2007.
[37] K.W. Chan, M. Ismail, Supercritical carbon dioxide fluid extraction of *Hibiscus cannabinus* L. seed oil: a potential solvent-free and high antioxidative edible oil, Food Chemistry 114 (3) (2009) 970−975.
[38] A. Rajaei, M. Barzegar, Y. Yamini, Supercritical fluid extraction of tea seed oil and its comparison with solvent extraction, European Food Research and Technology 220 (3) (2005) 401−405.
[39] M. Llompart, M. Celeiro, T. Dagnac, Microwave-assisted extraction of pharmaceuticals, personal care products and industrial contaminants in the environment, TrAC, Trends in Analytical Chemistry 116 (2019) 136−150.
[40] A. Delazar, L. Nahar, S. Hamedeyazdan, S.D. Sarker, Microwave-assisted extraction in natural products isolation, in: S.D. Sarker, L. Nahar (Eds.), Natural Products Isolation, Humana Press, Totowa, NJ, 2012, pp. 89−115.
[41] M. Vinatoru, T.J. Mason, I. Calinescu, Ultrasonically assisted extraction (UAE) and microwave assisted extraction (MAE) of functional compounds from plant materials, TrAC, Trends in Analytical Chemistry 97 (2017) 159−178.
[42] S. Périno-Issartier, Zill-e-Huma, M. Abert-Vian, F. Chemat, Solvent free microwave-assisted extraction of antioxidants from Sea Buckthorn (Hippophae rhamnoides) food by-products, Food and Bioprocess Technology 4 (6) (2011) 1020−1028.
[43] Y. Wang, R. Li, Z.T. Jiang, J. Tan, S.H. Tang, T.T. Li, et al., Green and solvent-free simultaneous ultrasonic-microwave assisted extraction of essential oil from white and black peppers, Industrial Crops and Products 114 (2018) 164−172.
[44] Z. Liu, B. Deng, S. Li, Z. Zou, Optimization of solvent-free microwave assisted extraction of essential oil from Cinnamomum camphora leaves, Industrial Crops and Products 124 (2018) 353−362.
[45] V. Mandal, Y. Mohan, S. Hemalatha, Microwave assisted extraction − an innovative and promising extraction tool for medicinal plant research, Pharmacognosy Reviews 1 (1) (2007).
[46] F. Chemat, N. Rombaut, A.G. Sicaire, A. Meullemiestre, A.S. Fabiano-Tixier, M. Abert-Vian, Ultrasound assisted extraction of food and natural products. Mechanisms, techniques, combinations, protocols and applications. A review, Ultrasonics Sonochemistry 34 (2017) 540−560.
[47] C.S. Dzah, Y. Duan, H. Zhang, C. Wen, J. Zhang, G. Chen, et al., The effects of ultrasound assisted extraction on yield, antioxidant, anticancer and antimicrobial activity of polyphenol extracts: a review, Food Bioscience 35 (2020) 100547.
[48] A. Carreira-Casais, P. Otero, P. Garcia-Perez, P. Garcia-Oliveira, A.G. Pereira, M. Carpena, et al., Benefits and drawbacks of ultrasound-assisted extraction for the recovery of bioactive compounds from marine algae, International Journal of Environmental Research and Public Health 18 (17) (2021) 9153.
[49] A. Mushtaq, U. Roobab, G.I. Denoya, M. Inam-Ur-Raheem, B. Gullón, J.M. Lorenzo, et al., Advances in green processing of seed oils using ultrasound-assisted extraction: a review, Journal of Food Processing and Preservation 44 (10) (2020) e14740.

[50] M.M. Rahman, B.P. Lamsal, Ultrasound-assisted extraction and modification of plant-based proteins: impact on physicochemical, functional, and nutritional properties, Comprehensive Reviews in Food Science and Food Safety 20 (2) (2021) 1457−1480.
[51] R. Indiarto, B. Rezaharsamto, A review on ohmic heating and its use in food, International Journal of Scientific & Technology Research 9 (2020) 485−490.
[52] T. Kumar, A review on ohmic heating technology: principle, applications and scope, International Journal of Agriculture, Environment and Biotechnology 11 (2018) 680−687.
[53] M. Gavahian, S. Sastry, R. Farhoosh, A. Farahnaky, Chapter 6 − Ohmic heating as a promising technique for extraction of herbal essential oils: understanding mechanisms, recent findings, and associated challenges, in: F. Toldrá (Ed.), Advances in Food and Nutrition Research, Academic Press, 2020, pp. 227−273.
[54] P. Sun, D.W. Armstrong, Ionic liquids in analytical chemistry, Analytica Chimica Acta 661 (1) (2010) 1−16.
[55] I. Billard, A. Ouadi, C. Gaillard, Liquid−liquid extraction of actinides, lanthanides, and fission products by use of ionic liquids: from discovery to understanding, Analytical and Bioanalytical Chemistry 400 (6) (2011) 1555−1566.
[56] R. Liu, J.F. Liu, Y.G. Yin, X.L. Hu, G.B. Jiang, Ionic liquids in sample preparation, Analytical and Bioanalytical Chemistry 393 (3) (2009) 871−883.
[57] A. Dhakshinamoorthy, A.M. Asiri, M. Alvaro, H. Garcia, Metal organic frameworks as catalysts in solvent-free or ionic liquid assisted conditions, Green Chemistry 20 (1) (2018) 86−107.
[58] N.H. Snow, Solid-phase micro-extraction of drugs from biological matrices, Journal of Chromatography A 885 (1) (2000) 445−455.
[59] A. Spietelun, A. Kloskowski, W. Chrzanowski, J. Namieśnik, Understanding solid-phase microextraction: key factors influencing the extraction process and trends in improving the technique, Chemistry Review 113 (3) (2013) 1667−1685.
[60] J. Płotka-Wasylka, N. Szczepańska, M. de la Guardia, J. Namieśnik, Miniaturized solid-phase extraction techniques, TrAC, Trends in Analytical Chemistry 73 (2015) 19−38.
[61] Y. Wen, 4 − Recent advances in solid-phase extraction techniques with nanomaterials, in: C. Mustansar Hussain (Ed.), Handbook of Nanomaterials in Analytical Chemistry, Elsevier, 2020, pp. 57−73.
[62] H. Sereshti, O. Duman, S. Tunç, N. Nouri, P. Khorram, Nanosorbent-based solid phase microextraction techniques for the monitoring of emerging organic contaminants in water and wastewater samples, Microchimica Acta (2020) 187.
[63] J. Laaks, M.A. Jochmann, T.C. Schmidt, Solvent-free microextraction techniques in gas chromatography, Analytical and Bioanalytical Chemistry 402 (2) (2012) 565−571.
[64] R. Castro, R. Natera, E. Durán, C. García-Barroso, Application of solid phase extraction techniques to analyse volatile compounds in wines and other enological products, European Food Research and Technology 228 (1) (2008) 1−18.
[65] G. Theodoridis, E.H.M. Koster, G.J. de Jong, Solid-phase microextraction for the analysis of biological samples, Journal of Chromatography B: Biomedical Sciences and Applications 745 (1) (2000) 49−82.
[66] F.M. Musteata, J. Pawliszyn, Bioanalytical applications of solid-phase microextraction, TrAC, Trends in Analytical Chemistry 26 (1) (2007) 36−45.
[67] C.H. Xu, G.S. Chen, Z.H. Xiong, Y.X. Fan, X.C. Wang, Y. Liu, Applications of solid-phase microextraction in food analysis, TrAC, Trends in Analytical Chemistry 80 (2016) 12−29.

[68] M. Faraji, Y. Yamini, M. Gholami, Recent advances and trends in applications of solid-phase extraction techniques in food and environmental analysis, Chromatographia 82 (8) (2019) 1207−1249.

[69] A. Kabir, K.G. Furton, A. Malik, Innovations in sol-gel microextraction phases for solvent-free sample preparation in analytical chemistry, TrAC, Trends in Analytical Chemistry 45 (2013) 197−218.

[70] F. Chemat, A.S. Fabiano-Tixier, M. Vian, T. Allaf, E. Vorobiev, Solvent-free extraction, in: Comprehensive Analytical Chemistry, 2017.

[71] Y. Li, A.S. Fabiano-Tixier, M.A. Vian, F. Chemat, Solvent-free microwave extraction of bioactive compounds provides a tool for green analytical chemistry, TrAC, Trends in Analytical Chemistry 47 (2013) 1−11.

[72] Z.A. Aziz, A. Ahmad, S.H.M. Setapar, A. Karakucuk, M.M. Azim, D. Lokhat, M. Rafatullah, M. Ganash, M.A. Kamal, G.M. Ashraf, Essential oils: extraction techniques, pharmaceutical and therapeutic potential − a review, Current Drug Metabolism 19 (2018) 1100−1110.

[73] K. Vilkhu, R. Mawson, L. Simons, D. Bates, Applications and opportunities for ultrasound assisted extraction in the food industry − a review, Innovative Food Science & Emerging Technologies 9 (2) (2008) 161−169.

[74] L. Anfossi, M. Calderara, C. Baggiani, C. Giovannoli, E. Arletti, G. Giraudi, Development and application of solvent-free extraction for the detection of aflatoxin M1 in dairy products by enzyme immunoassay, Journal of Agricultural and Food Chemistry 56 (6) (2008) 1852−1857.

[75] Á.I. López-Lorente, F. Pena-Pereira, S. Pedersen-Bjergaard, V.G. Zuin, S.A. Ozkan, E. Psillakis, The ten principles of green sample preparation, TrAC, Trends in Analytical Chemistry 148 (2022) 116530.

[76] GreenMedChem: the challenge in the next decade toward eco-friendly compounds and processes in drug design, Green Chemistry 25 (6) (2023) 2109−2169.

[77] A. Khaled, J.R. Belinato, J. Pawliszyn, Rapid and high-throughput screening of multi-residue pharmaceutical drugs in bovine tissue using solid phase microextraction and direct analysis in real time-tandem mass spectrometry (SPME-DART-MS/MS), Talanta 217 (2020) 121095.

Chapter 9

How to evaluate the greenness and whiteness of analytical procedures?

Ebaa Adnan Azooz[1], Farah Abdulraouf Semysim[3], Estabraq Hassan Badder Al-Muhanna[2,4], Mohammad Reza Afshar Mogaddam[5,6] and Mustafa Tuzen[7]

[1]*The Gifted Students' School in Al-Najaf, Ministry of Education, Najaf, Iraq;* [2]*Radiology Department, College of Medical Technology, The Islamic University, Najaf, Iraq;* [3]*Department of Chemistry, The Al-Mutafawiqat High School in Najaf, Ministry of Education, Najaf, Iraq;* [4]*Department of Pathological Analyses Sciences, Faulty of Science, Jabir Ibn Hayyan University for Medical and Pharmaceutical, Najaf, Iraq;* [5]*Food and Drug Safety Research Center, Tabriz University of Medical Science, Tabriz, Iran;* [6]*Pharmaceutical Analysis Research Center, Tabriz University of Medical Science, Tabriz, Iran;* [7]*Tokat Gaziosmanpaşa University, Faculty of Science and Arts, Chemistry Department, Tokat, Turkey*

1. Introduction

In order to promote sustainable development, green chemistry has taken center stage in both academia and industry. The green chemistry trends have 12 concepts. They provide a framework for actions that could be implemented to make chemical products and processes more environmentally friendly. These procedures are established by chemists from a variety of chemistry-related domains, including analytical chemistry, chemical engineering, and medicinal chemistry [1–3]. Most efforts to make chemical procedures more environmentally friendly focus on using safer, more benign, and cleaner liquids or on eliminating solvents altogether and minimizing chemicals and supplements. Other efforts include limiting derivatization, adopting milder process parameters, and looking for substrates with renewable energies in order to reduce energy usage. To increase the yield, highly selective methods should be used rather than more substrates. These solutions, which were purposefully used since 1998, were well understood before they were put into practice [4–6].

The evaluation of the greenness of chemical methods is one of the issues with green chemistry. It is common knowledge that procedures cannot be regulated if they cannot be measured. Green chemistry control can be seen as a motivation to select the greenest alternative. One can compare the greenness of existing and newly generated solutions by creating and using indicators [7]. Different indicators with varying degrees of sophistication are presently used to estimate the environmental effects of chemical procedures to improve any analytical procedure by using fewer liquids, chemicals, and other reagents. Also, it is possible to separate out the green alternatives. Analysis can be performed on a smaller scale. The amount of energy used can also have an impact on the environment [8]. Twelve Green Analytical Chemistry (GAC) principles were developed in 2013 because not all concepts of green chemistry are totally applicable to analytical chemistry. GAC has its own objectives for evaluating how environmentally friendly the analysis methods are in this approach [9]. All 12 GAC concepts provide the most environmentally friendly option for each of the various components of the analytical method. These concepts were abstracted in the following [10].

(1) Reduced sample volume and sample count are desired outcomes.
(2) Direct analytical methods ought to be used to prevent sample treatment.
(3) Assessments ought to be made in situ.
(4) The connectivity of analytical processes decreases reagent usage and consumes less energy.
(5) Miniaturized and automated procedures should be chosen.
(6) It's best to avoid derivatization.
(7) Prevent the production of a lot of waste and arrange for its effective management.
(8) Approaches using more than one analyte at a time, such as those using several parameters, are recommended.
(9) Energy usage ought to be kept to a minimum.
(10) It is best to use reagents derived from renewable sources.
(11) It is advisable to modify or remove toxic reagents
(12) The operator's safety should be improved.

According to the GAC concepts, chemists are expected to take safety, healthcare, and the environment into their duties. GAC was initially put forth in 1999 as a way to lessen the negative consequences that analytical techniques have on people and the environment. Throughout the previous 10 years, there have been an increasing number of scholarly papers on the subject of green chemistry [11,12], as shown in Fig. 9.1. On March 20, 2023, information will be retrieved from the Scopus and Web of Science archives.

Numerous indicators have been devised to evaluate how environmentally friendly analytical processes work. According to how the final report will be represented, the green measurement tools have been separated into qualitative and numerical categories. A green analytical tool is a design platform for assessing the greenness of an analysis method and assisting in the exclusion of possible environmental costs associated with its creation and use. It frequently

How to evaluate the greenness and whiteness **Chapter | 9** **265**

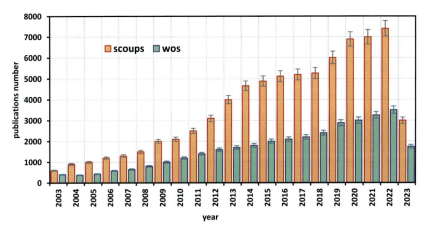

FIGURE 9.1 The number of documents there have been on the issue of "green chemistry" during the past 10 years, according to data from the Scopus and Web of Science indexes.

considers the quantity and kind of reagents used, energy consumption, possible dangers, and waste production in order to evaluate and aid in the decision-making process of such a process. The following greenness evaluation measurement tools will be discussed in this chapter.

1. National Environmental Methods Index (NEMI)
2. Developed NEMI
3. Eco-scale
4. Analytical Eco-Scale (AES)
5. Green Analytical Procedure Index (GAPI)
6. Complex Green Analytical Procedure Index (ComplexGAPI)
7. Analytical Method Greenness Score (AMGS)
8. Hexagon-CALIFICAMET
9. Analytical GREEness Scale (AGREE)
10. AGREE prep
11. White Analytical Chemistry (WAC)

This chapter offers a brief history, calcification, characteristics, and comprehensiveness assessment of the greenness tools in analytical chemistry. A comparison of their applications in the publications is also provided.

2. A brief history of greenness evaluation tools in analytical chemistry

To create less and cleaner "waste," the idea of "clean analytical chemistry" was first developed in 1993 [11]. The creator of the 12 green chemistry principles is Paul Anastas. As a result, he coined the phrase "green analytical chemistry" in 1999 [12,13]. Since then, there have been an increasing number of studies about green analytical chemistry each year. The 12 fundamentals of green analytical chemistry were developed in 2013 [10]. Then, in 2021, the 12 concepts of white analytical chemistry (WAC) were created. WAC also incorporated analysis, efficiency, and economic factors [14].

A number of tools and related indicators have been created to assess the environmental consequences of these analytical procedures, in line with the increasing interest in the development of green methodological approaches (Fig. 9.2). Greenness evaluation tools in analytical chemistry have been developed over the past decade in response to growing concerns about the environmental impacts of chemical analysis. The NEMI [15], is the most famous of these tools. It is an easy-to-use application that delivers a graphical evaluation based on four different metrics: waste, hazardous chemicals, procedures, and materials, as well as waste retention, bioaccumulation, and toxicity. Although it provides a brief summary of the analytical approach, it also demonstrates some of its weaknesses, especially its scope (e.g., no consideration of energy consumption). The NEMI [15] was the initial scale to be released in 2002. Although it has received minor upgrades over the decades [16], we chose the original model for the survey because it has not been fundamentally changed. The Eco-scale tool was introduced in 2006. The AES [17] was developed in 2012. It assigns a score out of 100 to the analytical procedure under consideration and works with penalty points deducted from the highest score of 100. The GAPI [18], which was established in 2018, assesses the "greenness" of a methodology from collecting through analysis and presents the results using five pentagram shapes with three specific colors, each pentagram representing a stage of the methodology.

The AMGS [19], created in 2019, was created by scientists working in the drug industry. It provides a formula-based score for the usage of the analytical tool and solvents. The hexagon-CALIFICAMET [20] was reported in 2019.

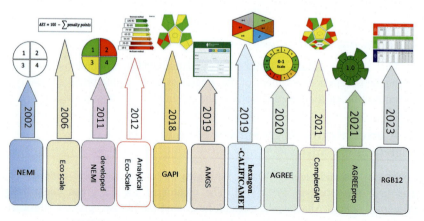

FIGURE 9.2 Historical development of green analytical tools.

Consistently performed, toxicology and safety, wastes, carbon emissions, and economic cost are only a few of the categories assessed. A hexagonal pictograph shows the qualification of these sections on a scale from 0 to 4. The GAPI tool was developed in 2021 to produce the Complex Green Analytical Procedure Index (ComplexGAPI) scale with many advantages. The AGREE scale [21], the most current metric, assigns a technique a score based on criteria that adhere to the 12 principles of green analytical chemistry. Although each selected tool has its own computations and requirements, these tools and indicators might point researchers in the direction of more environmentally friendly methods. Due to this, comparing them on an equal footing can be helpful in directing researchers toward one over the other, depending on their goals and the level of accuracy they are looking for from these indicators [22]. White analytical chemistry concepts have been added to the RGB (red-green-blue) additive color image as of 2021 [14]. The worker in this upgraded model is required to fully consider all 12 WAC parameters. The latest greenness evaluation tools, are designed to evaluate not only the environmental impact of chemical processes, but also their economic and social impact. These tools are becoming increasingly popular as scientists and industry strive to achieve more sustainable and responsible.

3. Classification of greenness evaluation tools

The outcomes or data needed, as well as whether they are qualitative or quantitative, can be used to classify green evaluation tools. The numerical scale, with or without a unit, is used to portray the quantitative data. It could be a count, a range, or a ratio. while the qualitative data's category, or non-numerical value. Tools like AMGS, Eco-Scale, WAC and AGREE are regarded as numerical evaluation tools, while the NEMI, modified NEMI, and GAPI are assumed to be qualitative assessment tools. Additionally, these tools were put to use while using particular green analysis techniques that were suggested.

There is another classification of green analytical tools, depending on the type of method used in calculating the greenness of the analytical method, into manual tools or software tools. The use of computational programmers for metrics (software, spreadsheets) has grown in popularity in recent years. Many works have made use of database designs and computing techniques to aid in the comprehension of green chemistry measurements. These modern programs, including AGREE, GAPI, ComplexGAPI, WAC, and AGREEprep, are available without charge. The AMGS tool have online program. Table 9.1. Shows summary of selected tools and each website for software.

TABLE 9.1 The principles, purposes, advantages, disadvantages, and ideal pictograms for selective tools.

Tool/debut year	Purposes	Parameters	Type	Advantages	Disadvantages	Ideal pictogram	References
NEMI 2002	Comparison of the parameters and getting the knowledge	1. Bio-accumulative, and hazardous materials. 2. Toxic waste 3. Corrosive conditions 4. Waste quantity	Qualitative	Simple pictograph and clear, direct information about the procedure's potential environmental effects.	Lack of energy, measurement of chemicals usage, waste generation quantity, and time-consuming searches for each chemical employed in the method in official listings		[15]
Developed NEMI 2011	Comparison of the parameters and getting the knowledge. It is available at the website http://www.nemi.gov	1. Operator danger 2. Chemicals 3. Energy 4. Waste	Semi-quantitative	Simple; speedy; take into account chemicals and energy use.	It does not provide a specific value.		[23–25]

Eco-scale 2006	Determine the quality of the organic production using six criteria	1. Yield 2. Cost 3. Safety 4. Technical system 5. Heating/time 6. Preparing/purifying	Semi-quantitative	Numerically simple; honest; calculated in less than 5 min; provides a general overview of the reaction conditions	Absence of energy and extra constants capable of distinguishing between technique applications at the micro, meso, and macroscales; lack of information in the case of negative environmental effects	**Ecoscale = 100 – the sum of each penalty**	[17,26,27]
Analytical Eco-scale 2012	Evaluation of a technique's green characteristics	1. Quantities 2. Hazardous 3. Energy 4. Occupational hazard 5. Waste	Quantitative	Semi-quantitative metrics of chemicals and waste levels; quantitative data on the effects of analytical methods on the surroundings; coverage of various environmental effect factors while evaluating the surroundings; the simplicity of comparing various	Absence of extra quantifiers able to distinguish between the nano-, micro-, meso-, and macroscales of technique usage; outcome is not instructive in case of adverse environmental effects; lack of knowledge in circumstances where the analytical process has an impact on the environment, such as when using		[28]

Continued

TABLE 9.1 The principles, purposes, advantages, disadvantages, and ideal pictograms for selective tools.—cont'd

Tool/debut year	Purposes	Parameters	Type	Advantages	Disadvantages	Ideal pictogram	References
				analytical techniques	solvents or other chemicals, posing a risk to workers, or producing waste.		
GAPI 2018	A semi-quantitative program to assess and measure the environmental effects associated with each stage of an analytical process for lab work and academic purposes. The free software is available at https://mostwiedzy.pl/en/justyna-plotka-wasylka,647762-1/gapi	1. Sample preparation
2. Method type
3. Reagents and solvents
4. Amounts of reagents
5. Instruments	Semi-quantitative/software	Measures environmental effects from samples taken to the end; offers qualitative data; makes it easier to compare various analytical techniques; takes a simple approach	Does not take into account the synthesis phase prior to sample collection, which leads to a lack of clarity regarding the overall greenness of the analysis procedure; the results do not always include the structure-related details of risks.		

ComplexGAPI 2021	For evaluating the greenness of analysis techniques, downloadable software is publicly available at mostwiedzy.pl/complexgapi	6. Sample preparation 7. Method type 8. Reagents and solvents 9. Amounts of reagents 10. Instruments 11. Yield, requirements 12. Green economic 13. Chemicals and reagents 14. Instrumentation 15. Final product 16. E-factor	Semi-quantitative/software	Includes all factors prior to and during the analysis procedure, allowing each step to be evaluated and controlled.	Difficulty of the modification	
AMGS 2019	Determine the effect of procedure model and instrument selection decisions. The free program is available on the ACS-GCI-PR website at https://www.acsgcipr.org/amgs	1. Instrument 2. Energy 3. Solvent energy 4. Solvent safety	Quantitative/online	Equation using a calculator and a reference SHE/SSG and CED values are among the criteria. Appropriate for the process of comparing	Need to search through many databases	

Continued

TABLE 9.1 The principles, purposes, advantages, disadvantages, and ideal pictograms for selective tools.—cont'd

Tool/debut year	Purposes	Parameters	Type	Advantages	Disadvantages	Ideal pictogram	References
Hexagon-CALIFICAMET 2019	Evaluating the effects of various analytical techniques on the environment	1. FM-1 (preparation of samples, method characteristics, calibration) 2. FM-2 (quality control, accuracy) 3. Toxic effects and safety evaluating 4. Residues development 5. General environmental 6. Economic cost	Quantitative	A perfect pictogram, a simple penalty score system that is suitable with the eco-scale, a clear reading, and a more thorough examination of a method's qualities are offered.	Many guidelines and knowledge are needed to give penalty points. Hexagon should be rescaled, as it now uses a five-level index with a qualification range of 0–4.		
AGREE 2020	Focused on the 12 principles of green analytical chemistry, assessing workplace and	1. Sample preparation 2. Sample size 3. In situ 4. Integration of analytical	Quantitative/ software	The overall result is instructive and pertinent according to the GAC's guiding principles.	It disregards the synthesis phase prior to sample preparation, leaving the total greenness of analytical methods in doubt;		[2,21]

	environmental risks connected to a certain analytical method. The free software is accessible at 1- https:// mostwiedzy.pl/ AGREE 2- git.pg.edu.pl/ p174235/ AGREE	procedure and used energy 5. Automation 6. Derivatization 7. Waste 8. Analyte numbers 9. Energy 10. Reagents 11. Hazardous materials 12. Safety of operator			outcomes do not always provide data on the structure of risks. The criteria for each part's weighting are unclear.	
AGREEprep 2021	The code of this tool available at https://git.pg. edu.pl/ p174235/ agreeprep It was designed using 10 classifications and the preparation guidelines for green samples. Additionally, a sub score	1. Sample preparation 2. Hazardous materials 3. Materials Renewability 4. Waste 5. Sample size 6. Sample throughput 7. Automation 8. Energy 9. Analytical instrument 10. Safety	Quantitative/ software	Covers the entire sample treatment stag	Does not support the effective use of the approach.	[29]

Continued

TABLE 9.1 The principles, purposes, advantages, disadvantages, and ideal pictograms for selective tools.—cont'd

Tool/debut year	Purposes	Parameters	Type	Advantages	Disadvantages	Ideal pictogram	References
	range from 0 to 1 is used, along with an overall qualitative computation value. The outcome is represented visually by a vibrant pictogram that shows the total sample preparation efficiency and the final assessment value located inside a colored sphere in the center of the image.						

| WAC 2021 | The WAC idea is an expansion and addition to the GAC concepts already supported in analytical chemistry, together with the RGB 12 algorithm as a specific method assessment tool. The supplementary data were available at https://ars.els-cdn.com/content/image/1-s2.0-S0165993621000455-mmc1.docx | 1. Four red parameters 2. Four green parameters 3. Four blue parameters | Quantitative/ software | A tool demonstrates the coherence and synergy of the analytical, ecological, and practical features with reference to the RGB color system. Whiteness can also be measured and used as a useful metric for comparing and choosing the best approach. | Include difficulty in use and the need for accuracy in the work. |

4. Summary of the chosen greenness evaluation tools

Numerous tools have been established over time to assess how environmentally friendly analytical techniques are. While many of them are specific and only apply to certain types of analytical methods, others are more generic and may be used with most analytical methods. Few of them are now particularly popular. Thus, this chapter will be focused on those particular measures.

NEMI, AES, GAPI, AMGS, hexagon-CALIFICAMET, AGREE, and WAC are the metrics chosen for this summary. Following are the justifications for including these tools.

(1) Analytical Eco-Scale, GAPI, and AGREE are the most commonly used tools since they apply to the vast majority of analytical methods. Some tools are restricted to specific types of analytical methods. HPLC-EAT and AMVI, for example, are only acceptable for liquid chromatography methodologies.
(2) Numerous criteria for reasonably green analytical chemistry principles are required for AES, AMGS, GAPI, and AGREE. They are provided with comprehensive descriptions of the entire process. Five characteristics contribute AES, ten green characteristics of AMGS, 15 characteristics cover GAPI, and 12 characteristics consist of AGREE.
(3) Due to the great number of Scopus, Springer, and WOS citations for the studies indicating AES, GAPI, and AGREE, these tools are extensively employed.
(4) NEMI and the developed NEMI were extremely simple.
(5) The WAC, ComplexGAPI, and AGREEprep software measurements are recent and have not yet been discovered.

Before delving into the specifics of the chosen metrics, Table 9.1 gives a quick summary of some of the metrics described in Fig. 9.1, along with descriptions of their goal, selection criteria, benefits, and drawbacks. Therefore, we have attempted to condense the basics of the many quantitative pictograms and compare them with each other.

4.1 National environmental method index (NEMI)

4.1.1 NEMI principles

The NEMI label is one of the earliest indicators for measuring the environmental friendliness of analytical techniques. A four-field wheel is used for the NEMI label. Each area corresponds to a distinct purpose of the analysis. When these conditions are met, the area is labeled green [23]. The following are the four principles of NEMI (Fig. 9.3).

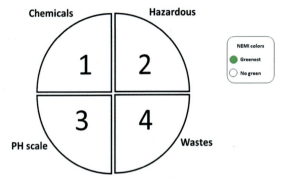

FIGURE 9.3 The graphic illustrates the evaluation score determined by the NEMI metric method. It has four quadrants (chemicals, toxic, corrosive, and acidity), each of which is colored green if the associated characteristic is met.

1. Area 1 was designated as "green," and the requirement is that none of the substances used in the experimental process relate to the list of (PBT) persistent, bio-accumulative, and hazardous materials.
2. Area 2 In order to be green, none of the substances used must be on the toxic waste D, F, P, or U lists.
3. Area 3 can be highlighted green if the pH of the test is between 2 and 12, avoiding extremely corrosive conditions.
4. Area 4 will be a green area if the quantity of waste produced does not exceed 50 g.

4.1.2 Strengths and weaknesses of NEMI

The fundamental benefit of NEMI as a method for evaluating greenness is that consumers may easily comprehend the procedure. It suffices to take a quick glance at the NEMI symbol to get the gist of how the operation affects the environment. The information gathered is relatively general, and filling out the NEMI symbol takes time, which are its two main downsides. Each risk is shown to be below a specified value by the NEMI symbol. As a result, it cannot be classified as semi-quantitative. The second drawback is related to the laborious creation of a sign, especially when numerous unusual compounds are utilized [23,24].

Furthermore, despite the fact that energy is one of the 12 GAC principles' fundamental terms, the NEMI tool didn't consider it a factor for evaluation. In addition, the evaluation's findings are only qualitative; no information is provided on the quantity of waste or risks, and each danger is only described as being over or below a particular threshold [25].

4.2 The developed NEMI tool

To make NEMI more quantifiable and accurate, the NEMI tool was developed in 2011(Fig. 9.4). The metrics of developed-NEMI are based on the amounts of hazardous substances, operator danger, wastes, and energy. Red, yellow, and green color additions to the fields' existing hues were recommended. Green represents an environmentally friendly technique, yellow represents a medium, and red represents a toxic analytical methodology [30–33]. The NEMI, an online database that may be viewed, is available at the website http://www.nemi.gov.

4.3 The Eco-scale tool

The Ecoscale was published in 2006 as an initial evaluation tool for measuring green organic reactions [17]. This method is based on the concept that a reaction qualifies as an "ideal process" with a penalty total score of 100. produces a full score of 100 if it uses inexpensive substances (reagents) at room temperature, a 100% yield and is not dangerous to workers or the surroundings. As a result, whenever a characteristic deviate from its optimum value, the penalty points are reduced and a lower score. The greener the procedure under investigation, the better the score. This model can be used to test whether research approaches are environmentally friendly [33]. Analytical Eco-Scale was developed in 2012 and is the most commonly used tool to assess the greenness of analytical methods [34].

4.3.1 The Eco-scale principles

On a laboratory scale, there are no limitations on the use of certain chemicals or liquids, although high-yield reactions may still be prohibited for larger-scale

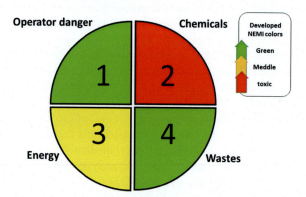

FIGURE 9.4 Developed NEMI pictogram for assessing greenness. It has four quadrants (chemicals, wastes, operator danger, and energy), each of which is colored by three colors: green, red, and yellow.

manufacturing, for instance by using a highly combustible or poisonous liquid or a costly reagent. Similar to this, since no energy is required for warming or cooling, processes at room temperature are significantly more significant on an industrial level. Additionally, the waste problem only matters slightly in a lab, but it can significantly increase the cost of a manufacturing phase. Therefore, it is crucial to emphasize that this Eco-Scale was created specifically for processes in the laboratory. More information on the foundation of these statistics in Ecoscale is provided below. In particular, since certain chemists could object to the suggested relative assignments, the subjective attribution of specific weights to different penalty points can be easily changed. It is intended for the Ecoscale to be a versatile tool.

4.3.1.1 Yield

One of the most crucial aspects is the yield. Indirectly, this showed that, on average, selectivity was higher since higher functional group compatibility results in better reactions. Although the pre-purification yield is ideal in theory, it is not feasible to use. First, this yield is frequently unknown (and not mentioned in the literature). Furthermore, the importance of reacting to situations in which the final product cannot be effectively filtered is debatable. Scores are assigned to the separated yield as well. The measurement of non-racemic synthesis can also be done using the Eco-Scale technique. In this situation, only the enantiomer's molecular composition is taken into account. The addition of effective asymmetric auxiliary species can greatly increase the Eco-Scale, but their quantity, quality, and low toxicity have a considerable impact on the final result [35].

4.3.1.2 Cost of used compounds

The costs are cumulative and account for every aspect of the reaction. The classification of the reaction's constituent parts as "cheap" is arbitrary. We classify a reaction component as affordable if it costs less than US$10 to use it to synthesize the final product on a scale of 10 mmol, and as very costly if it costs more than US$50. Solvents are frequently cheap and make up the majority of the reaction components found in reaction mixtures with more than 10 equivalents. However, typical solvents employed in completely anhydrous and/or high-dilution (big volume) situations should be reassessed as pricey components. The Eco-Scale score can be positively impacted by factors like a higher yield, a better safety profile, or an easier workup when an expensive solvent (like ionic liquid) is used. Two such unique examples can be mentioned. There is one less reaction component in a solvent-free reaction, and when the solvent is used as the reagent, no additional penalties are taken into account. It should be noted that improved selectivity (yield) and reduced energy requirements, which are considered in other areas of the eco-scale, are typically represented in the true benefits of integrating catalysts [16].

4.3.1.3 Safety

When performing organic chemistry experiments, safety must come first. There is always a risk involved when working with chemicals, thus it is important to completely comprehend any possible hazard. Among other things, organic molecules can be explosive, highly flammable, corrosive, mutagenic, teratogenic, carcinogenic, or lachrymatory. In addition, the hazard can rise over time, and photo oxidation of ether to form explosive peroxides are a good example. It must be stressed as well that it takes time for new goods' safety profiles to be completely described [36]. Finally, it's important to remember that specific combinations of different substances can lead to dangerous outcomes (e.g., exothermic reaction between acids and bases). A broad range of information, including the safety and health details in risk and safety terms, the material safety reports, and the danger warning marks on the containers, is easily accessible for evaluating these threats. The hazard warning signals are used as a guide to avoid performing a difficult computation. Each reaction element classified with T+ (highly toxic), F+ (highly flammable), or E (explosive) is penalized with 10 points, while compounds will be labeled with T (toxic), F (extremely flammable), or N (environmentally hazardous) and penalized with 5 points [26]. As shown in Table 9.2, the use of unsafe substances has the greatest influence on the overall synthesized quality when compared to other entries.

4.3.1.4 Technical system

There are no deductions for a simple system that includes a standard flask, a reflux condenser, and stirring. The general quality of the synthesis is reduced by any extras like specialized glassware, controlled chemical addition devices, pressurized vessels, the use of experimental methods like microwave heating, ultrasonic, or photocatalysis, and the requirement for an inert atmosphere, particularly in a glove box [37].

4.3.1.5 Reaction heat and time

The temperature and time of the reaction are inextricably linked. In an ideal case, a reaction would move forward quickly at room temperature. Yet, in order to complete formulation in a reasonable amount of time, heating is frequently necessary. Yet, cooling is more challenging than heating. When cooling conventionally (without the use of a cryostat), only repaired temperatures are useable (e.g., $0°C$ for an ice bath or $-78°C$ for acetone/sCO_2), and extreme care must often be used to avoid extra water in order to create repeatable results. Above room temperature, the heating scope is constant. The comparative penalty points take these characteristics into account. The penalties add up, so if heating and cooling are necessary for the process, both must be taken into consideration [31].

TABLE 9.2 The penalties used to determine the Eco-scale values.

Parameters	Degrees
1. Yield	(100-% yield)/2
2. Cost of used compounds to product (10 mmol) • cheap (>10$) • expensive (10–50$) • very expensive (<50$)	0 3 5
3. Safety • N (danger) • T (toxic) • F (flammable) • E (explosive) • F+ (highly flammable) • T+ (highly toxic)	5 5 5 10 10 10
4. Technical system • general system • addition instruments • alternative activation strategy • high pressure • special glassware • gas atmosphere (inert) • glove box	0 1 2 3 1 1 3
5. Reaction heat and time • 20–25°C, <1 h • 20–25°C, <24 h • heating, <1 h • heating, >1 h • cooling to 0°C • cooling less than 0°C	1 2 3 4 5
6. Preparing and purifying • none, cooling at room temperature, simple filtration, adding solvent, or remove solvent (boiling point <150°C). • filtration and crystallization • separation solvent (boiling point >150°C). • extraction by solid phase • extraction by liquid phase • distillation or sublimation • classic chromatography	0 1 2 2 3 3 10

4.3.1.6 Preparing and purifying

It can be difficult to prepare and purify the final product. The factor "a period of time to get the final products to a purity of above 98%" is chosen as the major criterion in giving the marks in order to prevent a complex computation. Standard purification processes are graded according to their runtime because using a clock during a lab workup operation is unnecessary. The penalty points are calculated by taking into account each step of the workup process [27].

4.3.2 Strengths and weaknesses

The ecoscale generally supports high-yielding, inexpensive, safe reaction mechanisms as well as simple purifying [17]. The tool meets the following criteria: (1) it is simple, clear, and quick; (2) it does not adopt a general viewpoint but considers the benefits and drawbacks of particular methodologies or auxiliary reagents; (3) it provides a general overview of the reaction conditions, and the areas for improvement are clearly indicated. This makes it a useful tool for understanding reaction protocols and useful for research as a way to compare several sets of preparations of the same substance [26]. Regrettably, the Ecoscale style has certain drawbacks because the score it produces does not take into account the identities of the toxic compounds. There is also a lack of information regarding the causes of unfriendly environmental impacts, such as the kinds of chemicals and liquids utilized as well as the waste that is produced.

Also, yield is usually given an excessive amount of weight. Only a tenuous relationship between reagent cost and greenness may exist. Not taken into account are the quantity and kind of reaction or workup solvents. Each category's penalty points are assigned at random (e.g., safety, technical set-up). The product's chemical and physical characteristics are not taken into account [27].

4.4 Analytical eco-scale tool

In order to assess how environmentally friendly a test procedure is, a novel comprehensive system called the analytical Eco-Scale was introduced in 2012. It is focused on deducting points for analytical process characteristics that do not align with the ideal green test. With this method, several analytical factors and processes are contrasted. The suggested analytical Eco-Scale can be an effective semi-quantitative instrument for laboratory application. It is quick and simple to use and can be combined with any established or novel approach. It also includes well-established assessment criteria. All of these characteristics will affect how widely the analytical Eco-Scale is used in labs. The suggestion has some drawbacks. Despite their obvious benefits in determining a specific analyte group, some analysis procedures and methods may not receive the highest score on the analytical Eco-Scale. For instance, determining organic molecules is typically far more difficult and multi-stage than determining trace amounts, resulting in a significantly larger number of penalty points and a lower ranking on the Eco-Scale [38].

The analytical Eco-Scale identifies the weakest points in analytical processes. Deployment will be crucial for the search for innovative, environmentally friendly approaches. The ideal green analysis includes the use of removal reagents, the use of less energy, and the generation of no waste. Only if these requirements are satisfied, the following equations should be used to calculate the Eco-Scale, considering the total penalty points for the entire procedure:

$$\text{Analytical Eco-Scale score} = 100 - \text{total penalty points} \quad \dots \quad (9.1)$$

It is necessary to explain how penalty points are assigned to energy and hazardous classes. Several classes of dangerous chemicals can be used to evaluate the risks associated with the agents used in analysis techniques. For instance, the EPA's Toxic Release Inventory, the Clean Water Act, and the Clean Air Act are all used by the NEMI system to evaluate environmental risks, while the NFPA's categorization is used to measure safety and health risks [38]. The Globally Harmonized Standard of Classification and Labeling of Substances (GHS), which is the most complete and current category of substances, will be used to inform users of their environmental and health effects [39].

4.4.1 The analytical Eco-scale principles
4.4.1.1 Chemical penalty points

Liquids or solvents can be given penalty points in a pretty straightforward manner. This process begins with a chemical box that has warning symbols on it. A single score is given for each danger symbol. Moreover, the number of hazardous icons is multiplied by two if the substance is defined with the term "danger." but it is multiplied by one if the substance is represented with the term "warning." Examples for calculating hazardous PPs for several regularly applied liquids and reagents are provided in Table 9.3. The penalty points are associated with the quantity of the chemicals [28]. The dosage of reagents or liquid is multiplied by one if it is just under 10 mL or 10 g, by two if it is in the range of 10–100 mL (g), and by three if it is greater than 100 mL or 100 g. The following generic equation can be used to perform these measurements:

Chemicals PPs = Number of hazard icons × Risk label (Warning = 1 or Danger = 2) × chemical amount (<10 mL or 10 g = 1, 10–100 mL or g 2, >100 mL or 100 g 3). .... (9.2)

4.4.1.2 Energy PPs

The following conditions are used to calculate the PPs for energy usage: Energy usage of less than 0.1 kWh per test means zero; energy usage of more

TABLE 9.3 The penalty points (PPs) values of many reagents and solvents.

Chemicals	Number of hazard icons	Symbol label (warning (1), danger(2))	Hazard PPs
Acetic acid (30%)	1	Danger	2
Acetone	2	Danger	4
Acetonitrile	2	Danger	4
Acetylene	2	Danger, warning	4
Ammonia solution (25%)	3	Danger	6
Benzene	3	Danger	6
Benzoic acid	1	Warning	1
Carbon tetrachloride	2	Danger	4
Chloroform	3	Danger	6
Dichloromethane	2	Danger	4
Dimethyl ether	2	Danger	4
Ethanol	2	Danger	4
Ethyl acetate	2	Danger	4
Glacial acetic acid	2	Danger	4
Hydrochloric acid (30%)	2	Danger	4
Hydrogen peroxide (30%)	2	Danger	4
Isooctane	4	Danger	8
Isopropanol	2	Danger	4
Methanol	3	Danger	6
n-hexane	4	Danger	8
Nitric acid (65%)	2	Danger	4
Potassium dichromate	5	Danger	10
Sodium hydroxide (30%)	1	Danger	2
Sulfuric acid (25%)	1	Danger	2
Toluene	3	Danger	6

than 1.5 kWh per test equals one [40]. Table 9.3 provides details on the list of several frequently employed equipment and systems, their classification in terms of energy consumption, and any related penalty points.

TABLE 9.4 The PPs value related on energy and instruments usage.

Instruments	Energy used (KWh/sample)	PPs
UV-VIS spectrometry, FTIR, titration, UPLC, immunoassay, spectrofluorometric, hot-plate solvent evaporation, (less than 10 min), needle evaporator, sonicator.	<0.1	0
AAS, Chromatography(GC, LC), ICP-MS	≤1.5	1
NMR, XRD, LC-MS, GC-MS, hot-plate solvent evaporation, (more than 2.5 h)	>1.5	2

The PPs for energy consumption are assigned based on the following criteria: Energy consumption ≤ 0.1 kWh per sample = 0, Energy consumption ≤ 1.5 kWh per sample = 1, Energy consumption ≥ 1.5 kWh per sample = Information on the list of some commonly used instruments and processes and their characterization in view of energy consumption, and associated penalty points are provided in Table 9.4.

4.4.1.3 Occupational hazard PPs

Occupational hazards have two values: zero or three. Three PPs are listed for any vapors or gases released into the air during the analysis process. However, if the analysis process is harmonized, no PPs are assigned [41].

4.4.1.4 Waste PPs

In the event that the process of analysis generates waste, the following PPs are granted according to the amount of the trash: <1 mL(g) = 1, 1−10 mL(g) = 3, >10 mL(g) = 5. If waste is produced, the disposal of the waste also was given points according to the following criteria: no treatment (3), passivation (2), degradation (1), recycling (0).

4.4.1.5 Description of the total score

The total PPs are subtracted from 100 to arrive at the final result. A "great green analysis" is one that has a score of over 75. A suitable green evaluation is defined as a value between 50 and 75. "Incomplete green analysis" is defined as a value of less than 50 [41].

4.4.2 Strengths and weaknesses of AES

The AES offers a numerical evaluation of the strategy's greenness that can be contrasted with a high score of 100. Furthermore, depending on their ultimate score, several approaches can be quickly contrasted in terms of how green they are. The AES doesn't really take into account the type of pictogram when calculating the PPs of danger for a substance, which is one of its limitations. It simply asks for the quantity of icons that include the word "warning" or "danger" multiplied. The pictogram's seriousness or danger factor are not considered. One cannot infer from the score alone which stage in the analysis process resulted in the environmental effects. Waste is allocated PPs depending on the amount and how it was treated; waste kind and hazards are not taken into account. Also, it would be ideal if the AES findings could be retrieved using free software. Although it has several drawbacks, it gives quantitative results by taking a number of analytical procedures into account (Table 9.5).

TABLE 9.5 The principles and measurements of analytical eco-scale.

Reagents

Penalty points (PPs) of analytical eco-scale		Sub-value of PPs
1. Quantities	Less 10 g (mL)	1
	10–100 g (mL)	2
	More than 100 g (mL)	3
2. Hazardous (Physical, environmental, health)	None	0
	Less dangerous risk	1
	More dangerous risk	2
PPs reagents value = quantities value of PPs × hazardous value		

Instruments

3. Energy	Less than 0.1 KWh/sample	0
	Less than 1.5 KWh/sample	1
	More than 1.5 KWh/sample	2
4. Occupational hazard	Analytical process hermetization	0
	Vapors or gas emissions into the atmosphere	3
5. Waste	None	0
	Less 1.0 g (mL)	1
	1–10 g (mL)	3
	More than 10 g (mL)	5
	Recycling	0
	Degradation	1
	Passivation	2
	No treatment	3
Instruments value = $\sum (3 + 4 + 5)$		
Total value = $\sum$(PPs reagents value + instruments value) = $\sum 1-5$		
Analytical eco-scale value = 100−total value (green value = ≥ 75)		

4.5 Green analytical procedure index (GAPI) tool

Potka-Wasylka created the GAPI as a semi-quantitative tool in 2018. At first, it was utilized to evaluate aromatic hydrocarbons in wine samples. The measuring tool produces a pictogram using a three-stage color scheme to evaluate an analytical method's impact on human health and the environment. This tool is constructed around a five-pentagon structure that has been broken into 15 parts. These cover all phases of the analysis procedure, including the technique type, chemicals, waste produced, instruments, and quantification. Green denotes a technique that is environmentally friendly, whereas red denotes a procedure that is not environmentally friendly. These colors are employed to evaluate the influence of each factor: green denotes a low, yellow denotes a medium, and red denotes a major impact on the environment.

4.5.1 GAPI tool principles

GAPI was an environmental measurement tool used to evaluate several analytical process phases, starting with the collection of samples and continuing through sample treatment, equipment, and procedure quantification [40]. The GAPI pictogram's five primary components are listed in Table 9.6, along with its 15 specific components.

As seen in Fig. 9.5, a five-pentagon symbol depicts a description of the degree to which each step of an analysis procedure is considered to be green using a color scale with three varying colors of evaluation for each step [2]. The area turns green if the given requirements have been satisfied, yellow for medium environmental effects, and red for a non-eco-friendly feature. Each of the 15 parts represents a unique part of the researched analytical technique. The key criteria suggested by the originally stated GAPI parameters are shown in Table 9.6.

4.5.2 Strengths and weaknesses of GAPI

The GAPI tool is used to evaluate all pertinent aspects associated with an analytical procedure. Sample collection, preparation, preservation, reagents, equipment, and waste are all included. Additionally, it provides insightful data with thorough descriptions of every step of the analytical procedure. This allows the identification of problems that the method was unable to resolve, allowing for further advancement in terms of greenness. This evaluation index is also one of the best tools for comparing various analytical methods and determining which is the greenest of them all. In comparison to other evaluation tools, the GAPI tool's main flaw is how complicated it is to manipulate. Moreover, the same results might be drawn by combining NEMI with ESA.

4.6 Complex green analytical procedure index (ComplexGAPI)

Several studies used software tools to aid in the comprehension of green chemistry measurements. Potka-Wasylka and Wojnowski released an easy

TABLE 9.6 Lists of the 15 parameters of the GAPI tool.

Section	Green	Yellow	Red
a – sample preparation			
1- Collection	In-line	On-line/ at-line	Off-line
2- Preservation	None	Physical/ chemical	Physical-chemical
3- Transport	None	Necessary	-
4- Storage	None	Under normal conditions	Special conditions
b- method type (direct/ indirect) A ring in the center of the graph refers to this procedure as qualification.			
5- General method	No preparation of sample	Filtration, decantation	Extraction
c- reagents and solvents			
6- Scale of extraction	Nano extraction	Micro extraction	Macro extraction
7- Solvents used	Free	Green solvents	Non-green solvents
8- Addition treatments	None	Simple treatments (clean up, remove of solvent)	Advanced (derivatization, mineralization)
d- amounts of reagents			
9- Quantities	<10 g (mL)	10-100 g (mL)	> 100 g (mL)
10- Health hazard	NFPA danger ranking (0 or 1), slightly poisonous, slightly irritating	NFPA (2 or 3), moderately poisonous, may lead to brief incapacity.	NFPA (4), Low exposure causes major harm; famous small animal cancer.
11- Safety hazard	No special dangers, the maximum NFPA flammability ranking is 0 or 1.	special dangers, the maximum NFPA flammability ranking is 2 or 3.	the maximum NFPA flammability ranking is 4.
e- Instruments			
12- Energy	Less than 0.1 KWh/sample	Less than 1.5 KWh/sample	More than 1.5 KWh/sample
13- workplace hazards	analysis procedure hermetic securing	-	Vapour emissions into the environment
14- Waste	Less 1.0 g (mL)	1-10 g (mL)	More than 10 g (mL)
15- Waste treatment	Recycling	Degradation Passivation	No treatment

NFPA, national fire protection association.

update of the work mentioned above under the title Complex Green Analytical Procedure Index (ComplexGAPI). In this tool, two or three stages of assessment are used along with a pentagram-pictogram and a color scheme. A hexagon-pictogram was also used, as seen in Fig. 9.6, to support the assessment by representing the yield, chemicals, liquids, parameters, instruments, workup, and finished product purification [42]. By including more factors relevant to the procedures carried out before the analytical method itself, the ComplexGAPI builds upon the well-known green analytical procedure index. The hexagonal that was created for the GAPI pictograph has six additional fields, each of which represents a separate step in the outlined procedure and is

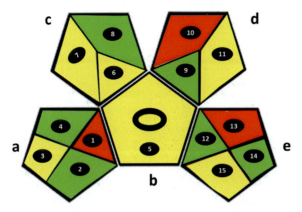

FIGURE 9.5 GAPI tool for assessing greenness. It has 15 quadrants each of which is colored by three colors: green, red, and yellow.

colored green when certain criteria are fulfilled. Fig. 9.10 shows the ComplexGAPI symbol, which is color-coded for clarity and has specific areas of the new hexagon symbol organized together. The basic GAPI schematic diagram is whited out in the backdrop. Table 9.7 explains all additional sections in ComplexGAPI.

The goal of the tool's advancement was to provide a more thorough assessment of the procedure's "greenness" by enabling it to evaluate as much information as possible about a specific analysis technique, covering operations carried out before analysis [42]. The ComplexGAPI is a model that includes every step of an analysis procedure, including collection of samples, transportation, preservation, and storage, as well as sample preparation and evaluation. It also includes those processes and procedures that are performed before the overall analysis techniques. In order to evaluate processes like the creation of novel ionic liquids (ILs), deep eutectic solvents (DESs), nanomaterials, etc., in addition to other components utilized during the separation process, such as phases for columns, numerous scientists who utilized GAPI in their lab work had questions about how to modify the initial GAPI software [43]. The same ideas that underpinned the creation of GAPI—the analytical eco-scale and the eco-scale—also served as the foundation for ComplexGAPI. Also, some specifications from the CHEM21 program were considered when creating ComplexGAPI. As a result, the new index will be easier to use. Due to the availability of the ComplexGAPI computer system, the evaluation procedure should be a lot simpler and quicker. By including a second hexagonal box at the bottom of the pictogram made for GAPI, the ComplexGAPI index broadens it [44]. This section reflects the "green" nature of pre-analysis procedures. It addresses topics including yield and circumstances, chemicals and liquids, equipment, work-up, and the purification of the final products (Fig. 9.10).

TABLE 9.7 Additional parameters in the ComplexGAPI scale (yield, and E-factor sections).

Type	Green	Yellow	Red
Initial procedures Yield, selection, and factors			
(I) Yield%	>89	70-89	<70
(II) Temperature / time	Without heating / < 1 h	Without heating / >1 h Heating / < 1 h Cooling to 0 °C	Heating / > 1 h Cooling < 0 °C
Regarding the green economy			
(III) Rules number met	5-6	3-4	1-2
(IV) Chemicals and reagents			
(IVa) Health risk	NFPA health concern value is 0 or 1; slightly poisonous or slightly irritating.	moderately poisonous; may temporarily incapacitate; NFPA = 2 or 3.	Brief exposure causes serious harm; minor animal cancer that is known or suspected; NFPA score of 4
(IVb) Safety risk	NFPA value of 0 or 1 for the highest flammability and instability. No specific risks	greatest NFPA combustibility or instability value is a 2, a 3, an associated special danger.	The NFPA's maximum score for combustibility or instability is 4.
(V) Instrumentation			
(Va) mechanical system	Common setup	additional arrangements /semi-advanced equipment	Gloves box; pressure apparatus >1 atm
(Vb) energy	Less than 0.1 KWh/sample	Less than 1.5 KWh/sample	More than 1.5 KWh/sample
(Vc) Workplace risk	analysis procedure hermetic securing	-	Vapour emissions into the environment
(VI) final product			
(VIa) Workup and purification	None	standard purification techniques	advanced purification techniques
(VIb) purity %	>98	97-98	<97

E-factor = *total waste mass/total product mass*.

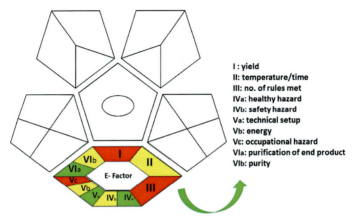

FIGURE 9.6 The ComplexGAPI pictograph, with specific fields of the new hexagon symbol organized and color-coded for clarity and the old GAPI symbols completely blacked out in the backdrop.

4.6.1 The ComplexGAPI principles

Legibility and ease are the basic requirements for the development of ComplexGAPI as an alteration of GAPI. Simultaneously, it must include the full range of factors that describe the analysis methodology and the pre-analysis methods (reagents, conditions, and techniques). Because the GAPI software has been thoroughly explained, the criteria were supported by an additional hexagonal pictogram explaining the processes that occur prior to the analytical protocol. All additional criteria were explained in Table 9.7.

Yield, requirements, and selection. Without a question, yield is one of the most significant components of synthesis. A high yield is considered a sign of success since it means that the limited reagents have almost completely been transformed into the intended product. To increase selective in the case of small yield, it is required to test the fundamental chemistry. Yet, in other cases the yield is small, but selective in the direction of the target chemical is great and the converting is likewise low [36]. Further inquiries are appropriate in this situation. Depending on the CHEM21 criteria, the applicable area will be colored green for yields >89%, yellow in the situation of yields in the range of 70%−89%, and red for yields <70% [42].

Time and temperature are taken into account together because of their interdependence. In an ideal process, the reaction would occur quickly and at room temperature, but in practice, longer synthesis times frequently necessitate the use of higher temperatures [37]. Cooling is considerably more difficult because there are frequently only predetermined temperatures available (for example, 0°C for an ice bath or 5°C for an acetone/ice bath). Moreover, eliminating moisture might be difficult and is occasionally advised in order to get repeatable results. In fact, such a phase needs the use of inert gases,

Schleck lines, gloveboxes, etc., which benefits the sales of the complete process but also its overall duration [43].

Green economic. In addition to protecting the environment by preventing pollution, green chemistry also improves manufacturing efficiency and lowers production costs. As a result, green chemistry and green economy are complementary approaches that, when applied together to complex situations, have a synergistic effect. Costs are minimized by the application of green chemistry. Because the combination of green chemistry and green economics helps address environmental problems and moves society closer to sustainability, green economy characteristics should be evaluated in relation to reactions and other procedures. To emphasize the significance of the combination of two strategies, parameters pertaining to both green chemistry and the economy were separated out as a separate ComplexGAPI category. Table 9.7 lists these specifications along with the corresponding marks. A system is designated green if it receives five to six points, yellow if it receives three to four points, and red if it receives less than 3 points since it is thought to be closer to the optimal green economy [44].

Chemicals and reagents. Aspects pertaining to wellness and security dangers are included in the category of reagents and solvents. Consideration must be given to every aspect of the reaction. For its evaluation, the criteria specified by the US National Fire Protection Association (NFPA) are employed (as is the case in GAPI). The most significant difficulties in this field are covered by these categories, including health risks, ignitability, reactive, and special dangers.

Instrumentation. Three factors are taken into account in the field of devices: the technological setup, energy usage, and occupational danger. If the setup only consists of a standard flask, reflux condenser, stirring, and other low-energy components (tools for controlled mixing of substances or specific glassware, for example), the section is colored green. The space is colored yellow if any supplementary, but not overly difficult, and often used sample preparation methods, such as ultrasound scans or photo- and microwave-irradiation, are applied. When a pressurized container is used and an inert environment is required, such as in a glove box, the section is colored red. These factors have an effect on the quantity of energy consumption, so they must be considered as well (Table 9.7). However, only two things are separated when it comes to occupational hazards: the area is colored green when the entire process is harmonized, and red when vapors are emitted into the environment [42—44].

Final product. The purification, workup, and generation of waste are all evaluated here, along with the end products. Green characteristics are those that do not use traditional purification methods such as quenching, filtering, centrifuging, low-temperature distilling, evaporating, or sublimation. The space is colored yellow when techniques like solvent exchange or quenching into an aqueous system are used. When sophisticated purifying processes are required (e.g., HPLC, multiple recrystallizations, ion exchange), the section is colored red. Green markers indicate process parameters that result in a high-purity (more than 98%) final. Yellow (97%−98%) and red (less than 97%) are used to indicate lower purity [42].

It was chosen to employ the **E-factor** parameter in order to account for the waste produced, which includes solvent loss, wasted catalysts and catalyst supports, and residual reactants in addition to waste by-products. As a result, it can be argued that the E-factors are generated from the quantity of liquids, chemicals, and consumables used per unit mass of product created, and the proper equation for its assessment may be utilized (Eq. 9.3). It can be simpler to compute the E-factor from a different angle because it can be challenging to account for losses and precise waste streams. In this situation, Eq. (9.4) must be applied [43].

$$\text{E-factor} = \frac{\text{total waste mass}}{\text{total product mass}} \quad \ldots \quad (9.3)$$

$$\text{E-factor} = \frac{\text{mass of raw materials-mass of product}}{\text{total product mass}} \quad \ldots \quad (9.4)$$

If the majority of the liquids and reagents are recycled, waste can be reduced. These chemicals aren't counted as by-products in this way. Moreover, water can be excluded from the overall amount of waste in the computation because it is a substantial by-product of many biochemical syntheses and other processes and is normally safe. For the purpose of calculating the E-factor, the mass of the water must be taken into account if it is significantly contaminated and difficult to recover in a state that is clean enough to apply to or discharge into a publicly owned wastewater treatment facility. The amount of waste produced, the negative influence on the environment, and the sustainability of a chemical process all increase with the E-factor. For the purpose of demonstrating the process' overall greenness, this factor was added to ComplexGAPI. In this instance, only the E-factor value is shown in the graph to make it easier to compare various approaches applied at the same chemical level [44].

4.6.2 Strengths and weaknesses of ComplexGAPI

The presence of software that will make using such a solution easier is a benefit of the ComplexGAPI model. By using the ComplexGAPI to evaluate the overall protocol's greenness, it is possible to see at a glance where the procedures under consideration differ and which areas need special attention to prevent specific problems. We advise adhering to the green chemistry tenets in all facets of laboratory activity. Because it includes 26 parameters both before and during the analysis procedure, the recently developed ComplexGAPI evaluation approach is unique in that it enables each stage to be examined and evaluated.

4.7 Analytical method greenness score (AMGS)

In 2019, Hicks and co-researchers published a tool to generate the AMGS as a green assessment scale in cooperation with the American Chemical Society Green Chemistry Institute Pharmaceutical Roundtable (ACS-GCI-PR). It is a mathematical index that takes into account solvents' environmental, health, and safety (EHS) evaluations as well as their cumulative energy demand (CED), apparatus energy use, and waste output (Fig. 9.7). This tool serves as a common metric reference for comparing development processes. It is not intended to be used as a strict yardstick for how green a methodology is or as a way to eliminate a certain methodology based on its finite score. The greener approach has a lower AMGS value.

This greenness calculator combines a focus on preservation into a single tool for determining the general quality of analytical procedures, thanks to a collaborative effort by eight drug companies. Input variables from each of the three points of concern—waste, environmental protection, and energy consumption—are counted by the AMGS using a single, simple instrument. The ACS-GCI-PR posts these instructions and makes the program accessible to anyone for usage on its website.

4.7.1 The AMGS principles

The AMGS tool was created to promote more environmentally friendly practices that ensure sustainability and efficiency. The Analytical Mass Volume Intensity (AMVI) for procedure solvent consumption is paired with other

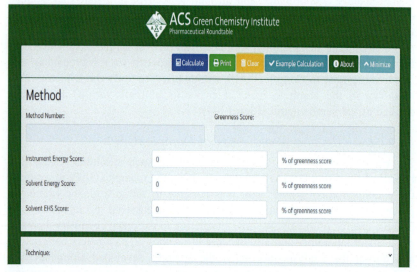

FIGURE 9.7 The AMGS free program is available on the ACS-GCI-PR website at https://www.acsgcipr.org/amgsprogram.

solvent EHS characteristics found in the solvent selection guide (SSG), which also contains the overall in-use CED, to create a more thorough evaluation tool. Now this tool incorporates equipment energy usage in addition to the SSG data from earlier solvent energy guides when evaluating the environmental friendliness of specific analytical or preconcentration methods. The AMGS is created using a differential assembly of many factors, such as the AMVI-described volume of wasted liquid, EHS/SSG values computed using the geometric mean for liquid safety, and CED for the choice of both the liquid and the equipment, as shown in Eq. (9.5).

$$\text{AMGS} = \frac{(mSn + mIn) \times (3\sqrt{Sn \times Hn \times En} + CEDn + (Ei \times R))}{\text{Number analytes of interest}} \quad \ldots$$

(9.5)

AMGS is evaluated for several solvents (1–n) by combining measures for AMVI, EHS, HPLC-EAT, and the energy consumed by the solvents used and the equipment selected.

- The mass m_s of the solvents used to prepare a sample or test is represented by mSn.
- Indicated by mIn is the mass m of the solvent n that the equipment employs to elute the gradient mobile phase at the stated flow rate.
- Depending on reactivity, explosion risk, and toxic effect at unit-less levels, Sn is the safe indicator of liquid n; values vary from 0.01 for methanol to 2.72 for acetonitrile.
- Depending on handle, irritation, and long-term hazard, the health score for solvent n, or Hn, has scores that vary from 0.01 for methanol to 1.2 for acetonitrile.
- En is the solvent environmental indicator rate for solvent n1 depending on the persistence, air, and water dangers of a specific solvent, e.g., 3.37 for methanol and 3.44 for acetonitrile (unit-less values).
- The term "CED" stands for "accumulated energy demand," which is calculated as the unit "kg-solvent/Mjoule-equivalent" for each specific solvent n depending on production and disposal.
- The term "Ei" refers to the energy usage of the equipment or the tabulated values obtained for each equipment type chosen, for example, 0.64 for HPLC and 1.22 for SFC-MS in terms of energy units of kW-hours.
- R is the total number of test injections needed for a particularly, which includes blanks, suitable standards, and reference standards.

In order to use the AMGS tool, the values of Sn, Hn, and En are entered into the AMGS calculator based on the current health, environmental, and safety indicators. Also, the choice of liquids as mobile phases or for standard dissolution is made from a pre-set drop-down menu. It is necessary to manually enter technique data such as flow rate, eluted time, and gradient factors. The slide-down menu on the tool allows users to choose from a number of instrument styles.

The AMGS is broken down into three parts in order to establish a clear point of view: The parameters to be taken into account are the energy of the device, the energy of the liquid, and the safety of the liquid. To highlight each value's significance to the whole AMGS, each of these data points is categorized and standardized by the total number of desired analytes at the top of the spread sheet. The liquid application part of the AMGS tool includes all the information required for the procedure: sample preparation, system suitability, sensitivity, mobile phase selection, and equipment settings. The three separate AMGS parts will be represented in the chart by a yellow to red color shift if they are close to or over 50% of the total AMGS data, while the lower AMGS score is attained by using eco-friendly liquids and more practical tools. The assessment of the analytical strategy under discussion produces a score that details the impact of each factor on the calculating value, along with color-labeling that denotes the key parameter that determines the marks.

4.7.2 The strengths and weakness of AMGS

The AMGS, a program that offers a novel strategy that combines a quantitative methodology to explain how methodology choices affect greenness, aids researchers in developing more integrated and sustainable methods. The three main factors that influence this impact are instrument selection as measured by energy consumption, solvent selection as measured by CED, and solvent use or waste formation as measured by environmental health and safety. As a result, the tool is designed to do a full review of the environmental friendliness of a system as well as a method-to-method comparison. However, the software is difficult for all users to handle and maintain due to the large amount of input data required.

4.8 Hexagon-CALIFICAMET scale

The five factors of a methodology are rated using the hexagonal numerical tool, which assigns penalty points (PPs). Analytical traits or figures of merit, related chemical and health issues, environmentally friendly nature, sustainability, and economic benefit are the five categories into which the factors are classified. The analytical effectiveness of the technique that is being evaluated is specifically included in the figures of merit, which are divided into the following categories: Figures of Merit 1 (FM-1) consider sample handling, method features, and calibration techniques, while Figures of Merit 2 (FM-2) consider quality assurance and accuracy. The globally harmonized system (SGA) assesses chemical toxicity, hazards, and safety issues. The residues resulting from the method's analysis and their potential for recycling are considered. The carbon footprint measures, which account for the energy demand of the tools used and the time required for the study, also quantify the environmental effects. The cost of the necessary equipment, its electricity usage cost, the cost of the chemicals and materials utilized, and the compensation paid to skilled individuals are taken into account when estimating the associated annual cost of the analytical determination. While penalty points are assigned to the other criteria, the carbon footprint and annual cost are measured in absolute terms. The total of the PPs, calculated carbon footprint, and cost quantities are then quantified overall for each parameter on a scale from zero to four, and the resulting pictogram is arranged in the shape of a hexagon. Higher scores are not always preferred. The final step involves calculating the arithmetic mean of the 0–4 score (S_{av}) to rank the analytical techniques and ultimately compare the expand when using the other provided programs. The scale corresponds to great, good, useful, bad, and refusing results of the evaluated analysis approach for the total score: 0, 1, 2, 3, and 4, respectively.

The Hexagon program has been used in a wide range of procedures that make use of various analytical methodologies. Atomic absorption spectroscopy (AAS), inductively coupled plasma mass spectrometry (ICP-MS) and optical emission spectrometry (ICP-OES), liquid and gas chromatography, and radioactivity are some of the methods that have already been assessed while examining the water sector. Recently, many techniques for analyzing ammonium in water, including ISE potentiometry, fluorescence, quimicoluminescence, and UV-Vis spectroscopy, have been evaluated (Fig. 9.8).

4.8.1 Hexagon-CALIFICAMET principles

The five distinct pieces that make up the suggested evaluation method are shown in Tables 9.8, 9.9, 9.10, 9.11, 9.12, 9.13, 9.14, 9.15, and 9.16.

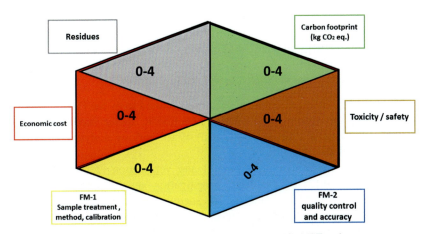

FIGURE 9.8 The scheme of hexagon-CALIFICAMET tool.

4.8.1.1 Figures of merit measurements

As shown below, there are two separate types of quality characteristics or figures of merit: FM-1 and FM-2. The adaptation of the figures of merit for handling the analytical issue is better when there are fewer penalty points involved. The FM-1 measure calibrated (Table 9.8), certain procedure features (Table 9.9), and criteria linked to sample treatment (Table 9.10). These tables include the factors taken into account for each factor as well as the penalty points allocated based on the many topical options as well as the optimal choice.

Depending on each parameter's occurrence in the overall position, the greatest number of penalty points has been assigned to it. In terms of parameters relating to sample treatment, Table 9.8 shows that the best option, free of penalties, corresponds to a method that does not use preservation and storage conditions for the sample, needing only a small amount of it, and does not use reagents or liquids, dilution, or concentration to achieve the required quantities for the chosen technique and pretreatment, and the number of samples for weekly analysis is greater than 50. Table 9.8 lists the penalty points for the additional factors that are different from the previously specified best choice. To make using the measuring tool easier, often no more than four scenarios have been taken into account. No penalty points are given to a technique that is in-line, automated, transportable, nondestructive, multicomponent, quick (analysis time under 10 min), and robust (Table 9.9). The optimum calibration technique is taken into account when the basic requirements for this phase are completed, as given in Table 9.10.

In particular, the Horwitz formula (ISO/IEC 17025 Testing Laboratory) in accordance with Eq. (9.6) is used to assess the precision (Table 9.10) of the procedure.

$$\text{RSDR \%} = 2^{-0.5 \times \log C} \ldots \quad (9.6)$$

Where,

RSDR: the relative standard deviation assessed from findings produced under reproducibility parameters.

C: the amount of the substance in the sample.

The repeatability of the findings (RSDr) is referred to by the HorRat score, which creates the presumption $RSDr = 0.66\ RSDR$. In the supporting data (SI), measurements for the precision analysis for various analyte concentrations are provided.

The factors linked to quality control and verification (Table 9.11) and accuracy (Table 9.12) of the under investigation analytical procedure are covered in Figures of Merit 2 (FM-2). The adequacy of the laboratory's chosen analysis technique is evaluated using the uncertainty methodology [45]. Eq. (9.7) (ISO/IEC 17025 Testing Laboratory) states that an uncertainty function

TABLE 9.8 The figures of merit for the treatment of the sample in the hexagonal tool. On the right side, the appropriate penalty points (PPs) are shown.

Preparation of samples		PPs
Preservation	None	0
	Physical	1
	Chemical	2
Storage	None	5
	Normal conditions	1
	Special conditions	2
Quantities	Micro	0
	Macro	1
No. of reagents/liquids	None	0
	≤3	1
	>3	2
Amounts of reagents/liquids	<1 g	1
	1–10 g	2
	10–50 g	3
	>50 g	4
Suitability to the procedure via instrumental	No need to dilute/concentrate	0
	Dilute/concentrate 5 time	1
	More than 5 time dilution	2
No. of sample (week)	≥50	0
	50–1	1
	<1	2
Pretreatment	None	0
	Filtration	1
	Stirring/drying on stove	2
	Acid digestion	3
		18

TABLE 9.9 The figures of merit for the assessment tool's methodological features The relevant penalty points (PPs) are shown on the right side.

Method characteristics		PPs
Method categories	In-line	0
	Off-line	1
Operational mode	Automatic	0
	Semi-automatic	1
	Manual	2
Portability	Yes	0
	No	1
Method/sample	No destructive	0
	Destructive	1
Analytes/sample	Multicomponent	0
	Unicomponent	1
Analysis time/sample	<10 min	0
	10–100 min	1
	>100 min	2
Robustness	Yes	0
	No	1
		10

(Uf) uses the detection limit of the method (LOD) and the amount of component (C) in the sample to create a maximal uncertainty measure:

$$Uf = \sqrt{\left(\frac{1}{2}LOD\right)^2 + (0.1 \times C)^2}\ldots\ldots \qquad (9.7)$$

The results from an appropriate method must have the maximum standard uncertainty specified by the aforementioned calculation.

4.8.1.2 Toxic effects and safety evaluating

The ranking and identification of substances in the internationally harmonized system (SGA) were taken into account. The penalty points are assigned according to the published information for these criteria. The assessment process entails collecting the pictograms found in the chemicals used in the

TABLE 9.10 Merit figures for the calibration stage as assessed by this tool and the related penalty points (PPs) on the right side.

	Calibration	PPs
Frequency	Annual	1
	Monthly	2
	Weekly	3
	Daily	4
Required time	≤30 min	1
	30 min-2 h	2
	2–8 h	3
	>8 h	4
No. of standard	5	1
	5–7	2
	>7	3
Linear adjustment R^2	$R^2 \geq 0.990$	0
	$R^2 < 0.990$	1
LOD/LOQ	LOD <0.1 legislation value	0
	$0.1 \leq$ LOD ≤ 0.33	1
	LOD >0.33	2
Linearity/working range	Suitable	0
	Partially suitable	1
	Inadequate	2
Precision	HorRat score (r/R) ≤ Horwitz	0
	HorRat score (r/R) > Horwitz	1
		17

analytical technique under consideration [46]. As shown in Table 9.13, the toxic effects analysis section considers both environmental and health risks. The penalty criteria for assessing safety are shown in Table 9.14. The higher the penalty, the worse the impact on health, the environment, and safety.

It is important to note that a quantitative evaluation of the substances used in the study is taken into account by the "Quantity of Reagent" item found in

TABLE 9.11 The FM-2 numbers associated with evaluation tool quality checks and the related penalty points (PPs) on the right side.

Quality control		PPs
Frequency	Working day	1
	0.5 working day	2
	Series of 5 samples	3
	Each sample	4
Required time	0–30 min	1
	0.5–1 h	2
	>1 h	3
Standards number	≤2	1
	3	2
	>3	3
		10

the examination template for toxicology and safety. Penalties are assigned based on Eq. (9.8), ranging from one to three points:

$$\sum (A_{reagents} + B_{calibration} + C_{quality\ control} + D_{accuracy}) \times \frac{3}{maximum\ PPs}.... \quad (9.8)$$

The quantity of reagents or liquids used during description is referred to as "reagents" in the prior equation (Table 9.8). The B term represents the total of the penalty points awarded to the frequency and number of standards used for calibration, whereas the C term represents the standards and frequency utilized throughout the method of quality control (Table 9.11). The final term D accounts for the penalty points connected to the appropriate frequency of accuracy validation (Table 9.12). The total of the penalties for all of the factors in Tables 9.13 and 9.14, respectively, the exception for the item relating to the quantity of chemical is then used to calculate the maximum penalty points in the denominator. As a result, tests for toxic effects and health might result in up to 24 and 22 penalty points, respectively [47].

4.8.1.3 Residues development

In Table 9.15, the concept of green chemistry measurements used to assess the amount of waste generated by analytical techniques is given in line with the values listed in the literature [2,5]. The amount of waste generated during the

TABLE 9.12 The FM-2 figures of merit in terms of the assessment hexagon-CALIFICAMET's accuracy.

	Accuracy	PPs
Frequency	Working day	1
	0.5 day for working	2
	5 samples	3
	Each sample	4
Required time	0–0.5 h	1
	0.5–1 h	2
	>2 h	3
Levels of quantities	≤2	1
	3	2
	>3	3
Magnitude–size	RSD_{method} (%) ≤ RSD (Uf %)	0
	RSD_{method} (%) > RSD (Uf %)	1
Selectivity	Yes	0
	No	1
		12

study, the likelihood of waste treatment, like recycling, and the type of disposable product made once the evaluation has ended are all taken into account when assigning penalties [48].

4.8.1.4 General environmental assessment

The term kilograms of CO_2 equivalent, also known as the carbon footprint [14], is employed to describe the factor used to measure the environmental influence of the analytical technique [49]. Its measurement is calculated using Eq. (9.9):

$$Kg\ CO_2 eq = \sum instrument\ power\ (kW) \times analysis\ time\ (h) \\ \times electricity\ emission\ factor\left(kg\ \frac{CO_2}{kWh}\right)\dots \quad (9.9)$$

4.8.1.5 Economic cost assessment

The sum of the metrics defined in Table 9.16 that is used to evaluate an estimate of the annual effective price (€).

TABLE 9.13 This figure shows the toxic effects penalty metrics for the products used in the evaluation.

Toxicity (health and environmental hazards)	PPs
Severe toxicity	3
Corrosive	2
Irritating	1
Irritating to eyes	2
Irritating to respiratory system and skin	3
Mutant	3
Carcinogen	3
Toxic—poisonous	3
System toxicity for target organ	3
Toxicity to the aquatic environment	1
Reagent amount	1—3
	27

TABLE 9.14 The safe work penalty metrics utilized during the analysis. The related penalty points (PPs) are noted on the side.

Safety physical hazards	PPs
Explosives	4
Flammables	1
Causes burns	2
Low pressure gases	1
Self- reactive chemicals	1
Pyrophoric	1
Materials undergo spontaneous heating.	4
Combustible gases activated by water	2
Organic peroxides	4
Corrosive for metals	2
Reagent amounts	1—3
	25

TABLE 9.15 Definition of penalty points (PPs) to evaluate an analysis procedure's waste production.

Residues		PPs
Amount	None	0
	<1 g	1
	1–10 g	2
	10–100 g	3
	100–500 g	4
	500–1000 g	5
	>1 kg	10
Treatment of waste	Recycling- unnecessary	0
	Degradation	1
	None treatment	3
Disposable substance	No	0
	Yes (Glass)	2
	Yes (Plastic)	5
		18

TABLE 9.16 The factors that are considered when calculating the annual expense.

Parameters	References values
No. of samples	Annual average
Analysis time (h)	Per sample
Cost of equipment	To be described for every analysis
Timeframe of amortization	10 years
Paying skilled personnel	15 €/h
Usage of electricity	0.5 €/kWh
Reagents	To be described for every analyzation

After calculating the penalty points for hexagons, a score of five dimensions is produced. The total score ranges from 0 to 4. The specification of the intervals of the penalty points is linked to the value assigned for each parameter of the examined technique in Table 9.17. It is important to note that [a, b] implies that a ≤ x ≤ b while [a, b] covers the interval a ≤ x < b. If the score is higher, that means the less adept one is at controlling experimental parameters, the greater the effects on the environment and health, and the less efficient analytical method [47]. A standard hexagon with six equilateral triangles constitutes the last qualified sign. excluding the toxic effects and safety parameters that are contained within the same triangle, each triangle relates to a factor of the analyzed procedure.

4.8.2 The strengths and weakness of hexagon-CALIFICAMET

The numerical scale is the hexagon-CALIFICAMET. By using the standard hexagonal pictograph, it enables a simple visual assessment of an analysis procedure's properties. This pictogram, which consists of six equilateral triangles, represents the analysis method's factors that are evaluated, including economic cost, toxic effects on health and wastes, and environmental and safety effects. This is a potent tool meant to help with the choice of the analytical approach that best balances sustainability and cost-effectiveness. But in order to assign penalty points, guidelines and understanding are required.

4.9 The analytical GREEness (AGREE) scale

The environment of an analytical approach must be measured using a metric. It includes many characteristics: Completeness of the offering The metric system needs to take many different factors into account because greenness is difficult to define and evaluate [2]. Input guidelines should take into account the requirements for materials (quality and quantity), production of waste, usage of energy, analyst security, and the overall strategy for the analytical process, such as the number of pretreatment stages and where the analytical instrument is located in relation to the topic of study, in order to accurately evaluate greenness [21]. The newest scale protocol, AGREE, was introduced in 2020. The 12 SIGNIFICANCE concepts are referenced in the input factors, which can be given various weights to provide some flexibility. According to the explanation in the Principles of Green Analytical Chemistry category, each of the 12 input parameters is converted into a standard scale with a 0−1 value. The sum of the evaluation outcomes for every concept produces the final assessment score. The result is a graph that resembles a clock, with the overall rating and color description in the center (Fig. 9.9). The weight of each concept is shown by the width of the part that corresponds to it, while the result of the operation on each basis is indicated by the simple red-yellow-green color scheme. This evaluation is simple and generates an automatically created diagram with a review report. The software's description is available on the website (https://mostwiedzy.pl/AGREE).

TABLE 9.17 The total certification of the method's factors based on penalty point scales.

Total value	FM-1	FM-2	Toxicity	Safety	Residues	Carbon footprint	Economic cost
					Penalty points scale		
0	[0,5]	[0,4]	[0,5]	[0,2]	[0,5]	[0,0.1]	[0,5.000]
1	[6,15]	[5,8]	[6,12]	[3,5]	[6,10]	[0.1,0.5]	[5000,15000]
2	[16,25]	[9,12]	[13,18]	[6,9]	[11,15]	[0.5,1]	[15000,30000]
3	[26,35]	[13,16]	[19,25]	[10,14]	[16,20]	[1,2]	[30000,50000]
4	[36,45]	[17,23]	[26,33]	[15,22]	[21,24]	[2,>30]	[50000,>50000]

AGREE Scale Parameters

1- Sample Preparation
2- Sample Size
3- In Situ
4- Integration of analytical procedure and used energy
5- Automation
6- Derivatization
7- Waste
8- Analyte Numbers
9- Energy
10- Reagents
11- Hazardous Materials
12- Safety of Operator

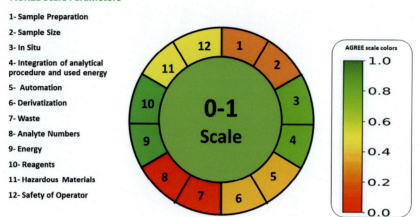

FIGURE 9.9 The AGREE parameters and the overall evaluation, as well as the appropriate color scale for references.

The total score is displayed in the center of the pictogram, and dark green values near one denote that the technique being evaluated is more environmentally friendly. The color in the section with a number relating to each factor reflects how well the procedure performed in each of the areas of evaluation [50]. The results for GAC concepts 1, 2, 7, and 8 in the example depicted in Fig. 9.9 are fairly poor, however, the result for ideas 9 and 10 is superb. Different weights for criteria are another significant piece of knowledge that is clear to understand from the pictograph. The evaluation does not include any other factors that do not pertain to greenness. An analytical process must be validated before it can be used, which necessitates the characterization of characteristics such as the limit of detection, accuracy, precision, and linear range by acceptable amounts. This makes it seem unnecessary to provide metrological criteria. Even though the eighth GAC principle takes into account the analytical throughput, a factor that is somewhat connected to economic growth, economic factors like the cost of chemicals or the price of analysis instruments are not directly taken into account.

4.9.1 The AGREE principles

There are 12 principles in AGREE scale.

1. **Sample preparation**: To prevent sample treatment, direct analytical methods should be used. Sample processing should be avoided by using direct analytical procedures. In reality, skipping the sample preparation and therapy processes might significantly reduce the environmental, health care, and safety risks connected with a certain procedure. Direct analysis is not usually feasible, though, as the samples must be in the right state of

matter and may need to have greater sensitivity and selectivity [50]. This approach allows for distinguishing between different greenness stages, as demonstrated in Table 9.18 by the greenest option of remote sensing without sampling damage and multistep procedures including pretreatment of the samples and batch assessment (the least green alternative). In order to translate the concept into numbers, a system for scoring was chosen, taking into account the widespread acceptance of online analysis as a friendly strategy. Also, the range of sample preparation stages needed before analysis was taken into account in the case of outside sample pretreatment to distinguish between treatments according to their difficulty [51].

2. **Sample size**: The objectives are a limited sample size and a small number of samples. The employment of quick and affordable vanguard analytical techniques was reduced. The number of samples that need to be examined by a rearguard analytical approach in a broader and more inclusive manner. With regard to the sample size, technological advancements in the reduction of analytical instruments have made it possible to conduct chemical tests with essentially negligible sample usage. All of these factors, as well as the division of (bio)chemical assessments based on the starting size of the sample, are taken into account when converting the second concept into a metric. Eq. (9.10) can be used to convert the mass (g) or volume (mL) of a sample to a score. In Table 9.19, ultra-micro analysis, microanalysis, semi-micro analysis, and macro analysis are rated as perfect.

$$\text{Score} = -0.142 \times \ln m \ (g \ or \ mL) + 0.65.... \qquad (9.10)$$

TABLE 9.18 The scores for several sample preparation procedures.

Sample preparation procedures	Score
Remote sensing with no contamination of samples	1.00
Distant sensing with minimal physical effect	0.95
Non-invasive analysis	0.90
Analysis directly and in-field collecting	0.85
On line analysis and in-field collecting	0.78
On line analysis	0.70
At line analysis	0.60
Off line analysis	0.48
External sample pre/treatment/batch analysis (reduced number of steps)	0.30
External sample pre/treatment/batch analysis (large number of steps)	0.00

TABLE 9.19 Score depending on sample size.

Analysis type	Sample size	Score
Ultra-microanalysis	<1	
Microanalysis	1–10	1.00
Semi-microanalysis	10–100	
Macro analysis	>100	According Eq. (9.10)

3. **In-situ**: It is important to take measurements in place. The third GAC concept aims for the most direct identification of target compounds. From the perspective of the GAC, it is crucial to position the instrument close to the assessment site since, in this case, the interval between analyses will be brief, as will the interval between collecting the samples and the acquisition of pertinent data for analysis. In this regard, field-portable devices and a minimized measuring site are important, since in such a circumstance the time between two tests is small and the time delay between sampling and collecting pertinent analytical data is likewise low. Portable tools for the field and miniature Additionally, the ability to do in-situ measurements has several other advantages, such as the ability to conduct chemical analysis with little to no sample pretreatment, better security for workers, and reduced reagent use. As a result, this approach takes into account the analytical device's placement in relation to the research item, as indicated in Table 9.20. In order to evaluate the third of the rules of GAC specifically, four possibilities—off line, at line, on line, and in line—have been taken into consideration as input information.
4. **Energy** is saved and reagent usage is decreased through the combination of analytical procedures and activities. Chemical analysis frequently involves a number of unitary processes, particularly when analyzing substances with medium-to high-level complexity. It saves time, money, and resources to reduce the number of analytical stages as much as possible. The synthesis

TABLE 9.20 Transformation of the location of the analytical device toward the investigated object to numerical scores.

Input information	Score
In line	1.00
On line	0.66
At line	0.33
Off line	0.00

of analytical procedures has received a lot of attention. The simultaneous performance of analytical processes and operations, such as testing with preconcentration by the use of passive sampling or sorption tubes or derivatization with removal or chromatographic determination (as pre- or post-column derivatization), is particularly noteworthy. A value of 1.0 was established for methods with three or fewer stages in order to evaluate the environment of a technique in accordance with the fourth GAC concept, while the scores were determined at 0.8, 0.6, 0.4, 0.2, and 0 for methods with 4, 5, 6, 7, and 8 stages or more, respectively.

5. **Automation** procedures should be chosen. Analytical procedures' automation and reduction both have positive effects on GAC since they use fewer substances, chemical solvents, and energy. Automating analytical operations lowers exposure at work, particularly to solvent fumes, and also lowers the chance of accidents. Table 9.21 displays the conversion of degrees of automation and reduction into points.

6. **Avoid derivatization**: The use of derivatization factors is problematic from the standpoint of the GAC since it requires additional procedures and the use of more substances, resulting in waste formation, and it usually has a detrimental impact on sample throughput. Preventing chemical derivatization is thus the best approach. Derivatization, on the other hand, is widely used in chemical analysis because micro-processes can enhance the extractability, analytical separation, and/or detection of specific analytes. The amount of hazard might vary greatly depending on the nature of the derivatization agent. A previously created derivatization agent selection guide is used to discriminate among derivatization chemicals. Safety of usage, environmental destiny, environmental persistence, and biological consequences are the assessment criteria. If no derivatization is used, a score of 1 is provided; alternatively, Eq. (9.11) is used to determine the score.

TABLE 9.21 Automated level and miniaturized sample processing step values converted to the numerical model.

Level of automation and miniaturization	Score
Automatic, miniaturized	1.0
Semi-automatic, miniaturized	0.75
Manual, miniaturized	0.50
Automatic, not miniaturized	0.50
Semi-automatic, not miniaturized	0.25
Manual, not miniaturized	0.0

$$\text{Score} = DA_1 \times DA_3 \times \ldots \times DA_n - 0.2 \ldots \quad (9.11)$$

where DA_i is the rating for the specific derivatization factor

For the purpose of evaluating the sixth GAC concept, the following premises are made.

a. To distinguish between the case in which no derivatization factor is utilized and the case in which the most environmentally friendly of derivatization factor is used, a value of 0.2 has been deducted. while the greenest derivatization compound would yield the same result as without an agent.
b. If more than one derivatization factor is utilized in the analytical process, the more troublesome one (with the smallest value) will have the biggest influence on the final result. It is uncommon to use more than one derivatization factor because only one is usually utilized.
c. If the derivatization agent appears in more than one chart, the rating with the lowest ranking is used.
d. In the program, enter the CAS numbers of the derivatization agent. The matching rating is subsequently added automatically.
e. When the resulting value is negative, it is changed to 0 as a substitute.

7. **Wastes**: There should be adequate handling of chemical waste provided, and there should be no large-scale production of analytical trash. An ideal solution from an environmental and financial standpoint would be to prevent the production of analytical waste. Sadly, in a great deal of instances, analytical waste is created. While avoidance should be the top priority in order to adhere to the seventh GAC concept, some strategies have been put forth to reduce the creation of waste and/or effectively handle analytical waste after analyte identification. These methods include online cleaning, reuse, and recycling of waste, among others. The mass of trash is computed in this work to score with Eq. (9.12):

$$\text{Score} = -0.134 \times \ln m \text{ (waste } g \text{ or } mL) + 0.6946\ldots \quad (9.12)$$

When this Eq. (9.12) is applied, a score of 1.0 is obtained for waste quantities less than 0.1 g (mL), 0.4 for waste quantities of 10 g (mL), 0.25 for waste of 25 g (mL), and 0.1 for waste of more than 100 g (mL).

8. **Analyte numbers**: In this principle, the number of chemicals that can be evaluated in a single hour is taken into account. The number of substances identified in a single run is multiplied by the analytical throughput, or the number of successive samples that can be examined in a single hour. This value serves as the input to Eq. (9.13) for the revision to the 0–1 scale:

$$\text{Score} = -0.2429 \times \ln n \text{ (no. of analytes per } h) - 0.0517\ldots \quad (9.13)$$

For 1, 10, 50, and 70 analytes evaluated throughout an hour, the resulting change yields scores of 0.0, 0.5, 0.9, and 1.0, respectively.

9. **Energy**: Energy consumption must be reduced. It might be difficult and hard to assess the energy used throughout the preparation, separation, and detection processes. It was suggested to base the traffic light energy value assessment on the overall kWh of each sample. According to this theory, there are two ways to calculate power usage. Thus, this modified and finished this method for an overall evaluation of energy usage per test, awarding the preparation of samples methods, quantitative separating processes, and devices with Table 9.22.

10. **Reagents**: The utilization of compounds produced from renewable resources is a very attractive and effective strategy in all fields of chemistry, not just analytical chemistry. There have been some attempts made in terms of analytical methods, mostly for the separation and identification of target chemicals, and more developments are anticipated. The treatment of the 10th GAC concept is simple. The result is 1 if no chemicals are used or if all are derived from bio-based origins. The result is 0.5 if some of them come from bio-based origins while others do not. The result is equivalent to zero if none of the chemicals come from bio-based origins [50].

11. **Hazard materials**: The elimination or substitution of harmful chemicals with more environmentally friendly ones is the goal of the 11th GAC concept. The quantity of harmful chemicals or solvents applied is equally important to consider as the sort of substances utilized. Indicate whether the evaluated analytical technique uses any harmful chemicals as the first point in evaluating a particular analytical approach in accordance with the 11th GAC concept. The result is 1 if no harmful substances are utilized.

TABLE 9.22 The AGREE score based on energy used and advices.

Score	Energy (KWh/sample)	Examples
1.0	<0.1	UV-VIS spectrometry, FTIR, titration, immunoassay, hot-plate solvent evaporation, (less than 10 min), needle evaporator, sonicator, SPE, microextraction methods, X-ray fluorescence
0.5	0.1–1.5	FAAS, EAAS, solvent extraction, Hot plate solvent evaporation (10–150 min), microwave-assisted extraction, LC, GC, ICP-MS, ICP-OES
0.0	>1.5	Hot plate solvent evaporation (>150 min), GC-MS, LC-MS, NMR, X-ray diffraction.

Alternatively, the substance's mass or volume is converted into the value of the score based on Eq. (9.14).

$$\text{Score} = -0.156 \times \ln m \text{ (chemicals amount in } g \text{ or } mL\text{)} + 0.5898\ldots \quad (9.14)$$

12. **Safety of operator**: The number of risks that are not averted is taken into account, along with the security of the worker and health hazards. Hazards come from the use of substances, in which case the material safety data sheets clearly indicate them, or from procedures where the risk of dangerous contact can be evaluated using specialist knowledge. The listing that follows should be used to choose the hazards that cannot be eliminated:
    - aquatic life poisonous
    - bio-accumulative
    - persistent
    - extremely flammable
    - very oxidizable
    - explosive
    - corrosive

If no risks are chosen, the result is 1. If there are 1, 2, 3, or 4 hazards, the results are 0.8, 0.6, 0.4, and 0.2, respectively. If ≥ 5 hazards are discovered, the result becomes zero [51].

4.9.2 The strengths and weakness of AGREE scale

AGREE is a metric scheme developed on the SIGNIFICANCE concepts for assessing the environmental impact of analytical operations. It is comprehensive (it incorporates all 12 rules), flexible (it allows you to apply weights), simple to comprehend (the result is a colored pictogram that shows the structure of both powerful and weak spots), and simple to use (it comes with a user-friendly GUI program). The publicly available software speeds up and simplifies the analysis. The analysis can be completed in a matter of minutes. The case studies demonstrate AGREE's full adaptability across different analytical approaches. The software can be downloaded from mostwiedzy.pl/AGREE, and the source code can be found at git.pg.edu.pl/p174235/AGREE.

While the overall greenness of the technique remains unknown because of the synthesis portion prior to preparing the sample, the results may not always convey knowledge of the structure of danger.

4.10 The AGREEprep scale

A new version of the previously developed AGREE tool called AGREEprep has just been introduced for 2022. Its goal is to assess the sample preparation process. Ten aspects are used to determine the total value; each is given a

How to evaluate the greenness and whiteness Chapter | 9 **315**

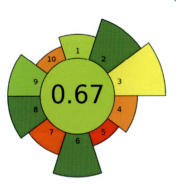

FIGURE 9.10 The outcomes of the AGREEprep evaluation of analytical techniques and all 10 principles with different weights for this tool.

sub-score with a 0−1 range and a weight to help differentiate between them. Substances, solvents, reagents, sample processing, energy, waste, and outcomes are a few of the assessment criteria. With free open-source programs, the most current edition of the tool is accessible at git.pg.edu.pl/p174235/agreeprep, and throughout the process of assessment, a pictogram is created with information on the total effect and composition of dangers [29]. Fig. 9.10 condenses the AGREEprep-related aspects in addition to the 12 GAC concepts.

4.10.1 The AGREEprep principles

The guidelines for evaluation depend on the 10 environmental preparation concepts for samples listed below.

4.10.1.1 Sample preparation

The initial principle proposes in situ sample preparation to reduce wastage of time, resources, and work. Also, difficulties with sample loss brought on by inappropriate storage during travel are prevented. In vivo sample preparation that is low- or even non-invasive as well as non-lethal avoids the need to take live things out of their natural environment [29]. Additionally, the process of sample preparation has a significant impact on numerous other factors that come after the use of solvents, substances, energy use, and waste production. Four different groups were taken into account when calculating this factor, and the results are shown in Table 9.23.

4.10.1.2 Hazardous materials

The addition of liquids and other supporting chemicals often increases the price, impact on the environment, and safety of sample preparation processes.

TABLE 9.23 The AGREEprep score value based on sample preparation.

Sample preparation method	Score	Explains
In line/in situ	1.00	Sample preparation is done inside the object under investigation. Sample preparation and sampling are typically combined. Using passive testers or applying SPME in vivo are two excellent examples.
On line/in situ	0.66	Sample preparation is carried out in situ, where sampling and sample preparation are carried out simultaneously in one location using fixed equipment, with the total process usually being wholly automatic.
On situ	0.33	Equipment is carried to the sampling location for on-site preparation of samples.
Ex situ	0.00	Following sample gathering and delivery, sample preparation is carried out at the lab.

The second concept of environmental sample preparation advises choosing safer liquids and chemicals with improved intrinsic characteristics that pose no or little danger to people and/or the surroundings. This strategy also strives to reduce the usage of harmful chemicals like acids and bases in derivation and digestion operations. The ideal situation to achieve in the second principle is the adoption of solvent-free and reagent-free sample preparation processes, and this scenario results in a score of 1 for this relevant criterion. The overall rating for this concept is zero in the worst-case scenario, where the preparation of samples requires techniques that consume more than 10 mL or 10 g of potentially dangerous solvents and chemicals [52]. Alternatively, Eq. (9.15) is used to determine the value. It is said that if a chemical is harmful through one of the routes of exposure or is classified as bio accumulative or constant, its mass must be taken into account (Table 9.24).

$$\text{Score} = -0.145 \times \ln m \text{ (hazardous amount in } g \text{ or } mL) + 0.3333... \quad (9.15)$$

TABLE 9.24 The application of AGREEprep scores based on the quantities of hazards.

Amount of hazards	Score
Free hazards (solvents, reagents)	1.00
>10 (g or mL)	0
≤10 (g or mL)	Applied Eq. (9.15)

4.10.1.3 Materials renewability

Substances must be stable for the duration of their (preferably extended) useful lives and decay after this point. The third factor to consider in this situation is how sustainable or recyclable substances are used in the sample preparation procedures. Components with a bio-based origin are preferred to those made of depleted compounds and fossil fuels. Also encouraged is the utilization of renewable and re-generable resources, such as trash, which has the added advantage of having a longer life cycle. Applying the third concept of green sample preparation, this rule also favors substances that can be repurposed over those that are only meant to be used once. When a substance is reusable, it can be utilized once more following a regeneration process, such as thermal desorption in cases involving solid sorbents. It is encouraged to consider substances as non-sustainable if knowledge regarding the sustainability of the substances employed to create them is unavailable [53]. Three factors are taken into consideration to determine the rating for this condition, as shown in Table 9.25.

4.10.1.4 Waste

Waste generation, handling, storage, and disposal use resources, time, and money. The fourth principle is focused on the idea that sample preparation procedures and equipment should be created in a way that minimizes waste production. Approaches for measuring the "greenness" of chemical production, like the "E-factor" or "atom economy," depend on the quantity of the substrates and the amount of the final product. All material inputs in chemical analysis, and specifically in preparing samples, can be treated as waste. Obviously, any item supplied with the sample should be discarded. Solvents used in extraction operations, derivatization substances, and acids or bases used for decomposition or pH correction are examples of waste. Disposable

TABLE 9.25 The AGREEprep score depending on the number of sustainable or recyclable substances.

Sustainable or recyclable substances	Score
100%, used several times	1.00
>75%	0.75
50%–75%, used once	0.50
0%, used several times	0.50
25%–50%	0.25
<25%, used once	0.00

items such as single-use glassware, SPE cartridges, sorbents, and filters are also included in the quantity of trash. Furthermore, if the sample becomes contaminated with dangerous compounds during the sample preparation stage, it should be discarded as waste. Eq. (9.16) are discussed in relation to the mass or volume of the trash created [29].

$$\text{Score} = -0.161 \times \ln m \text{ (sample in } g \text{ or } mL\text{)} + 0.6295... \quad (9.16)$$

Sample procedures for preparation that generate less than 1 g of trash receive a rating of greater than 0.5 in this impact section. This criterion simply evaluates waste mass, as the dangers of utilizing chemicals are addressed by other factors.

4.10.1.5 Sample size

The quantity of solvents, reagents, and additional substances to be employed in an analytical technique depends on the size of the sample. As a result, smaller sample sizes need less time, work, money, and resources while also having a higher chance of being automated or portable. Nevertheless, it is important to remember that sample accuracy must always be provided and that a drastic reduction in the sample size may degrade the analytical properties of the total analysis process [52]. Eq. (9.17) are used to determine the value for this principle:

$$\text{Score} = -0.145 \times \ln m \text{ (sample in } g \text{ or } mL\text{)} + 0.6667... \quad (9.17)$$

It should be observed that when just analytes are obtained (as with passive samplers), the amount of sample obtained (i.e., the amount of analytes gathered) is minimal, and an evaluation of 1 is given.

4.10.1.6 Sample throughput

The pace of the total sample preparation process, which is connected to energy usage and environmental issues, is covered in Guideline 6. There are two ways to get this principle's best scores. The first one has to do with using quick sample preparation methods to prepare numerous samples in a number of stages. The second one handles many samples concurrently [53]. The number of samples that can be made in a single hour (in series or parallel) is added to the score given by Eq. (9.18) to determine sample throughput.

$$\text{Score} = 0.2354 \times \ln n \text{ (no. of sample per hour)}... \quad (9.18)$$

4.10.1.7 Automation

The majority of sample processes involve multiple steps that can waste materials, use more substances and energy, and take longer. A development in sample preparation that has a beneficial effect on the approach's greenness is

TABLE 9.26 The sum of the two sub-scores determines the overall value for concept 7.

Sample preparation stages	Sub-score (1)	Sample preparation automation	Sub-score (2)
2	1.00	Fully automated	1.00
3	0.75	Semi-automated	0.50
4	0.50	Manual	0.25
5	0.25	–	
≥6	0.0	–	
Total value = sub-score (1) + sub-score (2)			

the search for simplicity of operation through the incorporation of stages. Automated systems also speed up the processing of samples, reduce waste production, the need for chemicals and fluids, and the amount of human mistakes and exposure risks associated with toxic compounds [21]. As shown in Table 9.26, the reduction and simplification of the stages needed are indicated in sub-scores.

4.10.1.8 Energy

In order to quantify the influence of this concept, the overall energy need is calculated in watt-hours (Wh) per sample. It should be mentioned that the energy need of the instrument is split by the number of tests running simultaneously if multiple specimens are handled in parallel or series mode using the same instrument [2]. The energy use is adjusted to the rating displayed in Table 9.27 based on whether the requirements are fully, partially, or not fulfilled. Eq. (9.19) were used to calculate the score for medium energy usage.

$$\text{Score} = -0.256 \times \ln\text{energy}\ (Wh\ \text{sample}/h) + 1.5886\ldots \quad (9.19)$$

TABLE 9.27 The score determines the overall value for energy usage.

Energy (Wh/sample)	Score
<10	1
10–500	Eq. (9.19)
>500	0

It should be noted that the usual numbers used to determine the energy needs of the utilized devices are those provided by the manufacturers. The results are nevertheless useful for comparison purposes, even though they only represent the highest possible scores and not the real power consumption of analytical devices [21].

4.10.1.9 Analytical instrument

The ninth concept advises cautiously choosing the greenest choice that is reasonably straightforward, energy-efficient, and uses the fewest substances. But it is understood that the ultimate decision is either based purely on availability or dependent on the analytical requirements for method performance [53]. According to the method used, the effect of the last choice phase may be substantial or small. Table 9.28 provides the ratings for the most popular final assessment approaches.

4.10.1.10 Safety

This concept contributes to the risks indicated by the safety symbols on the labels of the substances utilized during the process, such as poisoning of aquatic life (toxicity to people is not indicated by safety symbols), bioaccumulation potential, persistence, flammability, oxidazability, exploding, and corrosiveness. Additionally, this standard takes into account bio and physical risks such as gas compression. The multitude of distinct pictograms that can be used as input information for this evaluation allows for the easy extraction of the dangers from the MSDS of chemicals [29]. The overall rating for this requirement, which is provided in Table 9.29, is determined by the variety of the physical, chemical, or biological dangers that have been recognized.

TABLE 9.28 The total score depends on analytical instruments.

Instruments	Score
Simple and available (Smartphone, desktop scanner)	1.00
Optical spectroscopy (UV-vis spectroscopy, fluorimetric, chemiluminescence, etc.), Surface analysis methods, Voltammetry, Potentiometry	0.75
Gas chromatography, FAAS, capillary electrophoresis	0.50
LC, GC-MS	0.25
ICP-MS, ICP-OES	0.00

TABLE 9.29 The total score depends on number of hazards.

Number of hazards	Scope
None	0.10
1	0.75
2	0.50
3	0.25
≥4	0.00

There is critical data that has to be incorporated into this tool. The 10 factors that were used to judge how environmentally friendly the sample processing was demonstrate that they are not all of equal significance. Since chemicals and solvents have significant effects on how environmentally friendly a sample preparation process is, factor 2 was given the most weight. Although factor 1 has some bearing on how environmentally friendly sample preparation is, it is still possible to obtain a high level of greenness even when operations do not take place in situ [21]. Table 9.30 explains additional weights.

TABLE 9.30 The evaluation's standard weights of AGREEprep tool.

Principle	Description of principle	Weights
1.	In situ favorable principle	1
2.	Use green solvents and substances	5
3.	Choose materials that are recyclable and renewable.	2
4.	Reduce waste	4
5.	Reduce the amount of samples, chemicals, and materials	2
6.	Increasing sample throughput	3
7.	Incorporate steps and encourage automation	2
8.	Eliminate energy usage	4
9.	Select the most sustainable design for analyses' post-sample processing.	2
10.	Check the operator's safety protocols	3

4.10.2 The strengths and weakness of AGREE scale

By evaluating the pre-sample preparation steps, the technique's greenness score can be evaluated using the recently established AGREEprep tool. The shortcomings of the AGREE method might be fixed by giving the 12 assessment aspects different weights according to their significance and effects on the environment. Although AGREE provides a suitable equilibrium between procedure analytical efficiency and environmental effects as an ideal analytical technique, it has the disadvantage of not taking the analytical method's factors (accuracy, precision, and reproducibility) into consideration as an important factor for the strategy's final evaluation.

4.11 White analytical chemistry (WAC) scale

As an expansion of GAC, the idea of white analytical chemistry (WAC) was proposed. As a replacement for the well-known 12 GAC concepts, the 12 WAC ideas were submitted. Further to the green factors, WAC also considers the analytical (red) and practical (blue) factors as important variables impacting the quality of the approach [44]. A white analytical approach demonstrates the coherence and synergy of the analytical, ecological, and practical features with reference to the RGB color system, which holds that combining red, green, and blue light beams provides the perception of whiteness. According to the evaluation of individual rules, whiteness can also be measured and used as a useful metric for comparing and choosing the best approach [54,55].

4.11.1 White analytical chemistry (WAC) scale principles

The 12 GAC concepts were combined into four overarching "green" guidelines that address the most significant and mutually exclusive GAC (G1–G4) ideas. The four "red" concepts (R1–R4) and four "blue" concepts (B1–B4) referring to analytical effectiveness and practical/economic parameters, respectively, are then added to these four concepts to form the overall WAC concepts. As a result, every concept should be taken into account while evaluating the approach, but the ones that are particularly crucial for a certain application should be given a more thorough review that is suitable for the requirements. A quantitative term named "whiteness," which is a condensed measurement of how well the technique complies with the suggested criteria, is used to represent how generally the approach complies with them [54].

a. Red principles
   1. R1. The application's range The number of concurrently identified substances, the level of linearity of the determinations, compatibility with numerous kinds of tests, and resistance to the existence of potential interferences should all be as broad as possible in an analysis technique's application.

2. R2. Limits of detection as well as quantification (LOD and LOQ) for analytical procedures should be as low as feasible [14].
3. R3. Precision. The highest level of precision should be presented in the analysis. as shown by the repeatability and reproducibility of the findings.
4. R4. Accuracy. The accuracy of the approaches to analysis should be maximized (minimum relative error of evaluations, recovery as near to 100% as feasible) [44].

b. Green principles
1. G1. Chemical Hazard The use of as few harmful chemicals as possible and a high percentage of biodegradable or renewable chemicals and substances should define analytical processes.
2. G2. The quantity of waste and chemicals, analytical processes should produce the least amount of waste and use the fewest materials [55].
3. G3. Energy and other media, group. The least amount of electrical power and other resources should be used in analytical techniques. To save energy, on-site, automated, and high-throughput technologies are preferred.
4. G4. Direct influences, Analytical techniques shouldn't have an immediate impact on people, animals, or genetic nature. It is best to refrain from using animals, making genetic changes to humans (or other individuals), or exposing them to potentially dangerous substances [44].

c. Blue principles
1. B1. Cost effectiveness. Analytical techniques should be as economical as possible, taking into account staff, devices, supplies, media, and other costs.
2. B2. Time management. The best time efficiency (the shortest overall analysis time, taking into account the creation of methodologies and all phases of the analytical workflow) should be the hallmark of analytical techniques.
3. B3. Conditions. The minimum practical requirements for analytical procedures should include the volume of the sample used, accessibility to cutting-edge machinery, staff qualifications, and lab systems.
4. B4. Simplicity of work. The greatest degree of miniaturization, integration, transportation (on-site observations), and automation should be present in analytical processes [14].

The red, green, and blue data produced in the Excel template spread sheets must be completed in order to assess the approach utilizing the RGB 12 scheme (Fig. 9.11). Up to 10 approaches may be evaluated and compared concurrently using the template. The only step required is to input the scores in the correct gray columns, with 100 denoting that the approach is perfectly suited for a planned usage in terms of an idea and 0 denoting the poorest results. In justified circumstances (see the next paragraph's graph), it is also

			R1: Scope of application	R2: LOD and LOQ			R3: Precision			R4: Accuracy		
RED PRINCIPLES (analytical performance)	Method number	Method name	0-100	LOD	LOQ	0-100	RSD% (repeatability)	RSD% (reproducibility)	0-100	Relative error (%)	Recovery (%)	0-100
	1	The worst	0			0			0			0
	2	Appropriate	100			100			100			100
	3	Intermediate	50			50			50			50
	4		0			0			0			0
	5		0			0			0			0
	6		0			0			0			0
	7		0			0			0			0
	8		0			0			0			0
	9		0			0			0			0
	10		0			0			0			0

			G1: Toxicity of reagents (impact and biodegradation)		G2: Amount of reagents and waste			G3: Concumption of energy and other media	G4: Direct impacts (safety, use of animals and GMOs)			
GREEN PRINCIPLES (green chemistry)	Method number	Method name	Total number of pictograms	0-100	Reagent consumption	Waste production	0-100	1-100	Occupational hazards	Safety of users (0-100)	Use of animals (0 if no, 1 if yes)	Use of GMO (0 if no, 1 if yes)
	1	The worst		0			0	0		0	1	1
	2	Appropriate		100			100	100		100	0	0
	3	Intermediate		50			50	50		50	0	1
	4			0			0	0		0	1	1
	5			0			0	0		0	1	1
	6			0			0	0		0	1	1
	7			0			0	0		0	1	1
	8			0			0	0		0	1	1
	9			0			0	0		0	1	1
	10			0			0	0		0	1	1

			B1: Cost-efficiency		B2: Time-efficiency		B3: Requirements			B4: Operational simplicity		
BLUE PRINCIPLES (practical side)	Method number	Method name	Total cost	0-100	Speed of analysis	0-100	Sample consumption	Sample consumption (0-100)	Other needs: advanced instruments, skills, facilities (0-	Miniaturization (0-100)	Integration and automation (0-100)	Portability (0-100)
	1	The worst		0		0		0	0	0	0	0
	2	Appropriate		100		100		100	100	100	100	100
	3	Intermediate		50		50		50	50	50	50	50
	4			0		0		0	0	0	0	0
	5			0		0		0	0	0	0	0
	6			0		0		0	0	0	0	0
	7			0		0		0	0	0	0	0
	8			0		0		0	0	0	0	0
	9			0		0		0	0	0	0	0
	10			0		0		0	0	0	0	0

FIGURE 9.11 Method evaluation tables based on the red, green, and blue concepts using the RGB 12 methodology to complete the evaluation, suitable scores must be entered in the gray columns. Three model approaches were evaluated in the above case by giving them example ratings: The worst possible (1), acceptable (2), and medium (3).

feasible to award extra points of merit (measures over 100). The limitations for using animals and genetically modified organisms (GMOs) are an exception; in these cases, if they are employed at any point in the method, enter 1, and otherwise, enter 0. The most impartial and logical method possible should be used to assign these values. It is crucial that each individual criterion be consistently evaluated in light of the method's specified purpose [54,55].

The results of the assessment are generated and produced automatically in the form of tables when the required scores have been entered, and their interpretation is relatively straightforward (Fig. 9.12). The level of saturation of a specific main color (a value of 0 represents black, and 100 and greater is full saturation) is used to mathematically and visually represent how well the approach adheres to a particular WAC guideline. As can be seen below, the arithmetic mean scores for the three principal attributes R (%), G (%), and B (%) are each given separately, and a final score whiteness (%) is also provided as the average outcome of the evaluation of all the 12 concepts [14].

It is important to remember that the RGB 12 algorithm's choice of color is just one of several ways to display the outcomes. Based on the RGB color coding concept, the saturation utilized in the background style is based linearly on the numbers. As a result, it facilitates quicker interpretation and simpler outcomes in memory. This characteristic differs from the original RGB method, which distinguished the nine resultant colors as separate rating

How to evaluate the greenness and whiteness **Chapter | 9 325**

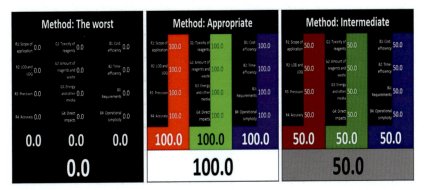

FIGURE 9.12 Presentation of the three design techniques' evaluation outcomes using the RGB 12 algorithm.

results. The RGB 12 system is streamlined; the entire assessment is defined solely by the "whiteness" component, formatted in black and white for expediency. As a result, when printed in black and white or for color-blind individuals, the analysis is simpler. Yet, the numbers of R (%), G (%), and B (%), along with the display of the related cells, still effectively convey the role of the distinct primary attributes symbolized by the colors red, green, and blue [44].

4.11.2 The strengths and weakness of WAC scale

Based on codified concepts and the same evaluation approach, the analytical (red) and practical (blue) metrics are compared in a manner similar to how greenness is evaluated. The WAC notion states that if a method satisfies many criteria, it becomes white, or appropriate, maintaining the completeness represented by the color white. The lack of evaluation criteria is one of the biggest problems, and it has a negative impact on the assessment's objectivity. Every scholar is free to assign his or her own assessment grade, which the authors conceded is a hard framework for judging their white analytical chemistry principles. To hitch-hike their idea for further development and provide a uniform framework for assessment, the project is deeply appreciated.

5. Applications of the studied greenness evaluation methods' publications

Table 2.6 contains a full literature analysis of the methods for evaluating greenness that were previously mentioned. It appears that the analysis tools mentioned were used to analyze multiple analytical techniques, involving electrophoresis, AA-DLLME-SFO, HPLC, UPLC, HNMR spectra, spectrophotometry, and spectrofluorometry, for a variety of analytes in multiple matrixes. Most papers used the AGREE technique, which was then used by

AES and GAPI, as shown in Table 9.31 (Fig. 9.13). This might be a result of their accuracy, sensitivity, and simplicity. Additionally, some articles used multiple tools of evaluation for a more accurate rating.

6. Conclusions

The use of hazardous solvents and reagents in the chemical and analytical sciences sectors has raised significant concerns about the effects of their practices on the environment. Analytical chemistry has played a key role in the utilization of greenness assessment systems following the development of the idea of green chemistry in recent years. There have been numerous attempts to create green techniques. It is only achievable, though, with efficient evaluation systems. In the past, a variety of green assessment methodologies have been employed to evaluate how environmentally friendly research projects are. The majority of research operations claiming greener analytical methods have not been adequately assessed by appropriate tools. Eleven various green profiling measurements, including NEMI, developed NEMI, analytical Eco-Scale, GAPI, ComplexGAPI, AMGS, AGREE, AGREEprep, and others, have been evaluated and discussed in this chapter. The process for determining each tool's greenness has been thoroughly explained, along with its operating concepts, problems, and advantages. The investigations and applications that utilized these tools have also been added.

7. Visions for the future

Going from broad concepts to practical applications and then spreading the model of environmentally friendly chemistry is important. Since their origin, green metrics have been extensively investigated and used, demonstrating their significance for pursuing sustainability. Teaching how to apply evaluation techniques to reduce the generation of dangerous hazardous waste is therefore crucial and provides a foundation for a healthier existence. The importance of the "sustainable" attitude" approach among scientists must also be emphasized. This circumstance will improve community perceptions of analytical science and support the incorporation of environmentally friendly chemistry projects. Overall, an evaluation of the analytical technique's greenness should be included in the procedure's acceptability criteria. Finally, in order to reduce the amount of pollutants and chemical risks released into the environment, a system for evaluating the sustainability level of analytical methodologies should be created before undertaking actual lab tests. Additionally, new legal steps should be taken to limit the damaging effects of chemical operations on the environment.

The suggested tools should help the analytical chemistry community select the most suitable approach for a particular application, assess recently produced methods, compare recently developed and established techniques, put

TABLE 9.31 Analytical Eco-Scale, WAC, AMGS, GAPI, Complex GAPI, AGREEprep, and AGREE programs are among the greenness evaluation metrics discussed in the published studies.

Type of research	Summary	Method	Sample/analyte	Number of citations	Applied tools	References
Tutorial	A guide to AGREEprep, a greenness index for analytical sample processing.	A Tutorial on AGREEprep	—	0	AGREEprep software	[29]
Review	A review of the AGREEprep software was successfully used to compare the differences in greenness, threat structures, and areas that needed improvement among six sample preparation processes for the detection of phthalate esters in water samples.	AGREEprep	—	69	AGREEprep software	[52]
Tutorial	The suggested analytical eco-scale can be an effective semi-quantitative tool for laboratory applications. It may be used for any established or novel approach and is quick and easy to use. It also includes well-defined evaluation guidelines.	Eco-scale principles	—	811	AES	[28]

Continued

328 Green Analytical Chemistry

TABLE 9.31 Analytical Eco-Scale, WAC, AMGS, GAPI, Complex GAPI, AGREEprep, and AGREE programs are among the greenness evaluation metrics discussed in the published studies.—cont'd

Type of research	Summary	Method	Sample/analyte	Number of citations	Applied tools	References
Research	analysis of greenness profile utilizing NEMI and eco-scale rating applications for a sustainable eco-friendly ultra-high-performance liquid chromatographic technique for simultaneous identification of caffeine and theobromine in commercial teas	HPLC	TEA/Caffeine, Theobromine	20	NEMI AES (88)	[32]
Research	The UASE-PMLS technology used to detect organic acids and polyphenols in wine collections. Organic acids and polyphenols in samples of Polish wine were assessed using the novel method in conjunction with GC-MS. The analytical eco-scale (88) and GAPI assessments provided positive findings and highlighted the supplied technique's green efficiency, indicating that it can be regarded as ecologically benign.	UASE-PMLS/GC-MS	Wines/Organic acids and Polyphenols	25	GAPI (3–9 green) AES (88)	[41]

Research	The ternary complex Tb^{+3}–8HQ–PRU is used in the proposed approach to effectively measure prucalopride succinate by spectrofluorometric.	Sensing probe	Prucalopride succinate	2	AES (91) Complex-GAPI (9 green)	[42]
Research	The first approach relies on directly determining the relative fluorescence intensity of Flibanserin at emissions and excitation wavelengths (λem/λex) (371 and 247 nm), whereas the second approach is a first-derived (D1) spectrofluorometric methodology that relies on peak amplitudes at 351 nm. For both approaches, linear regressions were seen in the 0.1–1.5 g/mL range. The new techniques' greenness was evaluated by NEMI, eco-scale, and GAPI.	Spectrofluorometric methods	Urine/Flibanserin	9	NEMI AES (88) GAPI (7 green)	[33]
Research	The analysis of nine water-soluble vitamins in honey is done using an environmentally safe, quick, and straightforward UPLC-ESI-MRM/MS approach. Analytical eco-scale and GAPI tools are used to compare the suggested approach to other approaches in terms of the preparation of samples, instrument energy use, utilization of dangerous substances, and waste creation.	UPLC-ESI-MRM/MS	Honey/vitamins	6	GAPI (7 green) AES (83)	[40]

Continued

TABLE 9.31 Analytical Eco-Scale, WAC, AMGS, GAPI, Complex GAPI, AGREEprep, and AGREE programs are among the greenness evaluation metrics discussed in the published studies.—cont'd

Type of research	Summary	Method	Sample/analyte	Number of citations	Applied tools	References
Tutorial	Utilizing greenness evaluation methods necessitates specialized instruments. This study suggests the analytical GREEnness calculator, a thorough, adaptable, and simple evaluation technique that yields understandable and instructive results.	Metric approach and software	—	437	AGREE software	[21]
Tutorial	The quality of the organic synthesis is assessed using a novel post-synthesis analysis tool according to yield, price, security, conditions, and simple workup or purification. The suggested method is based on giving these factors a variety of penalty points. Other chemical researchers who believe that specific factors should be given various relative penalty points may readily modify this semi-quantitative assessment. It is an effective tool for comparing different product preparations based on their safety, affordability, and environmental impact.	Eco scale summery	—	326	Eco-scale	[17]

Research	Analyses of four non-steroidal anti-inflammatory medications and diacerein using two flexible multi-analyte chromatographic approaches in comparison Assessment of greenness using the AGREE and analytical eco-scale measures	HPLC-DAD, HPTLC	Drugs	8	AES (85,81) AGREE (0.63, 0.7)	[50]
Research	This study quantifies the environmental effects of a recently developed electrochemical instrument that is entirely printed and helps with the colorimetric measurement of phosphate in saliva. Based on the concepts of "green analytical chemistry" and "white analytical chemistry", the analysis procedure's assessment was conducted.	Electrochemical instrument	Saliva/phosphate	3	AGREE (87) RGB (73.1)	[54]
Research	In order to choose the greenest analytical approach for the measurement of hyoscine N-butyl bromide (HNBB), a comparison of four greenness evaluation tools was conducted. The aqueous electrophoretic approach for HNBB evaluation created by Cherkaoui's group is the greenest approach, while the HNBB analysis of waste water by HPLC-MS/MS was the least green approach to analysis based on GAPI, AGREE, and ESA.	HPLC-MS/MS	Wastewater/HNBB	79	NEMI Eco-scale (96) GAPI (8 green) AGREE (0.71)	[56]

Continued

TABLE 9.31 Analytical Eco-Scale, WAC, AMGS, GAPI, Complex GAPI, AGREEprep, and AGREE programs are among the greenness evaluation metrics discussed in the published studies.—cont'd

Type of research	Summery	Method	Sample/analyte	Number of citations	Applied tools	References
Research	The primary goal of the proposed study is to provide a rapid, simultaneous UV approach for the simultaneous assessment of CVD and IBD in bulk (API) tablet dosage forms. The method of adaptation for quick screening in pharmaceutical corporations and commercial laboratories for these two medications will be greatly strengthened through the creation of an approach that makes use of eco-friendly solvents in UV spectrophotometers.	UV	Drugs/CVD, IBD	2	NEMI GAPI (10 green) AGREE (0.94)	[57]
Tutorial	By using ComplexGAPI to evaluate the overall protocol's greenness, it is possible to see at a glance where the procedures under consideration differ and which areas need special attention to prevent specific problems. We advise adhering to green chemistry tenets in all facets of lab activity.	Summary of complex GAPI	—	92	ComplexGAPI	[43]

Tutorial	A summary of the GAPI tool was introduced for the first time.	GAPI tool guidelines	—	651	GAPI	[18]
Research	In this research, an eco-friendly and accurate electrochemical carbon paste electrode for pyridoxine HCl and doxylamine succinate quantification utilizing square wave voltammetry is presented. This electrode has been altered chemically with zirconium dioxide and multi-walled carbon nanotubes.	Voltammetric/LC-MS	Drugs/pyridoxine HCl, Doxylamine succinate	2	AES (97)	[58]
Education letter	The eco-scale is the best paradigm for teaching and assessing green chemistry to undergraduates. Students gain knowledge of the eco Scale's form, use it to do a crude life cycle evaluation, and evaluate it as an academic model. Within a synthetic program where green chemistry and environmental responsibility are major concerns, the exam supports more conventional expository and independent research tasks.	Teaching of eco-scale tool		16	Eco-scale	[27]
Research	Co(II), Pb(II), and Pd(II) were isolated utilizing a cost-effective,	AA-DLLME-SFO/FAAS	Water/Co, Pd, Pd	0	GAPI (8 green) AGREEprep (0.78)	[53]

Continued

TABLE 9.31 Analytical Eco-Scale, WAC, AMGS, GAPI, Complex GAPI, AGREEprep, and AGREE programs are among the greenness evaluation metrics discussed in the published studies.—cont'd

Type of research	Summary	Method	Sample/analyte	Number of citations	Applied tools	References
	environmentally friendly microextraction method. Acetone was used as the dispersant, 1-undecanol served as the extractant, and folic acid served as the complexing reagent. The detection limits for Co (II), Pb (II), and Pd (II) were 0.042, 0.022, and 0.055 µg/L, respectively.					
Research	Six fluoroquinolone levels in chicken livers were determined using a straightforward HF-SLM/HPLC-FLD method. One-variable-at-a-time (OVAT) was utilized to optimize the HF-SLM process. The samples were isolated from the livers using a hollow fiber liquid phase microextraction method and a 100 mM Na_3PO_4 buffer at pH 7.4.	HF-SLM/HPLC-FLD	Chicken livers/6 fluoroquinolone	2	NEMI (3 green) GAPI (4 green) AGREE (0.46) AES (86)	[59]

| Research | By mathematically modifying UV absorption spectra, green analytical approaches with high sensitivity and selectivity were created for the quality control analysis of multi-component formulations containing REM and TEN. Last but not least, laboratory comparisons with the proposed HPLC approach were made after using proven UV spectroscopic techniques for the simultaneous measurement of REM and TEN from formulations. | UV spectroscopy | Pharmaceuticals/ REM, TEN | 4 | AGREE (0.91) Hexagon (0.0) WAC (97.5) | [60] |
| Research | Designing new, verified procedures that focus on several objectives, such as simplicity and sustainability, has become a major concern in quality control units. To estimate allopurinol with either benzbromarone or thioctic acid in their fixed dose combinations, sustainable analytical methods were developed and evaluated. The electrolyte solution of 50 mM borate buffer pH 8.5 was utilized, together with a 17-second injection period and a 30-kV voltage, to perform the electrophoresis of the capillary zone (CZE). | Capillary zone electrophoresis/UV | Drugs | 10 | AGREE (0.96) Hexagon (0.33) WAC (98.5) | [61] |

Continued

TABLE 9.31 Analytical Eco-Scale, WAC, AMGS, GAPI, Complex GAPI, AGREEprep, and AGREE programs are among the greenness evaluation metrics discussed in the published studies.—cont'd

Type of research	Summary	Method	Sample/analyte	Number of citations	Applied tools	References
Research	The main reason why SFC has been suggested as a substitute for conventional liquid chromatography is because it uses CO_2, a non-toxic, non-flammable solvent that is easily recyclable if desired. Although this calculator is a very helpful tool, it has to be modified in a number of ways, particularly in the way that CO_2 is taken into account in the AMGS calculator, in order to appropriately evaluate SFC's greenness.	SFC	Pharmaceutical industries	2	AMGS (83.92)	[62]
Research	This study suggests a brand-new, potent, white synergistic spectroscopy approach to assess their medicinal product with tiny substances (Paracetamol, Pseudoephedrine, and Loratadine). These spectroscopic approaches were effective in analyzing the combinations and doses, and a spectroscopic	Synergistic spectrophotometric strategy	Seasonal influenza/ spurious drug	3	Spider chart AES (82) AGREE (0.79) WAC	[63]

		similarity index was used to identify any false components and certify the purity of the recovered parent spectra for every medicine.			
Review	Application of WAC	The article discusses recent developments in biological food and environmental applications of white analytical chemistry. The ideas of this method have been effectively tested in the majority of these studies, even if the phrase "white analytical chemistry" has not been utilized. Here, the focus is primarily on using this method with published papers.	—	1	WAC	[55]
Review	Summary of WAC and GAC	As an expansion of GAC, the idea of WAC is put forth. As a substitute for the well-known 12 GAC rules, we suggest the 12 WAC ideas. Besides the green factors, WAC also considers the analytical (red) and operational (blue) characteristics as important factors impacting the quality of the approach.	—	137	WAC, AGREE	[14]

Continued

338 Green Analytical Chemistry

TABLE 9.31 Analytical Eco-Scale, WAC, AMGS, GAPI, Complex GAPI, AGREEprep, and AGREE programs are among the greenness evaluation metrics discussed in the published studies.—cont'd

Type of research	Summary	Method	Sample/analyte	Number of citations	Applied tools	References
Review	For the study of hydroxychloroquine in diverse biological mediums like human plasma, serum, whole blood, oral fluid, rat plasma, and tissues, numerous liquid chromatographic techniques have been published.	LC	Biological samples/hydroxychloroquine	4	NEMI AES (84) GAPI (6 green) AGREE (0.55)	[64]
Research	Lamotrigine in tablet dose form has been determined effectively using the suggested method without interference from excipients that are frequently used.	HNMR spectra	Drugs/lamotrigine	5	Spider diagram NEMI AES (91) AGREE (0.63)	[65]
Research	Magnetic effervescent tablet-assisted ionic liquid dispersive liquid–liquid microextraction employing the response surface method for the preconcentration of basic pharmaceutical drugs: Characterization, method development, and green profile assessment	META-IL-DLLME/HPLC-DAD	Water/Drugs	1	GAPI (6 green) AGREE (0.61)	[66]

Research	Various UV spectrophotometric and chemometric approaches have been used in this investigation to quantitatively analyze fluticasone propionate and azelastine in their pure forms, prepared mixtures, and medicinal dose forms without first separating them.	UV	Drugs/Fluticasone propionate, azelastine	0	AGREE (0.71) AES (88)	[67]
Research	In this research, we primarily concentrated on building environmentally friendly visible spectrophotometric techniques in order to establish a linkage between both theoretical and empirical uses. For accurate colorimetric measurement of ledipasvir in the presence of sofosbuvir, these techniques depended on charge-transfer combining (CTC) between ledipasvir and 2,3-dichloro-5,6-dicyano-1,4-benzoquinone (DDQ) or chloranilic acid (CA).	UV	Drugs/ledipasvir and sofosbuvir	3	NEMI GAPI (9 green)	[68]
Review	The AMGS tool is reviewed in this paper as an example of contemporary developments. More crucially, concerns for the AMGS effect on sample preparation, developing methodology, and energy usage are offered.	AMGS applications of pharmaceuticals		2	AMGS	[69]

Continued

TABLE 9.31 Analytical Eco-Scale, WAC, AMGS, GAPI, Complex GAPI, AGREEprep, and AGREE programs are among the greenness evaluation metrics discussed in the published studies.—cont'd

Type of research	Summary	Method	Sample/analyte	Number of citations	Applied tools	References
Review	Modernizing separations to be more environmentally friendly: a website and summary of the AMGS, a spreadsheet	Measurements of AMGS	—	65	AMGS	[19]
Research	This work is intended to establish a simple, accurate, and specific UPLC approach to calculating ISD and HDZ concurrently.	UPLC	Drugs/ISD/HDZ	17	NEMI GAPI (10 green) AMGS (44.61) AGREE (92)	[70]
Review	A hexagon-shaped symbol representing the processed data acts as a quick evaluation tool.	Summary of hexagon-CALIFICAMET	—	45	Hexagon-CALIFICAMET	[20]
Research	The suggested IT-SPME-CapLC-DAD method has been used to analyze synthetic dyes with a smaller impact on the environment by generating less waste and using fewer chemicals.	IT-SPME-CapLC-DAD	Chicken meat/dyes	1	Hexagon-CALIFICAMET (1 −2.71)	[71]
Research	The scientists concentrated on creating a UPLC-MS/MS technique to simultaneously calculate NER and NRN in rat plasma. Additionally, the established technique of analysis was used to evaluate the	UPLC-MS/MS	Rat plasma/NER and NRN	0	NEMI AES (90)	[72]

How to evaluate the greenness and whiteness **Chapter | 9** **341**

sustainability of both analytes and certification in accordance with recent regulatory criteria.					
Review	Two green criteria are utilized to rate and contrast the greenness of a solidified floating organic drop microextraction (SFODME) in different modes: AGREE and GAPI. The two metrics are incredibly straightforward and precise.	SFODME	—	4	AGREEprep GAPI AGREE AES	[2]
Research	The MSLE-DSPE approach made determining n-alkanes in sediments straightforward and fast. The recovery rates and the findings of the evaluation of real data both validated the efficacy of the process for extracting n-alkanes from a complicated sediment matrix.	MSLE-DSPE	Marine sediments/ n- alkanes	0	Eco-scale (80) GAPI (6 green) AGREE (0.66)	[73]
Research	The approach was successfully utilized on ascidian goods, and 10 PASHs with the greatest levels in the branchial basket were measured. This is the initial investigation to use a miniaturized green approach to determine PASHs from petrochemical sources in marine species.	MSLE-DµSPE/GC-MS/SIM	Phallusia nigra/ PASH	1	Eco-scale (82) GAPI (5 green) AGREE (0.6)	[74]

Continued

TABLE 9.31 Analytical Eco-Scale, WAC, AMGS, GAPI, Complex GAPI, AGREEprep, and AGREE programs are among the greenness evaluation metrics discussed in the published studies.—cont'd

Type of research	Summary	Method	Sample/analyte	Number of citations	Applied tools	References
Research	The pinggu peach of Beijing and its two neighboring regions in Heibei Province (China) were identified according to their geographical origin using extensively focused metabolomics using UHPLC-MS in conjunction with multivariate statistical modeling.	UHPLC-MS	Pinggu peach	0	AGREE (0.59)	[75]
Review	This paper gives an introduction to ferrofluids creation and highlights important factors for extraction procedures. The function of coating or composite substances, carrier fluids, and how they can help in analyte separation are also discussed in this review.	Summary of ferrofluids and applications of greenness tools	—	7	AES GAPI AGREE	[76]
Research	For the concomitant identification of two medicines, a green microemulsion electrokinetic chromatographic approach along with field-enhanced sample stacking was devised. Azelastine and budesonide concentration levels for the technique were 0.5–50.0 μg/mL and 1.0–20.0 μg/mL, respectively.	Microemulsion electrokinetic chromatographic method	Drugs/AZL, BUD	5	AES (0.84) GAPI (7 green) AGREE (0.71)	[77]

Research	For the accurate assay of VPA in raw materials and human plasma, a QbD methodology was used to construct a potentiometric screen-printed detector with the environment score to evaluate the process utilizing NEMI, ESA, GAPI, and AGREE evaluation tools.	Potentiometry with sensor	Sodium valproate/ Pharmaceutical and biological sample	4	NIME Eco-scale (91) GAPI AGREE (0.76) [78]
Research	Utilizing HPLC-RID, a brand-new analytical technique was created for the simultaneous finding of NAR in feed and its breakdown components.	HPLC-RID	Feed/narasin	1	GAPI (5 green) AGREE (0.52) [79]
Research	The created approach was successfully used to estimate three racetams with similar structural properties simultaneously in their pure form, synthesized mixes, and medicinal formulations.	RP-HPLC-UV	Plasma/three racetams	1	GAPI (9 green) AGREE (0.71) [80]
Research	A reliable analytical procedure using a mixed-micellar HPLC methodology. The separating effectiveness of meropenem and Ertapenem was increased by using a central composite design, which was utilized to examine the impacts of various chromatographic factors.	Mixed-micellar HPLC	Medicines/ Ertapenem and meropenem	3	GAPI (9 green) AGREE (0.75) AES (95) [81]

Continued

TABLE 9.31 Analytical Eco-Scale, WAC, AMGS, GAPI, Complex GAPI, AGREEprep, and AGREE programs are among the greenness evaluation metrics discussed in the published studies.—cont'd

Type of research	Summary	Method	Sample/analyte	Number of citations	Applied tools	References
Research	To separate the overlapped spectra of two compounds and measure them concurrently without prior isolation, suggested approaches depended on straightforward mathematical modification of scanning UV spectra.	UV spectroscopic	Bulk and formulation/ Chlorthalidone and Azelnidipine	3	AGREE (0.91) ComplexGAPI (12 green and E-factor less than 0.1) RBG (97.5)	[44]
Research	Dispersive liquid-liquid microextraction was used to remove and analyze four different sulfonamides from various water samples. Both natural monoterpenoids (menthol and thymol) were combined with various acidic substances, alcohols, and two separate sets of hydrophobic NDESs to create and experiment with them as microextraction solvents.	VA-DLLME/UHPLC-DAD	Water/sulfonamides	7	AES (73) GAPI (7 green) AGREE (0.64)	[82]

Research	In this study, a quick, environmentally friendly HPLC method for measuring IND, PER arginine, and ADB simultaneously in binary and ternary mixtures was reported. Comparing the described approaches, the overall chromatographic run time is the quickest (less than 3 min).	RP-HPLC	Drugs/IND, PER arginine, ADB	17	NEMI AGP (2 green) GAPI (7 green) AES (80)	[83]
Research	Two green HPLC methods were created and confirmed to separate the two antihypertensive combinations (Nebivolol hydrochloride/Valsartan mixture and Sacubitril/Valsartan mixture).	HPLC-PDA	Drugs/two antihypertensive binary mixtures	2	NEMI AES (91) GAPI (9 green) AGREE (0.84)	[84]
Research	For the concurrent measurement of MLK, BAM, and TER, the pharmacopeial-associated component of BAM, this research marks the creation of new environmentally friendly HPLC and HPTLC technologies.	HPLC, HPLTC	MLK, BAM, TER	0	AES (87, 81) GAPI (5 green) AGREE (0.71)	[85]
Research	For the quantification of SEP in the presence of its two contaminants in pure form and medicinal dose form, a straightforward, accurate, and environmentally friendly RPHPLC/UV approach was devised.	RP-HPLC/UV	Pharmaceutical dosage form/selexipag	1	NEMI GAPI (5 green) ESA (88) AGREE (0.54)	[86]

Continued

346 Green Analytical Chemistry

TABLE 9.31 Analytical Eco-Scale, WAC, AMGS, GAPI, Complex GAPI, AGREEprep, and AGREE programs are among the greenness evaluation metrics discussed in the published studies.—cont'd

Type of research	Summery	Method	Sample/analyte	Number of citations	Applied tools	References
Research	The present research investigates the planning and creation of a novel, rapid, and green UPLC approach for simultaneously detecting pitavastatin and ezetimibe. The Ishikawa fishbone model was utilized for combining the green analytical process with Perfection by design-based risk evaluation, and a rotatable center composite structure was employed for improvement.	UPLC-PDA	Drugs/Pitavastatin, ezetimibe	5	NEMI (4 green) GAPI (9 green) AES (97) AMGS (49.38) AGREE (0.89)	[87]
Research	Arsenic was extracted from botanical medicines using the UA-MSPD technique, MAE, and AADES.	UA-MSPD/ICP-MS	Drugs/As	0	AES (75) WAC (84.6)	[88]
Review	This study examined the AES, GAPI, and AGREE metrics, their fundamental ideas, and examples of when they were used with specific analytical methods. The advantages and	Summary for three tools	—	95	AES GAPI AGREE	[45]

	disadvantages of these measures are examined from the perspective of the typical reader or user.					
Review	The history of GAC and the current greenness evaluation metrics are thoroughly covered in this review, along with the criteria, ideas, advantages, and limitations of each tool. A comparison of the measurement tools is also provided, and their applicability for different analytical methodologies is investigated.	Summary for 10 tools	—	0	10 tools	[51]
Review	The green measurement programs were gathered and analyzed for this overview. Based on how the final report will be represented, the green tools were separated into two types: Quantitative and qualitative.	Summary for five tools	—	36	NEMI AES GAPI AMGS AGREE	[47]

348 Green Analytical Chemistry

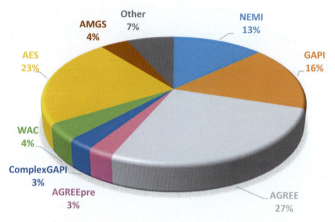

FIGURE 9.13 The rating of applications green tools in published papers.

recently made advancements into perspective, and make decisions about the future. In order to create a sustainable future, it should also help government and industrial experts decide which analytical tools to research further and which methods to accept.

Abbreviations

AA-DLLME-SFO Air agitation-dispersive liquid–liquid microextraction solidified floating organic drop
AADES Amino acids-based deep eutectic solvents
ADB Amlodipine besylate
AES Analytical Eco-scale
AGREE Analytical GREEness Scale
AMGS Analytical Method Greenness Score
AZL Azelastine
BAM Bambuterol hydrochloride
BUD Budesonide
ComplexGAPI Complex Green Analytical Procedure Index
CVD Carvedilol
ESIMRM/MS Electrospray ionization multiple reaction monitoring/mass spectrometry
GAC Green analytical chemistry
GAPI Green Analytical Procedure Index
GC-MS Gas chromatography-quadrupole mass spectrometry
GC–MS/SIM Gas chromatography coupled to mass spectrometry in selective ion monitoring
HDZ Hydralazine hydrochloride
HF-LPM Hollow fiber liquid phase microextraction
HPLC-FLD High performance liquid chromatography-fluorescence detection
IBD Ivabradine
IND Indapamide
ISD Isosorbide dinitrate

IT-SPME-CapLC-DAD In-tube solid-phase microextraction online coupled to capillary LC
LC Liquid chromatography
MAE Microwave-assisted extraction
META-IL-DLLME Magnetic effervescent tablet-assisted ionic liquid dispersive liquid−liquid microextraction
MLK Montelukast sodium
MSLE-DSPE Miniaturized device combined with cleanup via dispersive micro-solid-phase extraction
MSLE-DμSPE Microscale solid-liquid extraction method via dispersive micro-solid-phase extraction
NDESs Natural deep eutectic solvents
NEMI National Environmental Methods Index
NER Neratinib
NRN Naringenin
PASHs Polycyclic aromatic sulfur heterocycles
PER Perindopril
QbD Quality by design
REN Remogliflozin
SFC Supercritical fluid chromatography
TEN Teneligliptin
TER Terbutaline
UA-MSPD Ultrasound-assisted matrix solid-phase dispersion
UASE-PMLS Ultrasound-assisted solvent extraction of porous membrane-packed liquid samples
UPLC Ultra-performance liquid chromatography
VA-DLLME Vortex-assisted dispersive liquid−liquid microextraction
WAC White analytical chemistry

References

[1] A. Paul, N. Eghbali, Green chemistry: principles and practice, Chemical Society Reviews 39 (2009) 301−312, https://doi.org/10.1039/B918763B.

[2] A.R. Hussein, M.S. Gburi, N.M. Muslim, E.A. Azooz, A greenness evaluation and environmental aspects of solidified floating organic drop microextraction for metals: a review, Trends in Environmental Analytical Chemistry 37 (2023) e00194, https://doi.org/10.1016/j.teac.2022.e00194.

[3] P.T. Anastas, M.M. Kirchhoff, Origins, current status, and future challenges of green chemistry, Accounts of Chemical Research 35 (2002) 686−694, https://doi.org/10.1021/ar010065m.

[4] W. Wardencki, J. Curyło, J. Namieśnik, Green chemistry—current and future issues, Polish Journal of Environmental Studies 14 (2005) 389−395.

[5] B.D. Paul, A history of the concept of sustainable development: literature review, Ann. Univ. Oradea, Econ. Sci. Ser. 17 (2008) 581−585.

[6] E.A. Azooz, M. Tuzen, W.I. Mortada, N. Ullah, A critical review of selected preconcentration techniques used for selenium determination in analytical samples, Critical Reviews in Analytical Chemistry (2022), https://doi.org/10.1080/10408347.2022.2153579. In press.

[7] J. Płotka-Wasylka, H.M. Mohamed, A. Kurowska-Susdorf, R. Dewani, M.Y. Fares, V. Andruch, Green analytical chemistry as an integral part of sustainable education

development, Current Opinion in Green and Sustainable Chemistry 31 (2021) 100508, https://doi.org/10.1016/J.COGSC.2021.100508.
[8] E.A. Azooz, H.S.A. Al-Wani, M.S. Gburi, E.H.B. Al-Muhanna, Recent modified air-assisted liquid-liquid micro-extraction applications for medicines and organic compounds in various samples: a review, Open Chemistry 20 (1) (2022) 525–540, https://doi.org/10.1515/chem-2022-0174.
[9] M. Tobiszewski, M. Marć, A. Gałuszka, J. Namieśnik, Green chemistry metrics with special reference to green analytical chemistry, Molecules 20 (2015) 10928–10946, https://doi.org/10.3390/molecules200610928.
[10] A. Gałuszka, Z. Migaszewski, J. Namieśnik, The 12 principles of green analytical chemistry and the SIGNIFICANCE mnemonic of green analytical practices, Trends in Analytical Chemistry (Reference Ed.) 50 (2013) 78–84, https://doi.org/10.1016/J.TRAC.2013.04.010.
[11] W.I. Mortada, E.A. Azooz, Microextraction of metal ions based on solidification of a floating drop: basics and recent updates, Trends in Environmental Analytical Chemistry 34 (June 2022) e00163, https://doi.org/10.1016/j.teac.2022.e00163.
[12] P.T. Anastas, J.C. Warner, Green Chemistry: Theory and Practice, Frontiers, 1998.
[13] P.T. Anastas, Green Chemistry and the role of analytical methodology development, Critical Reviews in Analytical Chemistry 29 (1999) 167–175, https://doi.org/10.1080/10408349891199356.
[14] P.M. Nowak, R. Wietecha-Posłuszn, J. Pawliszyn, White analytical chemistry: an approach to reconcile the principles of green analytical chemistry and functionality, Trends in Analytical Chemistry 138 (2021) 116223, https://doi.org/10.1016/j.trac.2021.116223.
[15] NEMI (national, environmental methods index), 2002. https://www.nemi.gov/home/.
[16] E.A. Azooz, K.R. Rana, H. Ali Abdulridha, The fundamentals and recent applications of micellar system extraction for nanomaterials and bioactive molecules: a review, Nano Biomedicine and Engineering 13 (3) (2021) 264–278. http://nanobe.org/Data/View/710.
[17] K. Van Aken, L. Strekowski, L. Patiny, EcoScale, a semi-quantitative tool to select an organic preparation based on economical and ecological parameters, Beilstein Journal of Organic Chemistry 2 (3) (2006) 1–7, https://doi.org/10.1186/1860-5397-2-3.
[18] J. Płotka-Wasylka, A new tool for the evaluation of the analytical procedure: green Analytical Procedure Index, Talanta 181 (2018) 204–209, https://doi.org/10.1016/j.talanta.2018.01.013.
[19] M.B. Hicks, W. Farrell, C. Aurigemma, L. Lehmann, L. Weisel, K. Nadeau, H. Lee, C. Moraff, M. Wong, Y. Huang, P. Ferguson, Making the move towards modernized greener separations: introduction of the analytical method greenness score (AMGS) calculator, Green Chemistry 21 (2019), https://doi.org/10.1039/c8gc03875a, 1816e1826.
[20] A. Ballester-Caudet, P. Campíns-Falcó, B. Pérez, R. Sancho, M. Lorente, G. Sastre, C. González, A new tool for evaluating and/or selecting analytical methods: summarizing the information in a hexagon, Trends in Analytical Chemistry (Reference Ed.) 118 (2019), https://doi.org/10.1016/j.trac.2019.06.015, 538e547.
[21] F. Pena-Pereira, W. Wojnowski, M. Tobiszewski, Agree-analytical GREEnness metric approach and software, Analytical Chemistry 92 (2020) 10076–10082, https://doi.org/10.1021/acs.analchem.0c01887.
[22] P.M. Nowak, P. Kościelniak, M. Tobiszewski, A. Ballester-Caudet, P. Campíns- Falcó, Overview of the three multicriteria approaches applied to a global assessment of analytical methods, Trends in Analytical Chemistry (Reference Ed.) 133 (2020), https://doi.org/10.1016/j.trac.2020.116065, 116065.

[23] National environmental methods index (n.d.), https://www.nemi.gov/about/. (Accessed 28 July 2021).
[24] L.H. Keith, L.U. Gron, J.L. Young, Green analytical methodologies, Chemical Reviews 107 (2007) 2695−2708.
[25] M. De La Guardia, S. Garrigues (Eds.), Challenges in Green Analytical Chemistry, Royal Society of Chemistry, Cambridge, 2011. https://www.bookdepository.com/Challenges-Green-Analytical-Chemistry-James-H-Clark/9781849731324. (Accessed 25 October 2021).
[26] K. Van Aken, L. Strekowski, L. Patiny, EcoScale, a semi-quantitative tool to select an organic preparation based on economical and ecological parameters, Beilstein Journal of Organic Chemistry 2 (1) (2006) 3, https://doi.org/10.1186/1860-5397-2-3.
[27] A.P. Dicks, A. Hent, K.J. Koroluk, The EcoScale as a framework for undergraduate green chemistry teaching and assessment, Green Chemistry Letters and Reviews 11 (1) (2018) 29−35, https://doi.org/10.1080/17518253.2018.1431313.
[28] A. Gałuszka, P. Konieczka, Z.M. Migaszewski, J. Namieśnik, Analytical Eco-Scale for assessing the greenness of analytical procedures, Trends in Analytical Chemistry 37 (2012).
[29] F. Pena-Pereira, M. Tobiszewski, W. Wojnowski, E. Psillakis, A tutorial on AGREEprep an analytical greenness metric for sample preparation, Advances in Sample Preparation 3 (2022) 100025.
[30] M. Tobiszewski, Metrics for green analytical chemistry, Analytical Methods 8 (2016) 2993−2999.
[31] E.A. Azooz, G.J. Shabaa, E.H.B. Al-Muhanna, E.A.J. Al-Mulla, W.I. Mortada, Displacement cloud point extraction procedure for preconcentration of iron (III) in water and fruit samples prior to spectrophotometric determination, Bulletin of the Chemical Society of Ethiopia 37 (1) (2023) 1−10, https://doi.org/10.4314/bcse.v37i1.1.
[32] S. Heba, M. Ahmed, Sustainable eco-friendly ultra-high-performance liquid chromatographic method for simultaneous determination of caffeine and theobromine in commercial teas: evaluation of greenness profile using NEMI and eco-scale assessment tools, Journal of AOAC International 101 (6) (2018), https://doi.org/10.5740/jaoacint.18-0084.
[33] R. Ahmed, I. Abdallah, Development and greenness evaluation of spectrofluorometric methods for flibanserin determination in dosage form and human urine samples, Molecules 25 (2020) 4932, https://doi.org/10.3390/molecules25214932.
[34] A. Gałuszka, Z.M. Migaszewski, P. Konieczka, J. Namieśnik, Analytical eco-scale for assessing the greenness of analytical procedures, Trends in Analytical Chemistry (Reference Ed.) 37 (2012) 61−72, https://doi.org/10.1016/j. trac.2012.03.013.
[35] R. Ahmadi, E.A. Azooz, Y. Yamini, A.M. Ramezani, Liquid-liquid microextraction techniques based on in-situ formation/decomposition of deep eutectic solvents, TrAC, Trends in Analytical Chemistry 161 (2023) 117019, https://doi.org/10.1016/j.trac.2023.117019.
[36] F.A. Wannas, E.A. Azooz, R.K. Ridha, S.K. Jawad, Separation and micro determination of zinc(II) and cadmium(II) in food samples using cloud point extraction method, Iraqi Journal of Science 64 (3) (2023) 1046−1061, https://doi.org/10.24996/ijs.2023.64.3.2.
[37] K.R. Rana, E.A. Azooz, S.S. Taresh, Rapid palladium preconcentration and spectrophotometric determination in water and soil samples, Anal. Bioanal. Chem. Res. 9 (3) (2022) 251−258.
[38] L.H. Keith, L.U. Gron, J.L. Young, Chemical Reviews 107 (2007) 2695.
[39] United Nations, Globally Harmonized System of Classification and Labeling of Chemicals (GHS, Rev. 4), United Nations, New York, USA, 2011.
[40] A. El-Hawiet, F.M. Elessawy, M.A. El Demellawy, A.F. El-Yazbi, Green fast and simple UPLC-ESI-MRM/MS method for determination of trace water-soluble vitamins in honey:

greenness assessment using GAPI and analytical eco-scale, Microchemical Journal 181 (2022) 107625, https://doi.org/10.1016/j.microc.2022.107625.
[41] A.D. Robles, M. Fabjanowicz, J. Płotka-Wasylka, P. Konieczka, Organic acids and polyphenols determination in polish wines by ultrasound-assisted solvent extraction of porous membrane-packed liquid samples, Molecules 24 (2019) 4376, https://doi.org/10.3390/molecules24234376.
[42] M.S. Elshahed, S.S. Toubar, A.A. Ashour, R.T. El-Eryan, Novel sensing probe using Terbium-sensitized luminescence and 8-hydroxyquinoline for determination of prucalopride succinate: green assessment with Complex-GAPI and analytical Eco-Scale, BMC Chemistry 16 (2022) 80, https://doi.org/10.1186/s13065-022-00876-0.
[43] J. Płotka-Wasylka, W. Wojnowski, Complementary green analytical procedure index (ComplexGAPI) and software, Green Chemistry 23 (2021) 8657–8665.
[44] M. Attimarad, M.S. Chohan, V.K. Narayanaswamy, A.B. Nair, N. Sreeharsha, S. Shafi, M. David, A.A. Balgoname, A.I. Altaysan, E.P. Molina, P.K. Deb, Mathematically processed UV spectroscopic method for quantification of chlorthalidone and azelnidipine in bulk and formulation: evaluation of greenness and whiteness, Journal of Spectroscopy (2022), https://doi.org/10.1155/2022/4965138. Article ID 4965138, 13 pages.
[45] M. Sajid, J. Płotka-Wasylka, Green analytical chemistry metrics: a review, Talanta 238 (2022) 123046, https://doi.org/10.1016/j.talanta.2021.123046.
[46] E.A. Azooz, J.R. Moslim, S.M. Hameed, S.K. Jawad, E.A.J. Al-Mulla, Aspirin in food samples for separation and micro determination of copper(II) using cloud point extraction/solvation method, Nano Biomedicine and Engineering 13 (1) (2021) 62–71. http://nanobe.org/Data/View/687?type=100.
[47] K.P. Kannaiah, A. Sugumaran, H.K. Chanduluru, S. Rathinam, Environmental impact of greenness assessment tools in liquid chromatography—a review, Microchemical Journal 170 (2021) 106685, https://doi.org/10.1016/j.microc.2021.106685.
[48] E.A. Azooz, F.A. Wannas, S.K. Jawad, Developed cloud point extraction coupled with onium system for separation and determination cobalt in biological samples, Research Journal of Pharmacy and Technology 14 (2) (2021) 594–598.
[49] E. Adnan Azooz, F. Abd Wannas, R. Kadhim Ridha, S. Kadhim Jawad, E. Al-Mulla, A green approach for micro determination of silver(I) in water and soil samples using vitamin C, Analytical and Bioanalytical Chemistry Research 9 (2) (2022) 133–140, https://doi.org/10.22036/abcr.2021.277834.1609. http://www.analchemres.org/article_139800.html.
[50] D.S. El-Kafrawy, A.H. Abo-Gharam, M.M. Abdel-Khalek, T.S. Belal, Comparative study of two versatile multi-analyte chromatographic methods for determination of diacerein together with four non-steroidal anti-inflammatory drugs: greenness appraisal using Analytical Eco-Scale and AGREE metrics, Sustainable Chemistry and Pharmacy 28 (2022) 100709.
[51] M.S. Imam, M.M. Abdelrahman, How environmentally friendly is the analytical process? a paradigm overview of ten greenness assessment metric approaches for analytical methods, Trends in Environmental Analytical Chemistry 38 (2023) e00202, https://doi.org/10.1016/j.teac.2023.e00202.
[52] W. Wojnowski, M. Tobiszewski, F. Pena-Pereira, E. Psillakis, AGREEprep e Analytical greenness metric for sample preparation, Trends in Analytical Chemistry 149 (2022) 116553.
[53] E.A. Azooz, M. Tuzen, W.I. Mortada, Green microextraction approach focuses on air-assisted dispersive liquid–liquid with solidified floating organic drop for preconcentration and determination of toxic metals in water and wastewater samples, Chemical Papers 77 (2023) 3427–3438, https://doi.org/10.1007/s11696-023-02714-6.

[54] V. Mazzaracchio, A. Sassolini, K.Y. Mitra, D. Mitra, G.M. Stojanovi ć, A. Willert, E. Sowade, R.R. Baumann, R. Zichner, D. Moscone, F. Arduini, A fully-printed electrochemical platform for assisted colorimetric detection of phosphate in saliva: greenness and whiteness quantification by the AGREE and RGB tools, Green Analytical Chemistry 1 (2022) 100006, https://doi.org/10.1016/j.greeac.2022.100006.

[55] C.M. Hussain, C.G. Hussain, R. Keçili, White analytical chemistry approaches for analytical and bioanalytical techniques: applications and challenges, Trends in Analytical Chemistry 159 (2023) 116905.

[56] M. Gamal, I.A. Naguib, D.S. Panda, F.F. Abdallah, Comparative study of four greenness assessment tools for selection of greenest analytical method for assay of hyoscine N-butyl bromide, Analytical Methods (2020), https://doi.org/10.1039/d0ay02169e.

[57] H.K. Chanduluru, A. Sugumaran, Three spectrophotometric approaches for measuring ratio spectra of Ivabradine and Carvedilol in a binary mixture using green analytical principles, Current Chemistry Letters 11 (2022) 321−330.

[58] S.A. Atty, H.R. Abd El-Hadi, B.M. Eltanany, H.E. Zaazaa, M.S. Eissa, Analytical eco-scale for evaluating the uniqueness of voltammetric method used for determination of antiemetic binary mixture containing doxylamine succinate in presence of its toxic metabolite, Electrocatalysis 13 (2022) 731−746, https://doi.org/10.1007/s12678-022-00751-5.

[59] D. Moema, T.A. Makwakwa, B.E. Gebreyohannes, S. Dube, M.M. Nindi, Hollow fiber liquid phase microextraction of fluoroquinolones in chicken livers followed by high pressure liquid chromatography: greenness assessment using National Environmental Methods Index Label (NEMI), green analytical procedure index (GAPI), Analytical GREEnness metric (AGREE), and Eco Scale, Journal of Food Composition and Analysis 117 (2023) 105131, https://doi.org/10.1016/j.jfca.2023.105131.

[60] M. Attimarad, K. Narayanaswamy Venugopala, A. Balachandran Nair, N. Sreeharsha, E. Plaza Molina, R. Bhagavantrao Kotnal, C. Tratrat, A.I. Altaysan, A. Ahmed Balgoname, P.K. Deb, Environmental sustainable mathematically processed UV spectroscopic methods for quality control analysis of remogliflozin and teneligliptin: evaluation of greenness and whiteness, Spectrochimica Acta Part A: Molecular and Biomolecular Spectroscopy 278 (2022) 121303, https://doi.org/10.1016/j.saa.2022.121303.

[61] H.S. Elbordiny, S.M. Elonsy, H.G. Daabees, T.S. Belal, Sustainable quantitative determination of allopurinol in fixed dose combinations with benzbromarone and thioctic acid by capillary zone electrophoresis and spectrophotometry: validation, greenness and whiteness studies, Sustainable Chemistry and Pharmacy 27 (2022) 100684.

[62] R. Gray, B. Fitch, C. Aurigemma, M.B. Hicks, M. Beres, W. Farrell, S.V. Olesik, Improving the environmental hazard scores metric for solvent mixtures containing carbon dioxide for chromatographic separations, Green Chemistry 24 (2022) 4504, https://doi.org/10.1039/d1gc03749h.

[63] H.M. Lotfy, R.H. Obaydo, C.K. Nessim, Spider chart and whiteness assessment of synergistic spectrophotometric strategy for quantification of triple combination recommended in seasonal influenza Detection of spurious drug, Sustainable Chemistry and Pharmacy 32 (2023) 100980.

[64] H. Shaaban, The ecological impact of liquid chromatographic methods reported for bioanalysis of COVID-19 drug, hydroxychloroquine: insights on greenness assessment, Microchemical Journal 184 (2023) 108145, https://doi.org/10.1016/j.microc.2022.108145.

[65] H.A.-T. Noura, M. Nahed, E. El, T.E.-S. Dina, I.E.-S. Hussein, Spider diagram and Analytical Greenness metric approach for assessing the greenness of quantitative 1H-NMR determination of lamotrigine: taguchi method based optimization, Chemometrics and

Intelligent Laboratory Systems 209 (2021) 104198, https://doi.org/10.1016/j.chemolab.
2020.104198.

[66] S. Rajendran, S.H. Loh, M.M. Ariffin, W.M.A.W.M. Khalik, Magnetic effervescent tablet-assisted ionic liquid dispersive liquid–liquid microextraction employing the response surface method for the preconcentration of basic pharmaceutical drugs: characterization, method development, and green profile assessment, Journal of Molecular Liquids 367 (Part A) (2022) 120411, https://doi.org/10.1016/j.molliq.2022.120411.

[67] M.G. Fawzy, H. Saleh, A. Reda, E.A. Bahgat, A green spectrophotometric method for the simultaneous determination of nasal binary mixture used in respiratory diseases: applying isosbestic point and chemometric approaches as a resolving tool, greenness evaluation, Spectrochimica Acta Part A: Molecular and Biomolecular Spectroscopy 283 (2022) 121585, https://doi.org/10.1016/j.saa.2022.121585.

[68] A. Elsonbaty, A.W. Madkour, A.M. Abdel-Raoof, A.H. Abdel-Monem, A.M.M. El-Attar, Computational design for eco-friendly visible spectrophotometric platform used for the assay of the antiviral agent in pharmaceutical dosage form, Spectrochimica Acta Part A: Molecular and Biomolecular Spectroscopy 271 (2022) 120897, https://doi.org/10.1016/j.saa.2022.120897.

[69] M.B. Hicks, S. Oriana, Y. Liu, Assessment of analytical testing: the impact of metrics for the sustainable measurement of pharmaceuticals, Current Opinion in Green and Sustainable Chemistry 38 (2022) 100689, https://doi.org/10.1016/j.cogsc.2022.100689.

[70] H. Kumar Chanduluru, A. Sugumaran, Eco-friendly estimation of isosorbide dinitrate and hydralazine hydrochloride using Green Analytical Quality by Design-based UPLC method, RSC Advances 11 (2021) 27820.

[71] A. Ballester-Caudet, R. Navarro-Utiel, I. Campos-Hernández, P. Campíns-Falcó, Evaluation of the sample treatment influence in green and sustainable assessment of liquid chromatography methods by the HEXAGON tool: sulfonate-based dyes determination in meat samples, Green Analytical Chemistry 3 (2022) 100024, https://doi.org/10.1016/j.greeac.2022.100024.

[72] A. Altharawi, S.M. Alqahtani, S.S. Panda, M. Alrobaian, A.B. Alabbas, W.H. Almalki, M.A. Alossaimi, M.A. Barkat, R.A. Rub, S.N. Mir Najib Ullah, et al., UPLC-MS/MS method for simultaneous estimation of neratinib and naringenin in rat plasma: greenness assessment and application to therapeutic drug monitoring, Separations 10 (2023) 167, https://doi.org/10.3390/separations10030167.

[73] P.V.B. Bahia, M. Moreira Nascimento, Jailson Bittencourt de Andrade Maria Elisabete Machado, Microscale solid-liquid extraction: a green alternative for determination of n-alkanes in sediments, Journal of Chromatography A 1685 (2022) 463635, https://doi.org/10.1016/j.chroma.2022.463635.

[74] F.X.A. Sampaio, M.M. Nascimento, V.A. de Oliveira, S. Teixeira Martinez, J.B. de Andrade, M.E. Machado, Determination of polycyclic aromatic sulfur heterocycles in ascidians (Phallusia nigra) using a green procedure, Microchemical Journal 186 (2023) 108270.

[75] J. Zhao, A. Li, X. Jin, G. Liang, L. Pan, Discrimination of geographical origin of agricultural products from small-scale districts by widely targeted metabolomics with a case study on pinggu peach, Frontiers in Nutrition 9 (2022) 891302, https://doi.org/10.3389/fnut.2022.891302.

[76] M. Sajid, K. Kalinowska, J. Płotka-Wasylka, Ferrofluids based analytical extractions and evaluation of their greenness, Journal of Molecular Liquids 339 (2021) 116901, https://doi.org/10.1016/j.molliq.2021.116901.

[77] A.M. Zeida, A.A. El-Masryb, D.R. El-Wasseef, M. Eida, I.A. Shehata, Green microemulsion electrokinetic chromatographic method for simultaneous determination of azelastine and budesonide, Sustainable Chemistry and Pharmacy 29 (2022) 100795. https://doi.org/10.1016/j.scp.2022.100795.

[78] S.S. Soliman, G.A. Sedik, M.R. Elghobashy, H.E. Zaazaa, A.S. Saad, Greenness assessment profile of a QbD screen-printed sensor for real-time monitoring of sodium valproate, Microchemical Journal 182 (2022) 107859, https://doi.org/10.1016/j.microc.2022.107859.

[79] F. Abid, S.H. Youssef, Y. Song, P.A. Ankit, T.S.W.P. Darren, S. Garg, Development and validation of a new analytical method for estimation of narasin using refractive index detector and its greenness evaluation, Microchemical Journal 175 (2022) 107149, https://doi.org/10.1016/j.microc.2021.107149.

[80] N.M. Mansour, D.T. El-Sherbiny, F.A. Ibrahim, H.I. El Subbagh, Validation of a specific Reversed-Phase HPLC method for the quantification of three racetams; Piracetam, Levetiracetam, and brivaracetam in the presence of Co-administered drugs in their pharmaceuticals; greenness assessment and application to biological fluid and in-vitro dissolution testing, Microchemical Journal 181 (2022) 107703, https://doi.org/10.1016/j.microc.2022.107703.

[81] H.M. Hafez, S. El Deeb, M.M. Swaif, R.I. Ibrahim, R. Ali Kamil, S.A. Ahmed, A.E. Ibrahim, Micellar Organic-solvent free HPLC design of experiment for the determination of Ertapenem and meropenem; assessment using GAPI, AGREE and analytical Ecoscale models, Microchemical Journal 185 (2023) 108262, https://doi.org/10.1016/j.microc.2022.108262.

[82] A. Mostafa, H. Shaaban, A.M. Alqarni, M. Alghamdi, S. Alsultan, J.S. Al-Saeed, S. Alsaba, A. AlMoslem, Y. Alshehry, R. Ahmad, Vortex-assisted dispersive liquid−liquid microextraction using thymol based natural deep eutectic solvent for trace analysis of sulfonamides in water samples: assessment of the greenness profile using AGREE metric, GAPI and analytical eco-scale, Microchemical Journal 183 (2022) 107976, https://doi.org/10.1016/j.microc.2022.107976.

[83] S.S. Saleh, H.M. Lotfy, G. Tiris, N. Erk, Y. Rostom, Analytical tools for greenness assessment of chromatographic approaches: application to pharmaceutical combinations of Indapamide, Perindopril and Amlodipine, Microchemical Journal 159 (2020) 105557, https://doi.org/10.1016/j.microc.2020.105557.

[84] A. Hemdan, R. Magdy, M. Farouk, N.V. Fares, Central composite design as an analytical optimization tool for the development of eco-friendly HPLC-PDA methods for two antihypertensive mixtures containing the angiotensin receptor blocker Valsartan: greenness assessment by four evaluation tools, Microchemical Journal 183 (2022) 108105, https://doi.org/10.1016/j.microc.2022.108105.

[85] D.S. El-Kafrawy, A.H. Abo-Gharam, M.M. Abdel-Khalek, T.S. Belal, Eco-friendly chromatographic methods for concurrent estimation of Montelukast and Bambuterol with its pharmacopoeial related substance Terbutaline: greenness appraisal using analytical Ecoscale, GAPI and AGREE metrics, Microchemical Journal 176 (2022) 107236, https://doi.org/10.1016/j.microc.2022.107236.

[86] Y.M. Youssef, M.A. Mahrouse, E.A. Mostafa, Assessment of environmental impact of a novel stability-indicating RP-HPLC method and reported methods for the determination of selexipag in bulk and dosage form: a comparative study using different greenness assessment tools, Microchemical Journal 185 (2023) 108256, https://doi.org/10.1016/j.microc.2022.108256.

[87] H.K. Chanduluru, A. Sugumaranp, Estimation of pitavastatin and ezetimibe using UPLC by a combined approach of analytical quality by design with green analytical technique, Acta Chromatographica 34 (3) (2021), https://doi.org/10.1556/1326.2021.00949.
[88] T.G.S. Guimarães, F.S. Costa, I.M.N.R. Menezes, A.P.R. Santana, D.F. Andrade, A. Oliveira, C.D.B. Amaral, M.H. Gonzalez, Green approaches with amino acids-based deep eutectic solvents (AADES) for determining as in medicinal herbs by ICP-MS, Journal of Molecular Liquids 381 (2023) 121801, https://doi.org/10.1016/j.molliq.2023.121801.

Chapter 10

The CUPRAC method, its modifications and applications serving green chemistry

Reşat Apak[1], Mustafa Bener[2], Saliha Esin Çelik[1], Burcu Bekdeşer[1] and Furkan Burak Şen[2]
[1]*Istanbul University-Cerrahpaşa, Faculty of Engineering, Department of Chemistry, Avcilar, Istanbul, Türkiye;* [2]*Istanbul University, Faculty of Science, Department of Chemistry, Division of Analytical Chemistry, Fatih, Istanbul, Türkiye*

1. Importance of green analytical chemistry

Green chemistry (GC), evolved as a result of environmental concerns, aims to reduce the use of reagents, eliminate or minimize the use of solvents, and prevent pollution that harm the environment in chemical processes. In addition, its other goals include maximizing efficiency and reducing energy consumption in chemical processes [1,2]. Therefore, GC is based on reducing the use of chemicals that can harm the environment and humans, or on the synthesis and processing of chemicals that reduce risks [3]. GC designs all stages of the chemical-life cycle, includes monitoring to reduce the hazards of chemical processes at the source, and finally works systematically based on principles and designs [4]. In this context, after the first definitions, GC was based on 12 principles (Fig. 10.1) [5]. Chemists can achieve their sustainability goals with the help of these 12 principles [6].

Analytical chemistry briefly covers the science and art of qualitative and quantitative analysis of substances [7]. Analytical methods developed for the separation, recognition and quantification of real sample constituents involve the optimization of certain parameters such as sensitivity, selectivity, accuracy, repeatability, cost and time. Green analytical chemistry (GAC) is a field that is increasingly gaining ground within the field of green chemistry and attracting the attention of chemists. With GC applications, it is expected that the environment will not be negatively affected during chemical analyses and the quality of results will be high. After GC developments, the following goals emerged in GAC.

Green Analytical Chemistry. https://doi.org/10.1016/B978-0-443-16122-3.00007-X
Copyright © 2025 Elsevier Inc. All rights are reserved, including those for text and data mining, AI training, and similar technologies.

357

358 Green Analytical Chemistry

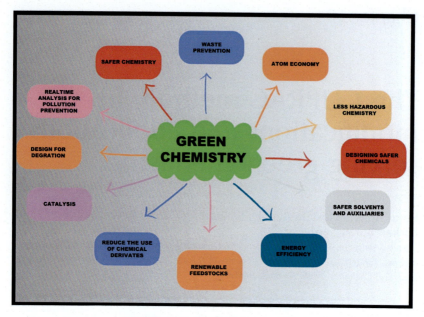

FIGURE 10.1 The 12 principles of GC.

- *Reducing or not using toxic solvents or reagents,*
- *Appropriate management of waste generated after analysis,*
- *Reducing energy consumption to a minimum level* [8–10].

In this context, damages to the environment can be minimized by reducing the solvent used through automation/miniaturization processes, by lessening the amount of solvent used in pre-treatments for analysis, and by direct analytical measurements that do not require solvents [9]. The fact that sample matrices are often complex and concentrations of analyzed substances are low enforces analytical chemists to perform pre-treatments such as extraction or pre-concentration before analysis. The amounts or chemical contents of solvents used during these pre-treatments can be very harmful for both the environment and the operator. In addition, the size and cost of instrumental devices used can be quite challenging. In the light of all this, in analytical laboratories within the scope of GAC; strategies such as processing waste, saving chemicals, reducing energy consumption, miniaturization and automation are at the forefront [11]. While traditional extraction techniques use large amounts of organic solvents with considerable toxicity, it has become essential to use alternatives within the scope of sustainable goals of green analytical chemistry [12]. While organic solvents are frequently used in analytical separation methods such as liquid chromatography and sample preparation methods such as liquid phase extraction, amphiphilic, ionic liquid and deep eutectic solvents have recently begun to be used instead [13]. Apart from these, it is known that reducing the need for reagents and performing analyses with less effort, time and people are incorporated in green chemistry. In fact, considering that the main purpose of analytical chemistry is to obtain information about the product, it is important to get fast, accurate and reliable results outside the laboratory, as well as to perform the analysis in the minimum equipped laboratory for the sake of greening while providing this information [14]. Nanosensors, paper sensors, and electrochemical sensors are frequently encountered in green analytical chemistry applications. From this perspective, sensors have been included in green analytical chemistry applications because they are portable, allow analysis outside the laboratory, provide rapid results, are simple in design, contain low amounts of solvent, enable analysis without sample preparation steps, and have high sensitivity [15,16].

2. Oxidative stress and antioxidants

Reactive oxygen and nitrogen species (RONS) consist of free radicals (hydroxyl radical ($\cdot$OH), superoxide anion radical ($O_2^{\cdot-}$), peroxyl radical (ROO$\cdot$), nitric oxide (NO$\cdot$) radical) and reactive molecules (hydrogen peroxide, hypochlorous acid, peroxynitrite anion). While in the human body these reactive species come from metabolic processes such as the respiratory chain, they are also produced under the influence of external sources such as smoking, exposure to X-rays, ozone, and air pollution [17]. Low-to-intermediate levels of RONS are necessary for normal cell functions such as intracellular signaling, regulation of gene expression and cellular defense against infective agents. However, there is a paradox in the biological functions of reactive species. If the balance between RONS production in the cell and the counteraction of defense system is disrupted, reactive species are produced uncontrollably and excessively. Under these conditions, reactive species can damage important biomolecules such as DNA, proteins, and membranes and play a role in the pathogenesis of various diseases, including chronic diseases, cancer, and inflammatory diseases.

Most living species have effective defense systems to protect themselves against this negative effect induced by RONS [18]. These defense systems can be produced in the body (antioxidant enzymes) or taken through diet (phenolic acids, flavonoids, thiol-type antioxidants). When the balance between RONS and antioxidants cannot be maintained by redox reactions involving electron transfer between the two chemical species, the cellular environment becomes oxidatively stressed [19].

Antioxidants are compounds that can delay or inhibit the oxidation of the substrate despite their low concentrations compared to the oxidizable substrate [20]. Antioxidants are basically divided into two classes according to their mechanism of action: primary (chain-breaking) antioxidants, which act mainly by ROS/RNS scavenging, and secondary (preventive) antioxidants, which act mainly by transition metal ion chelation [21].

Primary antioxidants act by scavenging reactive species through various mechanisms such as hydrogen atom transfer (HAT), single electron transfer (SET), or proton-coupled electron transfer (PCET), while secondary antioxidants inhibit the Fenton reaction due to their chelating ability of transition metal ions such as Fe(II) or Cu(I). In addition, endogenous antioxidant enzymes such as superoxide dismutase, catalase and glutathione peroxidase, known as "first-line defense antioxidants," can directly scavenge reactive species such as hydrogen peroxide and superoxide anion radicals [22].

3. CUPRAC method and its advantages

Antioxidants are analyzed for the purposeful comparison of foods regarding their antioxidant content and for screening variations within or between complex matrices. The chemical diversity of antioxidants complicates to separate and determine the amounts of individual antioxidants (i.e., polyphenols, vitamins, minerals, organosulfur compounds) from the food sample [23]. Total antioxidant capacity (TAC, measuring the combined power of antioxidants) is a more useful and meaningful parameter for an antioxidant sample, to reveal and compare beneficial effects of food/plasma antioxidants to health [24].

CUPric Reducing Antioxidant Capacity, developed and named by our research group as the "CUPRAC" method [25], is a simple, versatile, selective, low-cost assay for several types of phenolic compounds (e.g., flavonoids, phenolic acids), carotenoids, vitamins (e.g., ascorbic acid, α-tocopherol), plasma antioxidants (reduced glutathione, uric acid, and bilirubin) irrespective of chemical type or hydrophilicity. The light blue-colored bis(neocuproine) copper(II) complex (Cu(II)-Nc) is used as the chromogenic oxidant toward antioxidants. This is an electron-transfer based colorimetric assay that measures the antioxidant capacity of a sample by assessing its ability to reduce copper(II) ions to copper(I) in the presence of neocuproine ligand that provides significant elevation of Cu(II/I) reduction potential by selectively stabilizing the cuprous state in preference to the cupric state, i.e., cuprous-neocuproine is more stable than cupric-neocuproine by seven orders-of-magnitude [26]. Following the redox reaction with reducing antioxidants at pH 7, the absorbance of the stable yellow-orange colored product (Cu(I)-Nc chelate) is measured at a wavelength of 450 nm within 30 min. The intensity of color (as absorbance) is proportional to the reducing ability and concentration of antioxidants in the sample (Fig. 10.2).

The relationship between green chemistry and the CUPRAC assay lies in their shared focus on sustainability and environmental impact. Green chemistry, as a field, emphasizes the design and implementation of chemical products and processes that reduce or eliminate the use of substances

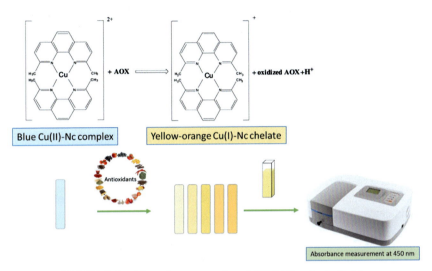

FIGURE 10.2 Schematic representation of CUPRAC method principle.

hazardous to human health and the environment [27]. CUPRAC assay uses environmentally friendly solvents with lower toxicity and lower environmental impact. As an additional feature of GAC, CUPRAC is probably the least solvent-dependent SET-assay, making it possible to analyze and compare the TAC results of antioxidant sample extracts prepared in different solvent media. Çelik et al. argued that the CUPRAC assay, involving a coordinatively saturated Cu(II)-complex reagent capable of outer-sphere electron-transfer, is relatively independent of solvent effects in alcohol−water mixtures of varying composition [28]. Short analysis period also makes the CUPRAC method valuable in terms of energy and time savings. In this procedure, an antioxidant-rich sample extract is mixed with aqueous copper(II) chloride solution (0.01 M), ethanolic neocuproine solution (7.5 mM), and ammonium acetate (NH_4Ac, 1 M) buffer solution (1:1:1, v/v/v). The typical CUPRAC buffer, NH_4Ac, can be prepared from aqueous ammonia and acetic acid, the latter known as edible vinegar acid. The reaction mixture is then incubated for a specific period (30 min). In this mixture, the reactive Ar−OH groups of polyphenolic antioxidants are oxidized to the corresponding quinones. However, fast reacting antioxidants may be oxidized in less time. The liberated protons from the CUPRAC reaction are buffered in NH_4Ac medium.

Standard redox potential of the Cu(II/I)-neocuproine is 0.6 V, much higher than that of Cu^{2+}/Cu^+ couple (0.17 V). This is why polyphenols are oxidized much more rapidly and efficiently when reacted with Cu(II)-Nc complex. For phenolics having relatively higher reduction potential close to that of Cu(II/I), reaction kinetics may be somewhat slower, requiring incubation in a 50°C-water bath for 20 min for optimal color development.

Calculation of individual antioxidant capacity coefficient of a phenolic antioxidant as Trolox (water-soluble vitamin E analogue) equivalents (TEAC, also known as "Trolox equivalent antioxidant capacity," defined as the reducing potency of 1.0 mM antioxidant solution in Trolox mM equivalents) in the CUPRAC method is calculated by dividing the molar absorption coefficient of antioxidant under investigation by that of Trolox under corresponding conditions. The antioxidant capacities measured in CUPRAC were found in accordance with theoretical expectations of structure–activity relationships, because the number and position of the phenolic hydroxyl groups as well as the degree of conjugation of the whole molecule are important for efficient electron transfer [25].

Total antioxidant capacity of a plant extract expressed in terms of Trolox equivalents (TE) is given below:

$$\text{TAC}\left(\frac{\text{mmol}}{\text{g}}\text{TE}\right) = \frac{\text{Absorbance at 450 nm}}{\varepsilon_{\text{TROLOX}}} \times \frac{\text{Total volume}}{\text{Sample volume}} \times \text{Dilution factor} \times \frac{\text{extract volume}}{m_{\text{dry sample}}}$$

3.1 Main advantages of the CUPRAC assay

Versatility: CUPRAC assay can be applied to a wide range of samples, including biological fluids, food products, and medicinal compounds. This versatility makes it useful in various fields such as biochemistry, food science, and pharmacology. This assay was successfully applied to biologically important antioxidants (e.g., ascorbic acid, α-tocopherol, β-carotene, reduced GSH, uric acid, and bilirubin) [29] and all types of food antioxidants (e.g., flavonoids, carotenoids, simple phenolic and hydroxycinnamic acids) [26]. CUPRAC method was successively adapted to different types of complex matrices, i.e., herbal plants [30–34], human serum [29], fruit and vegetable samples [35–37], cereals [38], edible oils [29,39]. Furthermore, ROS- such as hydroxyl radical-scavenging antioxidants can be indirectly measured with the CUPRAC assay by selecting a CUPRAC-responsive probe [40]. Thus, both antioxidants normally acting as reducing agents and reactive species normally acting as oxidizing agents can be determined with the same CUPRAC reagent after the necessary precautions. Versatility of a TAC assay is an essential part of its greenness, as different complex samples requiring different assay reagents would require the use of more chemical substances.

Sensitivity: The CUPRAC assay is sensitive to low concentrations of reducing agents. This allows researchers to detect and quantify substances with strong antioxidant properties even at low concentrations. Furthermore, the CUPRAC assay has been shown to produce higher values than other antioxidant assays, indicating its effectiveness in evaluating the potential environmental impact of chemical compounds and complex samples. For sensitivity stabilization, antioxidant capacities of phenolic mixtures were measured with CUPRAC, DPPH and ABTS assays in a time period of 60, 120 and 300 min, respectively. The results showed the kinetic advantage of CUPRAC over radical-based TAC assays [41].

Applicability: The method is easily and diversely applicable in conventional laboratories using standard colorimeters rather than necessitating sophisticated equipment and highly qualified operators. The method can simultaneously measure hydrophilic and lipophilic antioxidants (e.g., β-carotene and α-tocopherol) [35]. Moreover, methyl-β-cyclodextrin as solubility enhancer was utilized for the simultaneous determination of lipophilic and hydrophilic antioxidants as their inclusion complexes [42,43].

Simplicity and rapidity of procedure: The CUPRAC assay involves a relatively simple and fast procedure, making it a convenient choice for high-throughput screening of samples. This is particularly advantageous when dealing with a large number of samples or in situations where time is a critical factor.

Stability of reagents: The reagents used in the CUPRAC assay are relatively stable, contributing to the reproducibility of the results. The reaction product, Cu(I)-Nc chelate, is stable and relatively insensitive to a number of parameters adversely affecting radicalic reagents such as ABTS and DPPH, e.g., air, sunlight, humidity, and pH. Especially pH-insensitivity is an important feature of GAC, because the use of excessive acids and bases can be avoided in pH adjustments.

Realistic pH: The CUPRAC assay is applied at nearly pH 7 of ammonium acetate buffer which simulates physiological pH.

Additivity: The CUPRAC absorbances of mixture constituents monitored at 450 nm were additive, indicating lack of chemical deviations from Beer's law of optical density [26]. Additivity of absorbances gives rise to additivity of TAC values. For example, TAC values of antioxidants found with the CUPRAC assay are perfectly additive, that is to say, the TAC of a phenolic mixture is equal to the sum of TAC values of its constituent polyphenols.

Wide dynamic range: The CUPRAC assay has a broad dynamic range, enabling the assessment of reducing potential over a wide concentration range. The molar absorptivity for n–e$^-$ reductants $(8.5 \pm 1.0) \times 10^3$ n L mol^{-1} cm^{-1}, is sufficiently high to sensitively determine most phenolic antioxidants. This feature is beneficial when dealing with samples having varying antioxidant capacities.

Compatibility with other instrumental techniques: High-throughput methods produced from original CUPRAC assay are advantageous when there is a need to rapidly analyze a large number of samples. On-line HPLC (High Performance Liquid Chromatography)-CUPRAC assay was designed by Çelik et al. in order to separate constituents by HPLC column and simultaneously determine individual antioxidant capacities [23]. Moreover, CUPRAC method has been automated by Ribeiro et al. to microplate reader (96-well plates) within 4 min at 37°C and automated flow injection analysis [44].

4. Modifications and applications of CUPRAC method serving green chemistry

4.1 Optical sensors based on CUPRAC method

When compared to classical instrumental methods, optical sensors offer numerous advantages such as speed, remote control, flexibility, low cost, miniaturization, and on-site analysis. Due to these significant advantages, their use in many applications has become appealing today. Optical sensors evaluate analytical information using optical techniques (absorbance, reflection, fluorescence, etc.) and successfully perform analyte determination for complex samples where color interference and turbidity are possible [45].

Optical sensors play an important role in the context of green chemistry, aligning with the principles of sustainability and environmentally conscious practices. Their integration into analytical methodologies contributes significantly to the reduction of hazardous materials, energy consumption, and waste generation. For instance, the use of optical sensors allows for real-time monitoring, minimizing the need for extensive and resource-intensive sampling and sample preparation procedures [46]. This approach not only enhances analytical efficiency but also reflects a commitment to reducing the environmental impact of chemical analyses.

In summary, the incorporation of optical sensors into analytical processes not only enhances the precision and efficiency of chemical analyses but also contributes substantially to the overarching goals of green chemistry.

4.1.1 CUPRAC based optrode sensors

Optrodes are essential tools in analytical chemistry, particularly for detecting and measuring specific chemical components in solution. Optrodes are typically produced by immobilizing various reagents on suitable membrane surfaces through techniques such as covalent bonding, electrostatic interactions, and membrane incorporation [47,48]. Immobilization of small amounts of reagents on the optrode saves from excessive reagent consumption in solution. Analytical information is derived from changes in the properties of the interacting light with the coating. Optrodes play a crucial role in analytical detection and analysis by providing high sensitivity and selectivity. They are also effectively employed in small-scale and continuous monitoring applications. Optrode-based systems find widespread use in various analytical chemistry fields, notably in applications such as biochemistry, environmental analysis, and drug development. Studies on the adaptation of the CUPRAC method to optrodes have primarily focused on their attachment to polymeric membrane and paper-based substrate materials (Fig. 10.3) [49–51].

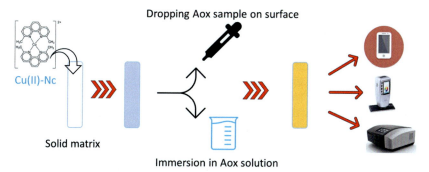

FIGURE 10.3 Schematic illustration of CUPRAC based optrode sensors.

In the study conducted by Bener et al. on the adaptation of the CUPRAC method to optical sensors, the authors immobilized the Cu(II)-Nc reagent onto a perfluorosulfonic acid-based cation-exchange polymer membrane matrix (Nafion) [49]. This sensor, considered the first optical sensor determining total antioxidant capacity in the literature, involves the in situ reduction of the highly colored Cu(II)-Nc chelate immobilized on the membrane with antioxidants, producing the highly colored Cu(I)-Nc on the membrane (which does not diffuse back into solution). Absorbance changes associated with this reduction are recorded on the membrane at a wavelength of 450 nm for the determination of antioxidant capacity. In another study, the same sensor was used to develop a reflectometric sensor by measuring reflectance changes at a wavelength of 530 nm in the presence of antioxidant components using a fiber optic spectrophotometer [50]. Analytical performance parameters for both sensors were examined, and it was reported that indicator paper strips (immersed in solution) could be used for determining antioxidant capacities in food samples without pretreatment. In another study, Bener et al. adhered to the same detection mechanism, utilizing a biopolymer film made from carrageenan obtained from red seaweeds instead of the synthetic chemical Nafion for the solid matrix [51]. Carrageenan, a polysaccharide, imparts cation-exchange properties to the biopolymer film due to its anionic sulfate groups in its structure. Absorbance changes with the developed sensor were evaluated using both a sensitive spectrophotometer and a smartphone. The chosen biopolymeric film as a support material not only results in 95% cost savings compared to the Nafion film but is also biodegradable and environmentally friendly, contributing to green analytical chemistry.

In the study conducted by Hidayet et al., paper microzone plates (PμZP) based on the CUPRAC method were designed for evaluating the total antioxidant capacities of plant extracts [52]. PμZPs were prepared by immobilizing Cu(II)-Nc onto paper with a 70-well pattern. Adhering to the mechanism of the traditional CUPRAC method, a color change from light blue to yellow (due to Cu(I)-Nc) occurred in a concentration-dependent manner in the presence of antioxidants. Using the developed sensor, the color intensity linearly increased 8 min after the addition of rutin, an antioxidant flavonoid, in the concentration range of 1−10 mM. The total antioxidant capacity values for the analyzed plant extracts exhibited a strong correlation (r = 0.9887) with those obtained using the traditional CUPRAC method. The short analysis duration of 8 min, the capability to analyze 70 samples simultaneously, and a significant reduction in the consumption of reagent and sample solutions make a valuable contribution to green analytical chemistry.

Puangbanlang et al. have developed a paper-based sensor for the simultaneous determination of antioxidant activity and total phenolic content in food samples [53]. This sensor enables the concurrent application of 2,2′-azino-bis(3-ethylbenzthiazoline-6-sulfonic acid) radical cation (ABTS), CUPRAC, and Folin-Ciocalteau reagent (FC) methods. The device is designed in an X shape, with three regions allocated for the respective methods and one for a blank sample. The sample is placed in the central sample zone of the device and progresses to the test zones. Different colors obtained for the respective methods are analyzed using ImageJ software to determine antioxidant concentrations. The simultaneous measurement with the three tests using this system is completed in 5 min. Results obtained for various food samples were compared with those of classical methods, showing good agreement. The paper-based sensor reduces sample and reagent consumption, analysis time, and analysis cost. With all these advantages, this sensor serves in the realm of green analytical chemistry. In the sensor developed by Szerlauth et al., the Cu(II)-Nc complex was immobilized onto a layer of alginate-functionalized layered double hydroxide (dLDH) nanosheets through electrostatic interactions [54]. The resulting composite mixture was constructed on cellulose paper. The paper-based sensor provided significant advantages over sensors prepared without dLDH support, attributed to the low LOD values and a widened linear working range. The sensor's notable superiority is attributed to the presence of solid dLDH particles, providing adsorption regions for dissolved antioxidant molecules and consequently reducing diffusion limitations.

In the study conducted by Akar and Burnaz, the CUPRAC method was transformed into a colorimetric sensor using thin layer chromatography (TLC) plates [55]. In the respective study, the CUPRAC reagent was dropped onto circular zones on TLC plates, dried, treated with the sample, and the resulting color changes were evaluated using the ImageJ program to determine antioxidant capacity. The developed TLC plate-based detection system determined the antioxidant capacities of standard antioxidants and plant extracts. The obtained results were found to be consistent with those obtained using the traditional CUPRAC method. The proposed detection system enables rapid, cost-effective, simple, and on-site analysis without the need for sophisticated equipment, similar to other sensor systems.

4.1.2 CUPRAC based optical nanosensors

Nanoscience is a multidisciplinary field that investigates the dimensions, properties, and interactions of matter at the nanometer scale, encompassing the fundamental principles and applications of nanotechnology [56]. In recent years, nanomaterial-based methods utilizing optical or electrochemical signal transducer platforms (nanosensors and nanoprobes) have begun to replace traditional methods for various analytes. Their minuscule size and increased surface area enable them to interact with analytes at the molecular and even atomic levels, providing a unique sensitivity in detection [57]. The integration of nanosensors has significantly enhanced the speed and accuracy of analyses, making them invaluable tools in various fields, from environmental monitoring to biomedical research. Thanks to the unique optical and electronic properties of nanomaterials, the determination of analytes at the nano level can occur in both solution and on solid surfaces (such as colorimetric kits, paper-based devices, or chemosensors). Furthermore, their real-time and on-site capabilities facilitate dynamic monitoring, offering insights into complex chemical processes. Serving as agents of innovation in analytical chemistry, nanosensors continue to open new frontiers, promising a future of more efficient, rapid, and environmentally sustainable analytical methodologies.

Nanosensors have emerged as pivotal tools within the realm of green chemistry, fostering sustainable and environmentally conscious analytical practices. The integration of nanotechnology into sensor design not only enhances sensitivity and selectivity but also contributes to minimizing the environmental footprint of analytical processes [58]. Nanosensors, due to their reduced size and increased surface area, often require smaller quantities of reagents and samples, leading to a reduction in waste generation. Additionally, the use of nanomaterials in sensor construction can facilitate rapid and efficient analyses, further promoting the principles of green chemistry by reducing energy consumption. The implementation of nanosensors in various analytical applications, from environmental monitoring to pharmaceutical analysis, underscores their versatility and relevance in advancing eco-friendly approaches within the field of chemistry.

The CUPRAC colorimetric sensor, based on heparin-stabilized gold nanoparticles (AuNPs), was developed by Bener et al. for the determination of total antioxidant capacity in food. In this method, negatively charged AuNPs stabilized with heparin are added to the Cu(I)-Nc solution obtained from the reaction of the CUPRAC reagent with antioxidants [59]. The absorbance of Cu(I)-Nc-AuNPs formed by the electrostatic adsorption of the cationic chelate Cu(I)-Nc onto gold nanoparticles is measured at 455 nm for the determination of total antioxidant capacity. This colorimetric sensor has exhibited both selectivity for antioxidant detection and an unusual inert electrolyte tolerance against aggregation. This method has increased the oxidation rate of compounds like thiourea due to the affinity of thiol antioxidants to the gold surface.

Additionally, this colorimetric sensor serves GAC by both significantly reducing reagent consumption in the determination of TAC and using natural anticoagulant heparin as a reducing/stabilizing agent instead of toxic chemicals for the synthesis of AuNPs (Fig. 10.4) [59].

Akyüz et al. utilized chicken egg white protein-protected gold nanoclusters (CEW-AuNCs) to measure the Cu(II)-induced prooxidant activity of certain antioxidant compounds [60]. These antioxidants reduced Cu(II) to Cu(I), with Cu(I) primarily binding to thiol groups in the structure of CEW-AuNCs. The ability of antioxidants to reduce Cu(II) is considered an indirect measure of their prooxidant powers since Cu(I) bound to the protein can potentially serve as a catalytic center for the generation of reactive oxygen species. The absorbance of the Cu(I)-Nc chelate, formed with neocuproine for the measurement of bound Cu(I), was recorded at 450 nm. The developed method allows for prooxidant ability determination without the need for a separate synthesis and pre-separation step. The developed biosensor is robust, reliable, easy to apply, cost-effective, and possesses a wide linear range, enabling the determination of prooxidant activities of natural antioxidant samples with high repeatability. The characteristics such as the green synthesis of gold nanoclusters with CEW and reduced reactive waste contribute to green analytical chemistry.

4.2 CUPRAC microplate- and flow injection-based methods

Microplate-based methods are one of the most used tools to perform laboratory analyses for clinical and biological practices [61]. Flow injection (FI) is a widely employed sample introduction technique that can significantly reduce sample consumption with easy automation [62]. The use of microplate-based methods and flow injection analysis (FIA) makes existing methods significantly green (Green analytical chemistry has brought new goals in the development and application of analytical methods and these methods meet some of these goals). Microplate-based methods can process multiple samples (usually 96 individual wells, but some can go up to 3456 wells), have low reagents/solvents consumption, minimize operation time and support methods based on different measurement techniques (i.e., colorimetry, fluorescence, and luminescence) [61]. Flow injection methods can achieve minimization of chemical waste generation, reduction of the use of harmful reagents and solvents, and increase the security of analysts [63]. Flow injection analysis and microplate-based methods fall within the scope of green chemistry in terms of automation and miniaturization of procedures, which are among established GAC strategies. Thanks to all these features, these methods are comparatively fast, cost-effective, reduce labor, provide high throughputs, and allow for a simpler and greener approach than the conventional method on which they are based.

Ribeiro et al. modified the CUPRAC method as a flow injection analysis and microplate-based method, and used it to determine the antioxidant capacities of biological samples [44]. Thus, not only a high throughput method

The CUPRAC method **Chapter | 10** 369

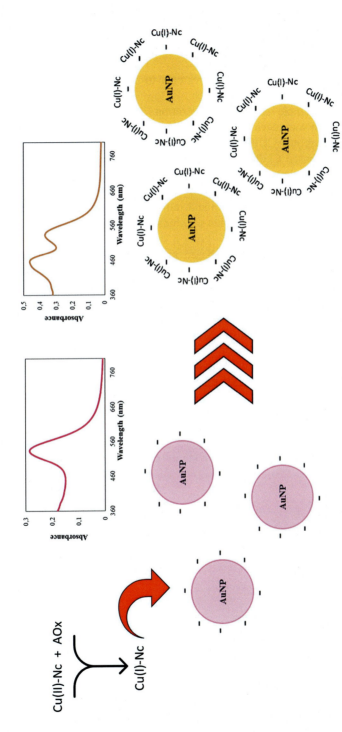

FIGURE 10.4 Schematic illustration of heparin-stabilized AuNPs based CUPRAC sensor.

was achieved, but also the reaction time was reduced by 7.5 times (30 min for the original method, compared to 4 min for microplate-base method) under the microplate format (Fig. 10.5) [44].

The CUPRAC-FIA method enables easy-to-implement and cost-effective analysis, owing to its features such as low reagent consumption, labor reduction and sample amount minimization. Total antioxidant capacity in urine and serum was successfully determined by both methods, in accordance with the conventional CUPRAC method. Similarly, Campos et al. developed the microplate-based CUPRAC−BCS assay to determine the antioxidant capacities of human plasma and urine [64]. Although this developed microplate-based method has the advantages mentioned above, it is different from CUPRAC in that BCS (bathocuproine disulfonate) is used instead of Nc as the chelating agent, the cupric complex of which may have a weaker response to lipophilic antioxidants.

The superiority of microplate-based methods and their service to green chemistry cause them to be frequently preferred in literature. The microplate-CUPRAC method has been used to determine the antioxidant capacity of many extracts as well as biological samples. Some of these can be listed as *Artemisia annua* L., *Asparaus africanus* Lam., *Ferula elaeochytris* and *Sideritis stricta*, *Celosia cristata* L. seeds, native edible berries of South Patagonia and red wines, dry sea cucumber. In addition to TAC determination, there are also microplate-CUPRAC methods developed for the determination of reducing sugars in food samples and GSH-Px enzyme activity determination.

4.3 Online HPLC-CUPRAC methods

It is possible to make analytical processes more sustainable and aligned with green chemistry practices by incorporating CUPRAC assay into an on-line

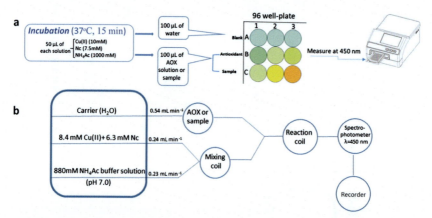

FIGURE 10.5 Schematic representation of (A) microplate CUPRAC and (B) CUPRAC-FIA methods. *AOX*, antioxidant; *Nc*, neocuproine; *NH₄Ac*, ammonium acetate.

chromatographic system. Conducting online HPLC coupled with a CUPRAC assay allows rapid, versatile and sensitive determination of individual antioxidant constituents in complex matrices. After quantification by UV detection, antioxidant capacity of a single constituent can be individually measured, and its contribution to the total antioxidant capacity can be monitored simultaneously (Fig. 10.6) [23].

The on-line HPLC-CUPRAC chromatogram is much more informative than the corresponding UV chromatogram, because it can be directly observed which chromatographic bands arise from antioxidant constituents providing the advantage of added specificity to antioxidant detection. As the non-antioxidant constituents are not detected, clear and simple chromatograms of complex samples can be obtained. In this assay, HPLC-separated compounds react with the CUPRAC reagent within 60 s time in a post-column reaction module equipped reaction coil containing 1 mL volume [23]. This approach provides detailed information about the antioxidant profile by identifying individual compounds and their concentrations. As the cupric-neocuproine complex is a powerful colorimetric oxidant, the CUPRAC assay is relatively fast, with all of the antioxidant compounds reaching almost 80% of their maximal molar absorptivities within ≤ 1 min (spent in the reaction coil). So, the reaction period for a precise calculation of antioxidant amounts is sufficient after contact with the CUPRAC reagent [33]. By our research group, phenolics in green and black tea (*Camellia sinensis*), mint (*Mentha*), sweet marjoram (*Origanum marjorana*) [33], flavonoid glucosides in elderflower (*Sambucus nigra*) were analyzed by on-line HPLC-CUPRAC method. This method was optimized by Rimkiene et al. in terms of flow rate of the reagent and reaction temperature for the assessment of potential

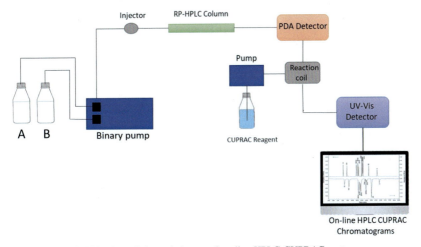

FIGURE 10.6 Schematic image of on-line HPLC CUPRAC system.

reducing compounds existing in the extracts of *Ginkgo biloba* leaves [65]. Otherwise, usage of concentrated and large volumes of derivatizing reagent in the reaction coil may lead to a deterioration in the observed separation efficiency and give rise to clogging of the reaction coil. The authors stated that CUPRAC method is more robust, has higher reagent selectivity, shows a wide concentration range, and post-column chemical reaction is generally faster than that of the ABTS method [66]. Suktham et al. also utilized reaction flow HPLC columns for post-column derivatization with CUPRAC reagent [67]. In reaction flow (RF) HPLC columns, reaction loop was not required, thus preserving the efficiency of separation. The authors compared the performance of the CUPRAC assay to that of ferric reducing antioxidant power (FRAP) assay, both in RF mode. CUPRAC assay with RF mode demonstrated less baseline noise interference, greater sensitivity, wider linear dynamic range and better assay precision than the FRAP assay with RF mode [67]. In another study, online HPLC-CUPRAC assay combined with chemometric analysis had distinct advantages including simultaneous analysis such as constituent (marker) analysis, quantification of individual tocopherol isomers and antioxidant capacity, and authentication of argan oil [68]. Consequently, this versatile approach not only benefits green chemistry but can also lead to authentication and quality detection of food samples, cost savings, low solvent consumption, improved efficiency and better precision.

4.4 Electroanalytical CUPRAC methods

In the voltammetric study conducted by Tufan et al. using the cupric-neocuproine complex, the total antioxidant capacities of polyphenolic compounds, real samples and ascorbic acid were measured [69]. First of all, the electrochemical behavior of the Cu(II)-Nc complex was examined by cyclic voltammetry. Then, using differential pulse voltammetry, the reduction current of the remaining Cu(II)-Nc complex after reaction with antioxidants was measured. This developed method saved time and material for the direct determination of turbid solutions that require pretreatment in the traditional spectrophotometric CUPRAC method, and also showed acceptable sensitivity (Fig. 10.7).

A voltammetric and chronoamperometric TAC method was developed by Cárdenas et al. taking the CUPRAC method as a reference [70]. This electrochemical TAC method, applied to commercial tea samples, is not affected by the interference of turbid solutions and does not require any pre-treatment. Additionally, its analytical linear range is quite wide.

In another study conducted by Trofin et al., the total antioxidant capacity of food samples was determined [71]. This study using screen printed microelectrodes is based on differential pulse voltammetry as well as cyclic voltammetry against open circuit potential. The results were compared with those of the traditional CUPRAC method and were found to be highly compatible.

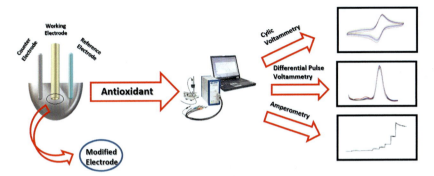

FIGURE 10.7 Schematic illustration of CUPRAC-based electroanalytical sensors.

Since screen-printed microelectrodes are resistant to organic substances, they are suitable for many solvent uses. During the analysis, very low volumes in microliter scale were used. In addition, the developed method had many advantages that serve green chemistry, such as short analysis time, low cost and ease of operation.

The TAC determination made by Cardenas and Frontana is based on the linear correlation of the anodic peak current intensity with antioxidant concentration in the cyclic voltammogram of the Cu(II)-Nc complex [72]. In the study, a chemically modified electrode with carbon ink was prepared using Cu(II)-Nc. A number of antioxidants were determined using the cyclic voltammetry technique. In addition, the TAC of tea samples was determined and found to be compatible with traditional CUPRAC results. This developed electrode provides significant savings in terms of both cost and time as it allows multiple analyses with the same electrode without the need for precursor solutions.

In the method developed by Ayaz et al., carbon nanotubes were functionalized with Nafion and dropped onto the glassy carbon surface [73]. Then, the sensor electrode was formed by dropping CUPRAC reagent onto the surface. A flow injection amperometric TAC determination method was developed using this sensor electrode. The essence of the method is based on the amperometric measurement of the anodic peak current of the [Cu(II)-Nc/Cu(I)-Nc redox couple in the presence of antioxidants. As the oxidation current of Cu(I)-Nc complex formed after reaction with antioxidants was measured, this approach provided higher sensitivity and selectivity than analogic electrochemical methods measuring reduction current of the remaining Cu(II)-Nc complex. The results of the method that lowered the detection limit were compatible with those of the conventional CUPRAC method. This developed sensor has many features that serve green chemistry, such as being fast, simple, selective and portable for antioxidant capacity.

4.5 Green solvents used in the CUPRAC method

To qualify as an environmentally harmless medium, a solvent must meet certain criteria such as availability, low toxicity, biodegradability and low cost [74]. Deep eutectic solvents (DES) are known as a new group of solvents obtained by mixing a hydrogen bond donor with a hydrogen bond acceptor [75]. Interest in DES is increasing nowadays, especially due to its environmental and economic features. In fact, DES is classified as "green" solvents because they consist of molecules that are considered environmentally friendly, although their ecotoxicological profiles are still not fully known [76].

DESs are used as green alternatives to traditional solvents due to their application-specific tunability, non-flammability, variable viscosity (according to the hydrogen bond donor/hydrogen bond acceptor structures or the molar ratio of these constituents), low vapor pressure, and chemical and thermal stability. In addition, compared to ionic liquids, natural DESs stand out in the field of green chemistry with their important features such as biodegradability, low toxicity, low production cost, simple preparation and easy gain from nature [77–79].

Besides their advantageous environmental and economic properties, DESs are more efficient than conventional organic solvents, leading to higher yields of bioactive compounds [80]. For this reason, DESs are also included among the applications of the CUPRAC method. Total antioxidant capacities of various food and plant extracts found using natural DES were successfully determined using the CUPRAC method.

Doldolava et al. examined the TAC and curcumin contents of turmeric extracts prepared in different DESs using the microwave-assisted extraction method [81]. CUPRAC method was used to determine the antioxidant capacities of extracts prepared in DES.

Bener et al. developed an environmentally friendly method using natural DES and microwave-assisted extraction to recover bioactive compounds from hazelnut pomace [36]. The antioxidant capacities of the extracts prepared using natural DES (choline chloride:1,2-propylene glycol with a molar ratio of 1:4), found most suitable for extraction, were determined by the CUPRAC method. Antioxidant capacities of extracts of all hazelnut pomaces were 2–3 times higher than those of ethanolic extracts [36]. Renjith and Lakshminarayanan showed that enhanced solvation of a hydrophobic species actively taking part in electron-transfer in DES solvents (relative to that in water) can shift redox potentials to more positive potentials compared to that in water; a favorable stabilization of redox-active species by the decreased polarity of DES solvent drives the oxidation of phenolic species to the positive potential range, thereby enhancing oxidation of phenolics and generally increasing the TEAC coefficients of antioxidant polyphenols, as found by the CUPRAC method [82]. Kinetic effects of hydrogen bonding on proton-coupled electron donation from phenols were investigated by Sjödin et al. who found that hydrogen-bonded concerted electron-proton transfer (CEP) mechanism may allow for a low energy barrier path that can operate efficiently at low driving forces such that even the weak oxidant $[Ru(bpy)_3]^{3+}$ can oxidize phenols via a CEP reaction in a single step [83]. This can also be valid for the CUPRAC oxidation of phenols that can establish hydrogen bonds with the intensive H-bonding DES solvents. Alasalvar and Yildirim investigated the green extraction of antioxidant phenolic compounds from Lavandula angustifolia using ultrasound-assisted extraction coupled with natural deep eutectic solvents [84]. Similar to the study mentioned above, the natural DES composed of choline chloride and glycerol (molar ratio of 1:2) showed higher extraction performance than other DESs and benchmark solvents (70% ethanol and 80% methanol) [84].

Thanks to the monovalent charge of the bis(neocuproin)-copper(I) chelate and the hydrophobicity of the chelate rings, the CUPRAC method was successfully used to determine the antioxidant capacities of extracts prepared with aqueous and organic solvents and alcohol-water mixtures [19]. As can be seen from the studies mentioned above, the CUPRAC method has been successfully applied to various NADES extracts in recent years. No interference effects between CUPRAC reagents and DES components based on choline chloride, organic acids, and sugars have been reported. Moreover, there is no difference in the kinetics of the electron transfer-based reaction between antioxidant compounds and CUPRAC reagent in DES medium. The conventional CUPRAC method was applied to determine the antioxidant capacities of extracts prepared with DES/natural DES (Fig. 10.8).

Another study that falls into the green chemistry perspective is the CUPRAC method modifications using cyclodexrins [42,43]. Cyclodextrins are cyclic oligosaccharides with a lipophilic interior cavity and a hydrophilic outer surface [85] that can increase the solubility of various hydrophobic organic molecules by forming inclusion complexes [86].

While the CUPRAC method is applied to hydrophobic antioxidants such as butylated hydroxyanisole, butylated hydroxytoluene, *tert*-butylhydroquinone, propyl gallate, lauryl gallate and vitamin E (α-tocopherol) by dichloromethane

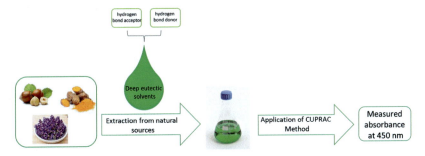

FIGURE 10.8 Usage of deep eutectic solvents in the CUPRAC method.

extraction, Çelik et al. improved the CUPRAC methodology by simultaneously analyzing lipophilic and hydrophilic antioxidants in aqueous media [43]. Creating "host-guest" complexes of these antioxidants with methyl β-cyclodextrin made it possible to analyze lipophilic and hydrophilic antioxidants simultaneously in aqueous media. Similarly, the antioxidant capacities of lipophilic and hydrophilic antioxidant compounds were determined by the CUPRAC method using 2% methyl β-cyclodextrin (w/v) in an acetone–H_2O (9:1, v/v) mixture [42]. The 2% methyl β-cyclodextrin mixture was found to adequately solubilize β-carotene, lycopene, vitamin E, vitamin C, synthetic antioxidants and other phenolic antioxidants. Here, it is worthy to mention that the widely used fluorescence quenching-based antioxidant assay, ORAC, can only solubilize vitamin E compounds and other lipophilic phenolic antioxidants in phosphate buffer by using randomly methylated β-cyclodextrinat a concentration as high as 7% [87]. Thus, the use of organic solutions harmful to both the environment and human health and laborious pre-treatments have been eliminated, and a greener and practical method has been put forward that gives a more realistic total antioxidant capacity value with simultaneous analysis.

4.6 Quencher CUPRAC methods

The QUENCHER approach abbreviated from "QUick, Easy, New, CHEap and Reproducible" involving the solubilization of matrix-bound phenolics incubated with oxidizing TAC reagents, is a time and cost-saving extraction-free procedure measuring in vitro antioxidant capacity. The QUENCHER approach facilitates the solubility of bound phenols through redox reactions that form intertwined equilibria with TAC reagents. Thus, solubilization of bound phenolics having a relatively smaller equilibrium constant is coupled to redox reactions having a larger equilibrium constant, thereby increasing thermodynamic favorability of overall reaction and enhancing dissolution. With the aid of the QUENCHER approach, the antioxidant capacity of compounds can be measured without preliminary extraction steps [88]. The authors showed the possibility that antioxidant functional groups, bound to the insoluble components of different food matrices, may better exert their antioxidant capability with the use of QUENCHER methods. With this approach, higher total antioxidant capacities were obtained compared to those of traditional procedures (such as alkaline hydrolysis of phenolics following multiple extraction steps) to evaluate the antioxidant capacity of bound phenolic compounds in cereals. Moreover, sulfhydryl-groups containing antioxidants may be prone to oxidation during preliminary hydrolysis in alkaline solution.

The CUPRAC assay was successfully adapted to QUENCHER strategy by Tufan et al. to determine the total antioxidant capacities of cereals (i.e., barley, wheat, rye, oat) and optimized in terms of reaction time and solvent composition parameters in which cereal samples were treated with the CUPRAC reagent in 1:1 (v/v) ethanol−water medium for 30 min (Fig. 10.9) [38]. The total antioxidant capacities of soluble and insoluble fractions of selected cereals were simultaneously measured. The order of TAC values of test samples was found as barley > rye > oat > wheat. The TAC values of cereals measured by QUENCHER-CUPRAC were higher than those of original QUENCHER method using free radical reagents such as $ABTS^+$ and DPPH. One of the applications of QUENCHER-CUPRAC assay was the determination of the TAC values of herbal plants, i.e., *Cistus incanus* (cistus), *Verbena officinalis* (vervain), and *Scutellaria baicalensis* (Chinese skullcap), and *Scutellaria lateriflora* (blue skullcap) [89]. Karadirek et al. modified the QUENCHER-CUPRAC method to make it applicable to humic substances (HS) [90]. From agriculture to green chemistry, humics are widely used as soil stabilizing agent and natural fertilizer. Humic substances have antioxidant and free radical scavenging properties. Besides free and soluble phenolics, humic acid-bound phenolics can be simultaneously extracted and oxidized with the CUPRAC reagent in $DMSO:H_2O$ (1:3,v/v) solvent medium. It was emphasized that water enables the dissolution of TAC reagents and facilitates electron-transfer reactions through enhanced phenol ionization while DMSO improves the solubilization of hydrophobic compounds in humic substances. DMSO is one of the solvents having a relatively high dielectric constant which increases ionization of phenolic protons [90]. In an another study, Csicsor and coworkers [91] also determined antioxidant capacities (gallic acid equivalents per gram) of different humic substances fractions obtained from leonardite by the aid of QUENCHER-based CUPRAC method procedure given in the study of Karadirek et al. [90]. The obtained results showed that the antioxidant potential of the HS fractions was affected due to the surface reaction

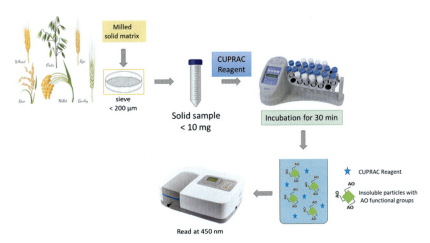

FIGURE 10.9 QUENCHER protocol adapted to the CUPRAC method.

phenomenon occurring between solid humic substances and liquid reagents and therefore the antioxidant capacity values of leonardite samples were higher than those measured by the traditional CUPRAC method.

References

[1] S.I. Kaya, A. Cetinkaya, S.A. Ozkan, Green analytical chemistry approaches on environmental analysis, Trends in Environmental Analytical Chemistry 33 (2022) e00157.
[2] M. Sajid, J. Płotka-Wasylka, Green analytical chemistry metrics: a review, Talanta 238 (2022) 123046.
[3] P.T. Anastas, Green chemistry and the role of analytical methodology development, Critical Reviews in Analytical Chemistry 29 (3) (1999) 167–175.
[4] P. Anastas, N. Eghbali, Green chemistry: principles and practice, Chemical Society Reviews 39 (1) (2010) 301–312.
[5] P.T. Anastas, J.C. Warner, Green Chemistry: Theory and Practice, Oxford University Press, 2000.
[6] W. Abdussalam-Mohammed, A.Q. Ali, A.O. Errayes, Green chemistry: principles, applications, and disadvantages, Chemical Methodologies 4 (4) (2020) 408–423.
[7] M. Koel, Do we need green analytical chemistry? Green Chemistry 18 (4) (2016) 923–931.
[8] A. Gałuszka, Z. Migaszewski, J. Namieśnik, The 12 principles of green analytical chemistry and the SIGNIFICANCE mnemonic of green analytical practices, TrAC Trends in Analytical Chemistry 50 (2013) 78–84.
[9] S. Armenta, S. Garrigues, M. de la Guardia, Green analytical chemistry, TrAC Trends in Analytical Chemistry 27 (6) (2008) 497–511.
[10] M. Shi, X. Zheng, N. Zhang, Y. Guo, M. Liu, L. Yin, Overview of sixteen green analytical chemistry metrics for evaluation of the greenness of analytical methods, TrAC Trends in Analytical Chemistry (2023) 117211.
[11] S. Armenta, S. Garrigues, F.A. Esteve-Turrillas, M. de la Guardia, Green extraction techniques in green analytical chemistry, TrAC Trends in Analytical Chemistry 116 (2019) 248–253.
[12] F. Pena-Pereira, A. Kloskowski, J. Namieśnik, Perspectives on the replacement of harmful organic solvents in analytical methodologies: a framework toward the implementation of a generation of eco-friendly alternatives, Green Chemistry 17 (7) (2015) 3687–3705.
[13] I. Pacheco-Fernández, V. Pino, Green solvents in analytical chemistry, Current Opinion in Green and Sustainable Chemistry 18 (2019) 42–50.
[14] M. del Valle, Sensors as green tools in analytical chemistry, Current Opinion in Green and Sustainable Chemistry 31 (2021) 100501.
[15] N. Wongkaew, M. Simsek, C. Griesche, A.J. Baeumner, Functional nanomaterials and nanostructures enhancing electrochemical biosensors and lab-on-a-chip performances: recent progress, applications, and future perspective, Chemical Reviews 119 (1) (2018) 120–194.
[16] G. Maduraiveeran, M. Sasidharan, V. Ganesan, Electrochemical sensor and biosensor platforms based on advanced nanomaterials for biological and biomedical applications, Biosensors and Bioelectronics 103 (2018) 113–129.
[17] K. Bagchi, S. Puri, Free radicals and antioxidants in health and disease: a review, EMHJ-Eastern Mediterranean Health Journal 4 (2) (1998) 350–360, 1998.
[18] H.E. Seifried, D.E. Anderson, E.I. Fisher, J.A. Milner, A review of the interaction among dietary antioxidants and reactive oxygen species, The Journal of Nutritional Biochemistry 18 (9) (2007) 567–579.

[19] R. Apak, M. Özyurek, K. Guclu, E. Capanoglu, Antioxidant activity/capacity measurement. 1. Classification, physicochemical principles, mechanisms, and electron transfer (ET)-based assays, Journal of Agricultural and Food Chemistry 64 (5) (2016) 997–1027.

[20] J.M. Gutteridge, Free radicals and aging, Reviews in Clinical Gerontology 4 (4) (1994) 279–288.

[21] D.L. Madhavi, S.S. Deshpande, D.K. Salunkhe, Food Antioxidants: Technological: Toxicological and Health Perspectives, CRC Press, 1995.

[22] J. Flieger, W. Flieger, J. Baj, R. Maciejewski, Antioxidants: classification, natural sources, activity/capacity measurements, and usefulness for the synthesis of nanoparticles, Materials 14 (15) (2021) 4135.

[23] S.E. Çelik, M. Özyürek, K. Güçlü, R. Apak, Determination of antioxidants by a novel on-line HPLC-cupric reducing antioxidant capacity (CUPRAC) assay with post-column detection, Analytica Chimica Acta 674 (1) (2010a) 79–88.

[24] R. Apak, Current issues in antioxidant measurement, Journal of Agricultural and Food Chemistry 67 (2019) 9187–9202.

[25] R. Apak, K. Güçlü, M. Özyürek, S.E. Karademir, Novel total antioxidant capacity index for dietary polyphenols and vitamins C and E, using their cupric ion reducing capability in the presence of neocuproine: CUPRAC method, Journal of Agricultural and Food Chemistry 52 (26) (2004) 7970–7981.

[26] R. Apak, K. Güçlü, M. Özyürek, S.E. Çelik, Mechanism of antioxidant capacity assays and the CUPRAC (cupric ion reducing antioxidant capacity) assay, Microchimica Acta 160 (2008) 413–419.

[27] M. Karpudewan, W.M. Roth, Z. Ismail, The effects of "Green Chemistry" on secondary school students' understanding and motivation, The Asia-Pacific Education Researcher 24 (2015) 35–43.

[28] S.E. Çelik, M. Özyürek, K. Güçlü, R. Apak, Solvent effects on the antioxidant capacity of lipophilic and hydrophilic antioxidants measured by CUPRAC, ABTS/persulphate and FRAP methods, Talanta 81 (4–5) (2010b) 1300–1309.

[29] R. Apak, K. Güçlü, M. Özyürek, S.E. Karademir, M. Altun, Total antioxidant capacity assay of human serum using copper (II)-neocuproine as chromogenic oxidant: the CUPRAC method, Free Radical Research 39 (9) (2005) 949–961.

[30] R. Apak, K. Güçlü, M. Özyürek, S. Esin Karademir, E. Erçağ, The cupric ion reducing antioxidant capacity and polyphenolic content of some herbal teas, International Journal of Food Sciences & Nutrition 57 (5–6) (2006) 292–304.

[31] L. Yıldız, K.S. Başkan, E. Tütem, R. Apak, Combined HPLC-CUPRAC (cupric ion reducing antioxidant capacity) assay of parsley, celery leaves, and nettle, Talanta 77 (1) (2008) 304–313.

[32] K. Alpinar, M. Özyürek, U. Kolak, K. Güçlü, Ç. Aras, M. Altun, S.E. Çelik, K.I. Berker, B. Bektaşoğlu, R. Apak, Antioxidant capacities of some food plants wildly grown in Ayvalik of Turkey, Food Science and Technology Research 15 (1) (2009) 59–64.

[33] S.E. Çelik, M. Özyürek, K. Güçlü, E. Çapanoğlu, R. Apak, Identification and antioxidant capacity determination of phenolics and their glycosides in elderflower by on-line HPLC–CUPRAC method, Phytochemical Analysis 25 (2) (2014) 147–154.

[34] S.E. Çelik, A.N. Tufan, B. Bekdeşer, M. Özyürek, K. Güçlü, R. Apak, Identification and determination of phenolics in Lamiaceae species by UPLC-DAD-ESI-MS/MS, Journal of Chromatographic Science 55 (3) (2017) 291–300.

[35] S.E. Çelik, M. Özyürek, K. Güçlü, R. Apak, Differences in responsivity of original cupric reducing antioxidant capacity and cupric–bathocuproine sulfonate assays to antioxidant compounds, Analytical Biochemistry 423 (1) (2012) 36–38.
[36] M. Bener, F.B. Şen, A.N. Önem, B. Bekdeşer, S.E. Çelik, M. Lalikoglu, Y.S. Asci, E. Çapanoglu, R. Apak, Microwave-assisted extraction of antioxidant compounds from by-products of Turkish hazelnut (*Corylus avellana* L.) using natural deep eutectic solvents: modeling, optimization and phenolic characterization, Food Chemistry 385 (2022) 132633.
[37] Ş. Karaman, E. Tütem, K.S. Başkan, R. Apak, Comparison of total antioxidant capacity and phenolic composition of some apple juices with combined HPLC–CUPRAC assay, Food Chemistry 120 (4) (2010) 1201–1209.
[38] A.N. Tufan, S.E. Çelik, M. Özyürek, K. Güçlü, R. Apak, Direct measurement of total antioxidant capacity of cereals: QUENCHER-CUPRAC method, Talanta 108 (2013) 136–142.
[39] M. Altun, S.E. Çelik, K. Güçlü, M. Özyürek, E. Erçağ, R. Apak, Total antioxidant capacity and phenolic contents of Turkish hazelnut (*Corylus avellana* L.) kernels and oils, Journal of Food Biochemistry 37 (1) (2013) 53–61.
[40] B. Bektaşoğlu, M. Özyürek, K. Güçlü, R. Apak, Hydroxyl radical detection with a salicylate probe using modified CUPRAC spectrophotometry and HPLC, Talanta 77 (2008) 90–97.
[41] L.M. Magalhães, L. Barreiros, M.A. Maia, S. Reis, M.A. Segundo, Rapid assessment of endpoint antioxidant capacity of red wines through microchemical methods using a kinetic matching approach, Talanta 97 (2012) 473–483.
[42] M. Özyürek, B. Bektaşoğlu, K. Güçlü, N. Güngör, R. Apak, Simultaneous total antioxidant capacity assay of lipophilic and hydrophilic antioxidants in the same acetone–water solution containing 2% methyl-β-cyclodextrin using the cupric reducing antioxidant capacity (CUPRAC) method, Analytica Chimica Acta 630 (1) (2008) 28–39.
[43] S.E. Çelik, M. Özyürek, K. Güçlü, R. Apak, CUPRAC total antioxidant capacity assay of lipophilic antioxidants in combination with hydrophilic antioxidants using the macrocyclic oligosaccharide methyl β-cyclodextrin as the solubility enhancer, Reactive and Functional Polymers 67 (12) (2007) 1548–1560.
[44] J.P. Ribeiro, L.M. Magalhaes, S. Reis, J.L. Lima, M.A. Segundo, High-throughput total cupric ion reducing antioxidant capacity of biological samples determined using flow injection analysis and microplate-based methods, Analytical Sciences 27 (5) (2011) 483–488.
[45] O.S. Wolfbeis, Fiber-optic chemical sensors and biosensors, Analytical Chemistry 78 (12) (2006) 3859–3874.
[46] C. Dincer, R. Bruch, E. Costa-Rama, M.T. Fernández-Abedul, A. Merkoçi, A. Manz, G.A. Urban, F. Güder, Disposable sensors in diagnostics, food, and environmental monitoring, Advanced Materials 31 (30) (2019) 1806739.
[47] D.T. Newcombe, T.J. Cardwell, R.W. Cattrall, S.D. Kolev, An optical redox chemical sensor based on ferroin immobilised in a Nafion® membrane, Analytica Chimica Acta 401 (1–2) (1999) 137–144.
[48] A. Kawalec, K. Jasek, M. Pasternak, Measurements results of SAW humidity sensor with nafion layer, European Physical Journal: Special Topics 154 (1) (2008) 123–126.
[49] M. Bener, M. Özyurek, K. Guclu, R. Apak, Development of a low-cost optical sensor for cupric reducing antioxidant capacity measurement of food extracts, Analytical Chemistry 82 (10) (2010) 4252–4258.

[50] M. Bener, M. Özyurek, K. Guclu, R. Apak, Novel optical fiber reflectometric CUPRAC sensor for total antioxidant capacity measurement of food extracts and biological samples, Journal of Agricultural and Food Chemistry 61 (35) (2013) 8381−8388.

[51] M. Bener, F.B. Şen, A. Kaşgöz, R. Apak, Carrageenan-based colorimetric sensor for total antioxidant capacity measurement, Sensors and Actuators B: Chemical 273 (2018) 439−447.

[52] M.A. Hidayat, R.I. Chassana, I.Y. Ningsih, M. Yuwono, B. Kuswandi, The CUPRAC-paper microzone plates as a simple and rapid method for total antioxidant capacity determination of plant extract, European Food Research and Technology 245 (9) (2019) 2063−2070.

[53] C. Puangbanlang, K. Sirivibulkovit, D. Nacapricha, Y. Sameenoi, A paper-based device for simultaneous determination of antioxidant activity and total phenolic content in food samples, Talanta 198 (2019) 542−549.

[54] A. Szerlauth, L. Szalma, S. Muráth, S. Sáringer, G. Varga, L. Li, I. Szilágyi, Nanoclay-based sensor composites for the facile detection of molecular antioxidants, Analyst 147 (7) (2022) 1367−1374.

[55] Z. Akar, N.A. Burnaz, A new colorimetric method for CUPRAC assay with using of TLC plate, LWT 112 (2019) 108212.

[56] F.C. Adams, C. Barbante, Nanoscience, nanotechnology and spectrometry, Spectrochimica Acta Part B: Atomic Spectroscopy 86 (2013) 3−13.

[57] S.P. Usha, H. Manoharan, R. Deshmukh, R. Álvarez-Diduk, E. Calucho, V.V.R. Sai, A. Merkoçi, Attomolar analyte sensing techniques (AttoSens): a review on a decade of progress on chemical and biosensing nanoplatforms, Chemical Society Reviews 50 (23) (2021) 13012−13089.

[58] R. Apak, S. Demirci Çekiç, A. Üzer, S.E. Çelik, M. Bener, B. Bekdeşer, Z. Can, Ş. Sağlam, A.N. Önem, E. Erçağ, Novel spectroscopic and electrochemical sensors and nanoprobes for the characterization of food and biological antioxidants, Sensors 18 (1) (2018) 186.

[59] M. Bener, F.B. Şen, R. Apak, Heparin-stabilized gold nanoparticles-based CUPRAC colorimetric sensor for antioxidant capacity measurement, Talanta 187 (2018) 148−155.

[60] E. Akyuz, F.B. Şen, M. Bener, K.S. Baskan, E. Tutem, R. Apak, Protein-protected gold nanocluster-based biosensor for determining the prooxidant activity of natural antioxidant compounds, ACS Omega 4 (1) (2019) 2455−2462.

[61] J.F. Bergua, R. Alvarez-Diduk, A. Idili, C. Parolo, C. Maymo, L. Hu, A. Merkoçi, Low-cost, user-friendly, all-integrated smartphone-based microplate reader for optical-based biological and chemical analyses, Analytical Chemistry 94 (2) (2022) 1271−1285.

[62] Y. He, L. Tang, X. Wu, X. Hou, Y.I. Lee, Spectroscopy: the best way toward green analytical chemistry? Applied Spectroscopy Reviews 42 (2) (2007) 119−138.

[63] A.M. Bevanda, S. Talić, A. Ivanković, Flow injection analysis toward green analytical chemistry, Green Analytical Chemistry: Past, Present and Perspectives (2019) 299−323.

[64] C. Campos, R. Guzmán, E. López-Fernández, Á. Casado, Evaluation of the copper (II) reduction assay using bathocuproinedisulfonic acid disodium salt for the total antioxidant capacity assessment: the CUPRAC−BCS assay, Analytical Biochemistry 392 (1) (2009) 37−44.

[65] L. Rimkiene, L. Ivanauskas, A. Kubiliene, K. Vitkevicius, G. Kiliuviene, V. Jakstas, Optimization of a CUPRAC-based HPLC postcolumn assay and its applications for Ginkgo biloba L. extracts, Journal of Analytical Methods in Chemistry (2015) 280167.

[66] R. Raudonis, L. Bumblauskiene, V. Jakstas, A. Pukalskas, V. Janulis, Optimization and validation of post-column assay for screening of radical scavengers in herbal raw materials and herbal preparations, Journal of Chromatography A 1217 (49) (2010) 7690−7698.

[67] T. Suktham, A. Jones, A. Soliven, G.R. Dennis, R.A. Shalliker, A comparison of the performance of the cupric reducing antioxidant potential assay and the ferric reducing antioxidant power assay for the analysis of antioxidants using reaction flow chromatography, Microchemical Journal 149 (2019) 104046.

[68] S.E. Celik, A. Asfoor, O. Senol, R. Apak, Screening method for argan oil adulteration with vegetable oils: an online HPLC assay with postcolumn detection utilizing chemometric multidata analysis, Journal of Agricultural and Food Chemistry 67 (29) (2019) 8279−8289.

[69] A.N. Tufan, S. Baki, K. Guclu, M. Ozyurek, R. Apak, A novel differential pulse voltammetric (DPV) method for measuring the antioxidant capacity of polyphenols-reducing cupric neocuproine complex, Journal of Agricultural and Food Chemistry 62 (29) (2014) 7111−7117.

[70] A. Cárdenas, M. Gómez, C. Frontana, Development of an electrochemical cupric reducing antioxidant capacity method (CUPRAC) for antioxidant analysis, Electrochimica Acta 128 (2014) 113−118.

[71] A.E. Trofin, L.C. Trincă, E. Ungureanu, A.M. Ariton, CUPRAC voltammetric determination of antioxidant capacity in tea samples by using screen-printed microelectrodes, Journal of Analytical Methods in Chemistry (2019) 8012758.

[72] A. Cárdenas, C. Frontana, Evaluation of a carbon ink chemically modified electrode incorporating a copper-neocuproine complex for the quantification of antioxidants, Sensors and Actuators B: Chemical 313 (2020) 128070.

[73] S. Ayaz, A. Üzer, Y. Dilgin, R. Apak, A novel flow injection amperometric method for sensitive determination of total antioxidant capacity at cupric-neocuproine complex modified MWCNT glassy carbon electrode, Microchimica Acta 189 (4) (2022) 167.

[74] L.B. Santos, R.S. Assis, J.A. Barreto, M.A. Bezerra, C.G. Novaes, V.A. Lemos, Deep eutectic solvents in liquid-phase microextraction: contribution to green chemistry, TrAC Trends in Analytical Chemistry 146 (2022) 116478.

[75] A.P. Abbott, G. Capper, D.L. Davies, H.L. Munro, R.K. Rasheed, V. Tambyrajah, Preparation of novel, moisture-stable, Lewis-acidic ionic liquids containing quaternary ammonium salts with functional side chains, Chemical Communications (19) (2001) 2010−2011.

[76] E. Durand, J. Lecomte, P. Villeneuve, From green chemistry to nature: the versatile role of low transition temperature mixtures, Biochimie 120 (2016) 119−123.

[77] X.X. Chang, N.M. Mubarak, S.A. Mazari, A.S. Jatoi, A. Ahmad, M. Khalid, R. Walvekar, E.C. Abdullah, R.R. Karri, M.T.H. Siddiqu, S. Nizamuddin, A review on the properties and applications of chitosan, cellulose and deep eutectic solvent in green chemistry, Journal of Industrial and Engineering Chemistry 104 (2021) 362−380.

[78] M. Zdanowicz, K. Wilpiszewska, T. Spychaj, Deep eutectic solvents for polysaccharides processing. A review, Carbohydrate Polymers 200 (2018) 361−380.

[79] Y. Dai, G.J. Witkamp, R. Verpoorte, Y.H. Choi, Natural deep eutectic solvents as a new extraction media for phenolic metabolites in Carthamus tinctorius L, Analytical Chemistry 85 (13) (2013) 6272−6278.

[80] D.C. Murador, L.M. de Souza Mesquita, N. Vannuchi, A.R.C. Braga, V.V. de Rosso, Bioavailability and biological effects of bioactive compounds extracted with natural deep eutectic solvents and ionic liquids: advantages over conventional organic solvents, Current Opinion in Food Science 26 (2019) 25−34.

[81] K. Doldolova, M. Bener, M. Lalikoğlu, Y.S. Asci, R. Arat, R. Apak, Optimization and modeling of microwave-assisted extraction of curcumin and antioxidant compounds from turmeric by using natural deep eutectic solvents, Food Chemistry 353 (2021) 129337.

[82] A. Renjith, V. Lakshminarayanan, Electron-transfer studies of model redox-active species (cationic, anionic, and neutral) in deep eutectic solvents, Journal of Physical Chemistry C 122 (2018) 25411−25421.

[83] M. Sjödin, T. Irebo, J.E. Utas, J. Lind, G. Merenyi, B. Akermark, L. Hammarström, Kinetic effects of hydrogen bonds on proton-coupled electron transfer from phenols, Journal of the American Chemical Society 128 (2006) 13076−13083.

[84] H. Alasalvar, Z. Yildirim, Ultrasound-assisted extraction of antioxidant phenolic compounds from Lavandula angustifolia flowers using natural deep eutectic solvents: an experimental design approach, Sustainable Chemistry and Pharmacy 22 (2021) 100492.

[85] T. Loftsson, M.E. Brewster, Pharmaceutical applications of cyclodextrins. 1. Drug solubilization and stabilization, Journal of Pharmaceutical Sciences 85 (10) (1996) 1017−1025.

[86] A. Braibanti, E. Fisicaro, A. Ghiozzi, C. Compari, G. Bovis, Host-guest interactions between β-cyclodextrin and piroxicam, Reactive and Functional Polymers 36 (3) (1998) 251−255.

[87] D. Huang, B. Ou, M. Hampsch-Woodill, J.A. Flanagan, E.K. Deemer, Development and validation of oxygen radical absorbance capacity assay for lipophilic antioxidants using randomly methylated β-cyclodextrin as the solubility enhancer, Journal of Agricultural and Food Chemistry 50 (7) (2002) 1815−1821.

[88] A. Serpen, E. Capuano, V. Fogliano, V. Gökmen, A new procedure to measure the antioxidant activity of insoluble food components, Journal of Agricultural and Food Chemistry 55 (19) (2007) 7676−7681.

[89] M. Dziurka, P. Kubica, I. Kwiecień, J. Biesaga-Kościelniak, H. Ekiert, S.A. Abdelmohsen, F.F. Al-Harbi, D.O. El-Ansary, H.O. Elansary, A. Szopa, In vitro cultures of some medicinal plant species (Cistus× incanus, Verbena officinalis, Scutellaria lateriflora, and Scutellaria baicalensis) as a rich potential source of antioxidants—evaluation by CUPRAC and QUENCHER-CUPRAC assays, Plants 10 (3) (2021) 454.

[90] Ş. Karadirek, N. Kanmaz, Z. Balta, P. Demirçivi, A. Üzer, J. Hızal, R. Apak, Determination of total antioxidant capacity of humic acids using CUPRAC, Folin−Ciocalteu, noble metal nanoparticle-and solid−liquid extraction-based methods, Talanta 153 (2016) 120−129.

[91] A. Csicsor, E. Tombácz, P. Kulcsár, Antioxidant potential of humic substances measured by Folin-Ciocalteu, CUPRAC, QUENCHER-CUPRAC and ESR methods, Journal of Molecular Liquids 391 (2023) 123294.

Chapter 11

Conclusion and future perspectives

Miryam Perrucci[1,2], Vincenzo De Laurenzi[2,3], Marcello Locatelli[4], Halil I. Ulusoy[5], Abuzar Kabir[6], Imran Ali[7], Fotouh R. Mansour[8] and Savas Kaya[9]

[1]*University of Teramo, Department of Biosciences and Agro-Food and Environmental Technologies, Teramo, Italy;* [2]*Department of Innovative Technologies in Medicine & Dentistry, University "Gabriele d'Annunzio" of Chieti-Pescara, Chieti, Italy;* [3]*Center for Advanced Studies and Technology (CAST), University "Gabriele d'Annunzio" of Chieti—Pescara, Chieti, Italy;* [4]*University "G. D'Annunzio" of Chieti-Pescara, Department of Pharmacy Chieti, Italy;* [5]*Department of Analytical Chemistry, Faculty of Pharmacy, Cumhuriyet University, Sivas, Turkey;* [6]*International Forensic Research Institute, Department of Chemistry and Biochemistry, Florida International University, Miami, FL, United States;* [7]*Department of Chemistry, Jamia Millia Islamia (Central University), Jamia Nagar, New Delhi, India;* [8]*Department of Pharmaceutical Analytical Chemistry, Faculty of Pharmacy, Tanta University, Tanta, Egypt;* [9]*Department of Chemistry, Faculty of Science, Sivas Cumhuriyet University, Sivas, Turkey*

The idea of this book stems from the fact that a complete book on sampling and sample preparation techniques in the field of green analytical chemistry (GAC) is not yet available.

Nowadays we are witnessing more and more the development/application of new techniques/procedures/devices that, following the concepts of the GAC, allow reducing the impact of anthropic activities related to the field of chemical analysis on the environment and at the same time increase analytical performances of the methods developed. This book wants to cover this gap, as well as wanting to be a reference for all colleagues/students/researchers who approach these issues.

In this book relating to "Green Analytical Chemistry" and from the contributions collected here, it is increasingly evident that Analytical Chemistry must increasingly evolve toward procedures, methods and mind-sets such as to minimize the anthropic impact of the research activity carried out in laboratory in order to preserve (and if possible improve) the environment and health for operators [1].

Over the years, especially from the birth of Green Chemistry (1990s) to Green Analytical Chemistry and Green Sample Preparation (the latter just a couple of years ago) [2], there has been an ever-increasing interest in the development of methods that responded to the principles of GAC and/or GSP. In particular, developing new devices [3—5], new methods [6—8], focusing attention on fast analyses [9] that allow us to increase productivity by reducing solvent consumption and waste. Another very important element is represented by the development of new materials [10,11], which allow not only sample treatment with the least impact possible, but also the possibility of developing instruments for in situ analysis [12,13].

In parallel, all these advances in the field of Analytical Chemistry and Sample Preparation must be "governed" by the main objective of increasing the sensitivity and selectivity [14] of the procedures but always responding to the motto "reduce the environmental impact, safeguard the health, increasing analytical performance."

It might seem like a very difficult task, but thanks to the (simultaneous) development of tools that allow the green profile to be assessed and the potential impact of anthropic activity to be assessed [2,15,16] we have at our disposal all the elements that should be present in the chemist's "toolbox." In this scenario, also the development of a specific tool that allow the evaluation of the applicability of the method [17] is a great achievement.

The chapters collected here are not intended to be exhaustive of a panorama that is constantly evolving, but rather want to "whet" interest and curiosity toward a field (the GAC and the GSP, chapters 1, 2, and 9). These chapters offers considerable challenges and great satisfaction and which, in recent years, has paved the way for unexpected developments in the Green Chemistry sector. An example of what has just been stated can in fact be found in the topics covered here. In this book are collected topics from the theoretical approach (chapter 3), to the development of new procedures and materials (chapters 4, 5, and 8), to the discovery of new solvents (chapters 6 and 7), up to the modification of less "green" methods so that they respond as much as possible to the principles of GAC and GSP (chapter 10).

In light of what has been reported, we can envisage ever greater attention in the development of procedures that have less and less impact, just as the need for the Analytical Chemist to interface, in an increasingly multidisciplinary way, with other sectors in order to be able to provide a complete "answer" to the principles of GAC and GSP. From this perspective, a possible development will certainly be the study and implementation not only of the procedures, but also of many technical aspects relating to the new instruments, the new physical structures in which to install the laboratories. The main target and the common objective is the achieving ideally a "zero impact" laboratory, i.e., a structure that is autonomous from an energy point of view and that does not produce waste resulting from the analysis activity (or that respects the concept of reducing the carbon footprint and the circular economy through reuse of as much material as possible).

Declaration of interests

The authors declare that they have no competing financial interests or personal relationships that could have appeared to influence the work reported in this paper.

Author contributions

All Authors contributed equally to Conceptualization; Investigation; Project administration; Resources; Supervision; Roles/Writing—original draft; and Writing—review and editing.

Funding

This research did not receive any specific grant from funding agencies in the public, commercial, or not-for-profit sectors.

Acknowledgments

This article is based upon the work from the Sample Preparation Study Group and Network, supported by the Division of Analytical Chemistry of the European Chemical Society.

References

[1] A. Gałuszka, Z. Migaszewski, J. Namieśnik, The 12 principles of green analytical chemistry and the SIGNIFICANCE mnemonic of green analytical practices, TrAC, Trends in Analytical Chemistry 50 (2013) 78–84, https://doi.org/10.1016/j.trac.2013.04.010.

[2] M. Locatelli, A. Kabir, M. Perrucci, S. Ulusoy, H.I. Ulusoy, I. Ali, Green profile tools: current status and future perspectives, Advances in Sample Preparation 6 (2023) 1–15, https://doi.org/10.1016/j.sampre.2023.100068.

[3] V. D'Angelo, F. Tessari, G. Bellagamba, E. De Luca, R. Cifelli, C. Celia, R. Primavera, M. Di Francesco, D. Paolino, L. Di Marzio, M. Locatelli, MicroExtraction by Packed Sorbent and HPLC-PDA quantification of multiple anti-inflammatory drugs and fluoroquinolones in human plasma and urine, Journal of Enzyme Inhibition and Medicinal Chemistry 31 (S3) (2016) 110–116, https://doi.org/10.1080/14756366.2016.1209496.

[4] M. Locatelli, A. Tartaglia, F. D'Ambrosio, P. Ramundo, H.I. Ulusoy, K.G. Furton, A. Kabir, Biofluid sampler: a new gateway for mail-in-analysis of whole blood samples, Journal of Chromatography B 1143 (2020), https://doi.org/10.1016/j.jchromb.2020.122055 article 122055.

[5] M. Locatelli, A. Tartaglia, H.I. Ulusoy, S. Ulusoy, F. Savini, S. Rossi, F. Santavenere, G.M. Merone, E. Bassotti, C. D'Ovidio, E. Rosato, K. Furton, A. Kabir, Fabric phase sorptive membrane array as non-invasive in vivo sampling device for human exposure to different compounds, Analytical Chemistry 93 (4) (2021) 1957–1961, https://doi.org/10.1021/acs.analchem.0c04663.

[6] A. Tartaglia, A. Kabir, F. D'Ambrosio, P. Ramundo, S. Ulusoy, H.I. Ulusoy, G.M. Merone, F. Savini, C. D'Ovidio, U. De Grazia, K.G. Furton, M. Locatelli, Fast off-line FPSE-HPLC-PDA determination of six NSAIDs in saliva samples, Journal of Chromatography B 1144 (2020), https://doi.org/10.1016/j.jchromb.2020.122082 article 122082.

[7] M. Locatelli, K.G. Furton, A. Tartaglia, E. Sperandio, H.I. Ulusoy, A. Kabir, An FPSE-HPLC-PDA method for rapid determination of solar UV filters in human whole blood, plasma and urine, Journal of Chromatography B 1118–1119 (2019) 40–50, https://doi.org/10.1016/j.jchromb.2019.04.028.

[8] A. Kabir, K.G. Furton, N. Tinari, L. Grossi, D. Innosa, D. Macerola, A. Tartaglia, V. Di Donato, C. D'Ovidio, M. Locatelli, Fabric phase sorptive extraction-high performance liquid chromatography-photo diode array detection method for simultaneous monitoring of three inflammatory bowel disease treatment drugs in whole blood, plasma and urine, Journal of Chromatography B 1084 (2018) 53–63, https://doi.org/10.1016/j.jchromb.2018.03.028.

[9] G.M. Merone, A. Tartaglia, S. Rossi, F. Santavenere, E. Bassotti, C. D'Ovidio, M. Bonelli, E. Rosato, U. de Grazia, M. Locatelli, F. Savini, Fast quantitative LC-MS/MS determination of illicit substances in solid and liquid unknown seized samples, Analytical Chemistry 93 (49) (2021) 16308–16313, https://doi.org/10.1021/acs.analchem.1c03310.

[10] F.R. Mansour, R.M. Abdelhameed, S.F. Hammad, I.A. Abdallah, A. Bedair, M. Locatelli, A microcrystalline cellulose/metal-organic framework hybrid for enhanced ritonavir dispersive solid phase microextraction from human plasma, Carbohydrate Polymer Technologies and Applications 7 (2024), https://doi.org/10.1016/j.carpta.2024.100453 article 100453.

[11] R. Abdelhameed, S.F. Hammad, I.A. Abdallah, A. Bedair, M. Locatelli, F.R. Mansour, A hybrid microcrystalline cellulose/metal-organic framework for dispersive solid phase microextraction of selected pharmaceuticals: a proof-of-concept, Journal of Pharmaceutical and Biomedical Analysis 235 (2023), https://doi.org/10.1016/j.jpba.2023.115609 article 115609.

[12] S.F. Hammad, I.A. Abdallah, A. Bedair, R.M. Abdelhameed, M. Locatelli, F.R. Mansour, Metal organic framework-derived carbon nanomaterials and MOF hybrids for chemical sensing, TrAC, Trends in Analytical Chemistry 170 (2024), https://doi.org/10.1016/j.trac.2023.117425 article 117425.

[13] F.R. Mansour, S.F. Hammad, I.A. Abdallah, A. Bedair, R.M. Abdelhameed, M. Locatelli, Applications of metal organic frameworks in point of care testing, TrAC, Trends in Analytical Chemistry 172 (2024), https://doi.org/10.1016/j.trac.2024.117596 article 117596.

[14] M. Locatelli, D. Melucci, G. Carlucci, C. Locatelli, Recent HPLC strategies to improve sensitivity and selectivity for the analysis of complex matrices, Instrumentation Science and Technology 40 (2–3) (2012) 112–137, https://doi.org/10.1080/10739149.2011.651668.

[15] F. Pena-Pereira, W. Wojnowski, M. Tobiszewski, AGREE—analytical GREEnness metric approach and software, Analytical Chemistry 92 (2020) 10076–10082, https://doi.org/10.1021/acs.analchem.0c01887.

[16] M. Sajid, J. Płotka-Wasylka, Green analytical chemistry metrics: a review, Talanta 238 (2022) 1–11, https://doi.org/10.1016/j.talanta.2021.123046, 123046.

[17] N. Manousi, W. Wojnowski, J. Płotka-Wasylka, V. Samanidou, Blue applicability grade index (BAGI) and software: a new tool for the evaluation of method practicality, Green Chemistry 25 (2023) 7598–7604, https://doi.org/10.1039/D3GC02347H.

Index

Note: Page numbers followed by *f* indicate figures and *t* indicate tables.

A

Acceptor phase (AP), 138
Acid-base chemistry, 46–47
Adjustable viscosity, 158–159
Advanced robotics automation, 91
AGREEprep principles, 315–321
AGREEprep scale, 314–322
Air-assisted solid phase extraction, 107–109
Air sampling, 73
Alginate-functionalized layered double hydroxide (dLDH), 366
American Chemical Society Green Chemistry Institute Pharmaceutical Roundtable (ACS-GCI-PR), 294, 294f
Amphiphilic solvents, 146–149, 153t–155t
Analyte numbers, 312–313
Analytical chemistry, 6–7, 386
Analytical eco-scale (AES), 285–286
  principles, 283–285
  tool, 282–286
Analytical GREEness (AGREE)
  parameters, 306, 308f, 313t, 315f, 316t–317t, 321t, 327t–347t
  principles, 308–314
  scale, 306–314, 322
Analytical instrument, 320
Analytical Mass Volume Intensity (AMVI), 294–295
Analytical method greenness score (AMGS), 294–296
Analytical process, 32–33
Atomy Economy (AE), 3

B

Bar coating technique, 96
Biocompatible extractions, 253
Biofluid sampler (BFS), 130–132, 133f
Biological samples, 93
Biomedical research, 79
Bisphenol A (BPA), 151–152
Blue principles, 323

Born-Oppenheimer approximation, 43
Bulk membrane extraction, 139

C

Capillary electrochromatography (CEC), 101–102
Capsule phase microextraction (CPME), 127–128, 129f
Carbon nanomaterial-based solid phase extraction, 130
Carboxen, 107
Cetyltrimethyl ammonium bromide (CTAB), 147, 215
Chemical derivatization, 311
Chemical measurement technique, 15–16
Chemical penalty points, 283
Chemical pollutants, 94
Chosen greenness evaluation tools, 276–325
Cold vapor ionic liquid assisted headspace single drop microextraction (CV-ILAHS-SDME), 159–165
Complex Green Analytical Procedure Index (ComplexGAPI), 266–267, 287–293, 291f
  principles, 291–293
  scale, 290t
Conceptual DFT-based applications, 52–55
Concerted electron-proton transfer (CEP), 374–375
Convection-powered agitation, 107
Conventional SPME, 82
Corona discharge (CD), 126
Cumulative energy demand (CED), 294
CUPric Reducing Antioxidant Capacity method (CUPRAC)
  advantages, 362–364
  assay, 360–361
  based optical nanosensors, 367–368
  based optrode sensors, 364–366
  green solvents, 374–376
  method, 361f, 365f, 375–376, 375f
    advantages, 360–364

389

CUPric Reducing Antioxidant Capacity method
    (CUPRAC) (*Continued*)
  serving green chemistry, 364–378
  microplate- and flow injection-based
    methods, 368–370
  optical sensors based, 364–368

D

Deep eutectic solvents (DESs), 165–169,
  172t–174t, 201–203, 204t–205t
Density Functional Theory (DFT), 50
Derivatization, 26–27, 311
Dialysis, 136–137
Didecyldimethylammonium bromide
  (DTAB), 151–152
Dimethyl benzylamine, 206–209
Dimethyl ether (DME), 218–219
Direct immersion (DI) modes, 70
Direct immersion (DI)-solid-phase
  microextraction (SPME), 77–79, 78f
Direct solid-phase microextraction, 77
Dispersive liquid-liquid microextraction
  (DLLME), 190–209
  deep eutectic solvents (DESs), 201–203,
    204t–205t
  extracting solvents, 192–193
  ionic liquids (ILs), 193–194
  reverse micelles based SUPRAS, 198–201
  supramolecular solvents (SUPRASs),
    197–198
  switchable solvents, for micro extraction,
    203–209
Dispersive micro solid phase extraction
  (DµSPE), 212–213
Dispersive solid phase extraction (DSPE),
  99–101, 211–213
Donor phase (DP), 138
Dry ashing, 26–27
Dynamic extraction, 86
Dynamic fabric phase sorptive extraction
  (DFPSE), 124
Dynamic stir bar sorptive extraction
  (DSBSE), 85–86

E

Economic cost assessment, 303–306
Eco-Scale method, 19–20
Eco-scale principles, 278–282
Eco-scale tool, 266, 278–282
Electroanalytical CUPRAC methods,
  372–373

Electrodialysis, 137
Electromembrane extraction (EME),
  138–139, 139f
Electron–electron interaction functional, 44
Electronic energy (E), 44
Electron transfer process, 45
Electrophilicity, 50–52
Electrophilicity index, 45, 50–52
Electrospinning method, 98
Electrothermal atomic absorption
  spectrometry (ETAAS), 159–165
Elution solvent, 68
Energy
  consumption, 10
  penalty points (PPs), 283–285
  sustainability, 248
Energy-intensive operations, 16–17
Enhanced sensitivity, 82
Enhanced solvent extraction (ESE), 209–210
Environmental, health, and safety (EHS), 294
Environmental analysis, 72, 79
Environmental benefits, 254–255
Environmental monitoring, 254
Environmental rehabilitation, 256
Environmental samples, 93–94
Equilibrium extraction techniques, 122
Euler equation, 44
Exogenous substances, 130
Extraction, 124
  kinetics, 71
  techniques, 191f, 207t–208t
  type, 68
Ex vivo SPME extraction, 70

F

Fabric phase sorptive extraction (FPSE),
  89–90, 104–106, 121–127
Fatty acid-based switchable solvents,
  222–223
Fenton reaction, 360
Ferric reducing antioxidant power (FRAP),
  371–372
Fiber, 83
Fiber solid phase extraction (SPME), 80–81
Filtration, 137
Flame atomic absorption spectrometry
  (FAAS), 213
Flavor analysis, 73, 79
Flow injection (FI), 368
Flow injection analysis (FIA), 368
Folin-Ciocalteau reagent (FC) methods, 366
Foods, 94

Index **391**

analysis, 73, 254
and beverage analysis, 79
processing, 256
Forensic analysis, 79, 254
Forensic toxicology, 73
Fragrance analysis, 79

G

Gas-assisted grinding (GAG), 248
Gas-assisted mechanical methods, 248
Gas chromatograph (GC), 223—224
General environmental assessment, 303
Genetically modified organisms (GMOs), 323—324
Graphitized carbon black, 108
Green analytical chemistry (GAC), 2, 2f, 6—11, 7f, 15—16, 20—24, 22f, 24f, 30—34, 31f, 59—60, 119, 169—175, 263—264, 357—359, 385
  evaluations, 32—33
  green sample preparation, 28—30
  methodology, 34—35
  metrics, 34—35
Green analytical methods, 30—34
Green analytical procedure index (GAPI) tool, 10, 287, 288t, 289f
Green analytical tools, 266f
Green chemistry (GC), 1—6, 2f, 15—20, 145—146, 247, 263, 265f, 357
  metrics, 19—20, 21f
Greener analytical techniques, 21—23
Greenest techniques, 34
Greenness assessment systems, 326
Greenness evaluation
  method, 325—326
  tools, 265—267
Green principles, 323
Green sample preparation method, 28—30, 29f, 34—35, 385—386
Green solvents, 26—27
Grinding, 248

H

Hard acid-base principle, 46—47
Hard acids, 48
Hard bases, 48
Hard molecules, 50—52
Hardness, 50—52
Hazard materials, 313—314
Hazardous materials, 315—316
Hazardous organic solvents, 248
Hazardous solvents, 11
Hazards, 314, 321t
Headspace solid phase microextraction (HS-SPME), 73—79, 253—254
Heat-stable extractive material, 96—97
Heavy metal ions, 49
Hexagon-CALIFICAMET, 266—267, 303t
  principles, 297—306
  scale, 296—306
  strengths and weakness, 306
Hexagon-pictogram, 287—289
Hexagon program, 297
High extraction efficiencies, 77
High-performance liquid chromatography (HPLC), 64—65, 370—371
Hohenberg-Kohn Theorem, 43
Hollow-fiber based liquid phase microextraction, 138
Homogeneous liquid-liquid microextraction (HLLME), 157
HPLC-CUPRAC chromatogram, 371—372
Humic substances (HS), 376—378
Hydrofluoric (HF) acids, 26—27
Hydrogen bond acceptor (HBA), 165, 169f, 201
Hydrogen bond donor (HBD), 165, 201
Hydrogen bonding, 66
Hydrophilic DES, 201
Hydrophilicity switchable solvent (SHS), 152
3-Hydroxy-2,2'-iminodisuccinic acid (HIDS), 49

I

Iczkowski-Margrave electronegativity, 44
Immobilization, 364
"In-column extraction" method, 81
Individual tocopherol isomers, 371—372
Inductively coupled plasma mass spectrometry (ICP-MS), 297
In-tube methods, 83
In-tube solid phase microextraction, 81—83
  principle, 81—82
Ionic liquids (ILs), 33—34, 157—165, 193—194, 194f, 195t—196t, 220—221, 252—253
Ionic strength, 69
Ionic surfactants, 148f
Ion mobility spectrometry (IMS), 126

L

"Lab on a disk" devices, 91
Life cycle assessment (LCA), 18−19, 189−190
Limit of detection (LOD), 63−64
Liquid chromatography (LC), 223−224
Liquid−liquid extraction (LLE), 59−60, 189−190, 221−223
  principle, 221
  solvent, 222−223
  traditional, 61−62
Liquid-liquid micro extraction (LLME), 135−136, 201−203, 221−222
Liquid phase micro-extraction (LPME), 146, 223−225
Liquid−solid extraction (LSE), 189−190
Lowered waste generation, 248

M

Magnetic dispersive solid phase (MDSPE), 212
Magnetic ionic liquids (MIL), 159
Magnetic nanoparticles (MNPs), 92−93
Magnetic solid phase extraction, 92−95, 168−169
Magnet-integrated fabric phase extraction (MI-FPSE), 124
Magnetizability, 50−52
Mass transfer kinetics, 86
Mass transfer process, 85−86
Material science, 256
Materials renewability, 317
Matrix-dispersed solid phase extraction, 83−84
Matrix solid-phase extraction (MSPE), 83−84, 92, 93f, 94−95
  applications, 92−94
  principle, 92
Maximum Hardness Principle (MHP), 50
Membrane-based extraction techniques, 119−120, 120f
  membrane-based liquid phase microextraction, 135−139
  membrane-based solid-phase microextraction, 121−135
Membrane-based liquid phase microextraction, 135−139
Membrane-based solid-phase microextraction, 121−135
Membrane extraction techniques, 138−139
Membrane-protected molecularly imprinted materials, 132−133
Membrane-protected solid phase microextraction, 133−135
Merit measurements, 298−300
Methyl β−cyclodextrin, 375−376
Micellar-assisted extraction (MAE), 214−215
Microextraction capsules (MECs), 127−128
Microextraction techniques, 168, 217t
Microplate-based methods, 368
Micro solid phase extraction, 103−104, 128−129
Microwave-assisted extraction (MAE), 251
Microwave energy, 251
Miniaturization, 82, 84
Minimum magnetizability principle, 50−52
Molecularly imprinted polymers (MIPs), 132−133
Molecular weight cut-off (MWCO), 133−135
Monolithic polymer stationary phases, 64−65
Multiretention mechanisms, 62

N

Nanoscience, 367
National environmental method index (NEMI), 8−9, 9f, 18−19, 266, 276−277, 277f−278f
  principles, 276−277
  tool, 278
National Fire Protection Association (NFPA), 292
Natural product extraction, 256
Non-ionic surfactants, 148f
Non-polar bio-based solvents, 175
Nonpolar compounds, 150−151
Non-polar solvents, 210−211
Nonporous membranes, 138
Non-racemic synthesis, 279
Normal-phase silica sorbents, 82

O

Occupational hazard PPs, 285
Octadecyl silane (ODS), 64−65
Ohmic heating extraction, 252
Online HPLC-CUPRAC methods, 370−372
Optrodes, 364
Organic solvents, 358−359
Organophosphorus pesticides (OPPs), 151−152
Oxidative stress and antioxidants, 359−360

P

Packed sorbent microextraction, 101−103
Packed sorbent microextraction (PSME), 102−103
  method, 102
  principles, 102
Penalty points (PPs), 282, 284t−285t, 299t−302t, 304t−305t, 307t
  energy, 283−285
  waste, 285
Perfluorosulfonic acid-based cation-exchange polymer, 365
Personal Protective Equipment (PPE), 6, 210−211
Petrochemical analysis, 79
pH, 69
Pharmaceuticals, 79, 256
Phycocyanin (PC), 220−221
Polarizability, 50−52
Polyacrylate, 80
Polyacrylonitrile (PAN), 98
Polycyclic aromatic hydrocarbons (PAHs), 126, 220−221
Poly dimethyl silane (PDMS), 71
Polyphenolic antioxidants, 360−361
Poly(styrene-codivinylbenzene) (PS-DVB), 64−65
Porous membrane−based liquid phase microextraction techniques, 136−137
Post-column derivatization (PCD), 371−372
Pressing process, 248
Pressurized fluid extraction (PFE), 209−211
Pressurized liquid extraction (PLE), 209−210, 249−250
Pressurized solvent extraction (PSE), 209−210

Q

Quantification techniques, 32−33
QUENCHER approach, 376
Quencher CUPRAC methods, 376−378
Quick-easy-cheap-effective-rugged-safe (QuEChERS), 59−60

R

RAMP, 5f
Reaction flow (RF), 371−372
Reaction heat and time, 280
Reactive oxygen and nitrogen species (RONS), 359
Reagents safety, 33−34
Red principles, 322−323
Reduced matrix effects, 82
Repeatability of the findings (RSDr), 298
Reported rotating disk solid extraction (RDSPE), 89−91
  applications, 90
  principles, 90
Residues development, 302−303
Reverse micelles based SUPRAS, 198−201
Robust analyte−matrix interactions, 83−84
Rotating disk solid phase extraction, 89−91
Rotating disk sorptive extraction (RDSE), 85−86

S

Safety, 280, 320−321
  hazard, of solvents, 11
Sample collection, 10, 31−32
Sample preparation, 11
  device, 131−132, 133f
  techniques, 25f, 26−27
Sample preservation, 10
Sample storage, 10
Sample throughput, 318
Sample transportation, 10
Scale of extraction, 11
Secure operation, 33−34
Semivolatile organic compounds (SVOCs), 73
Sodium dodecyl sulfate (SDS), 147
Soft acid-base principle, 46−47
Soft acids, 48
Soft bases, 49−50
Sol-gel technology, 128
Solid adsorbent selection, 212−213
Solidification of floating organic drop (SFOD), 192−193
Solidified floating organic drop microextraction (SFODME), 63
Solid phase extraction (SPE), 59−60, 61f−62f, 64−70, 89−90, 102−103, 211
  efficiency-related parameters, 67−70
  principle, 65−67
Solid-phase microextraction (SPME), 70−79, 119−120, 133−135, 146, 224−225, 253−254
  applications, 72−73
  classification, 73−79
  conventional, 82
  fiber, 80
  principle, 70−71

Solid-phase microextraction (SPME) (*Continued*)
  sorbents, 73
Solvent based extraction procedures, 189–190
  dispersive liquid-liquid microextraction (DLLME), 190–209
  dispersive solid phase extraction (DLPE), 211–213
  liquid–liquid extraction (LLE), 189–190, 221–223
  liquid phase microextraction (LPME), 223–225
  micellar-assisted extraction (MAE), 214–215
  pressurized fluid extraction (PFE), 209–211
  supercritical fluid extraction (SFE), 215–221
Solvent free extraction procedures, 247–254
  advantages, 254–255
  applications, 256
  considerations, 255–256
  solvent-free extraction, 248–254
Solvent-free sample preparation technique, 70
Solvent selection guide (SSG), 294–295
Sorbent
  amount, 67–68
  type, 67, 80–81
Sorbent-based dynamic extraction, 85–87
Sorbent-based extraction procedures, 59–109
  air-assisted solid phase extraction, 107–109
  dispersive solid phase extraction, 99–101
  fabric phase sorptive extraction, 104–106
  fiber solid phase extraction, 80–81
  in-tube solid phase microextraction, 81–83
  magnetic solid phase extraction, 92–95
  matrix-dispersed solid phase extraction, 83–84
  micro solid phase extraction, 103–104
  packed sorbent microextraction, 101–103
  rotating disk solid phase extraction, 89–91
  solid phase extraction, 64–70
  solid phase microextraction, 70–79
  sorbent-based dynamic extraction, 85–87
  stir bar sorptive extraction, 87–89
  thin film microextraction (TFME), 95–99
Standard Operating Procedures (SOP), 6
Stir-bar fabric phase sorptive extraction (stir bar-FPSE), 124

Stir bar sorptive extraction, 87–89
Stir-fabric phase sorptive extraction (stir-FPSE), 123–124
Subcritical water extraction (SWE), 33–34, 249, 249f
Supercritical fluid (SCF), 215–216, 216f
  applications, 220–221
  properties, 218
Supercritical fluid extraction (SFE), 33–34, 215–221, 250–251
Supported liquid membrane (SLM), 138
Supramolecular solvent (SUPRAS), 149–152, 191–192, 197–198, 199t–200t
Sustainable analytical procedures, 19–20
Sustainable green analysis, 30
Switchable hydrophobicity solvents (SHSs), 160t–162t, 191–192
Switchable solvents, 203–209

T

Technical system, 280
Teflon-coated bar magnet, 124
Temperature, 69–70
Tenax, 107
Tetrabutylammonium (TBA), 198
Tetracycline antibiotics (TC), 151–152
Tetrahydrofuran (THF), 150–151, 198–201
Tetra-n-butyl ammonium hydroxide, 198–201
2-Thenoyltrifluoroacetonate (TTA), 52
Thin film microextraction (TFME), 95–99, 135
  comparison and applications, 98–99
  materials devices, 98
  principles, 95–98
  solvent desorption, 97–98
Thin layer chromatography (TLC), 366
Total antioxidant capacity (TAC), 370, 373
Toxic effects and safety evaluating, 300–302
Traditional coating techniques, 98
Traditional extraction techniques, 146, 358–359
Triazole pesticides, 54–55
Triethanolamine (TEA), 152–156
Trolox equivalents (TE), 362

U

Ultra-high performance liquid chromatography (UHPLC), 215

Ultrasound-Assisted Extraction (UAE), 251–252
Ultra-thin extractive phases, 96

V
Vesicular coacervate phase micro extraction, 198
Volatile organic compounds (VOCs), 190
Vortex-assisted switchable hydrophilic solvent liquid phase microextraction (VA–SHS–LPME), 157

W
Warning pictograms, 9–10
Waste, 317–318
    generation, 34
    management, 255
    penalty points (PPs), 285
Water-immiscible organic solvents, 225
White analytical chemistry (WAC), 265–266, 322–325

Y
Yield, 279

Printed in the United States
by Baker & Taylor Publisher Services